Pergamon International Library
of Science, Technology, Engineering and Social Studies
The 1000-volume original paperback library in aid of education,
industrial training and the enjoyment of leisure
Publisher: Robert Maxwell, M.C.

Introductory Dynamical Oceanography

SECOND EDITION

THE PERGAMON TEXTBOOK INSPECTION COPY SERVICE

An inspection copy of any book published in the Pergamon International Library will gladly be sent to academic staff without obligation for their consideration for course adoption or recommendation. Copies may be retained for a period of 60 days from receipt and returned if not suitable. When a particular title is adopted or recommended for adoption for class use and the recommendation results in a sale of 12 or more copies, the inspection copy may be retained with our compliments. The Publishers will be pleased to receive suggestions for revised editions and new titles to be published in this important International Library.

D0036325

Related Pergamon Titles

Books

BEER
Environmental Oceanography (An Introduction to the Behaviour of Coastal Waters)

FEDOROV
The Thermohaline Fine Structure of the Ocean

*GORSHKOV
World Ocean Atlas

Volume 1: Pacific Ocean
Volume 2: Atlantic and Indian Oceans
Volume 3: Arctic Ocean

KRAUSS
Modelling and Prediction of the Upper Layers of the Ocean

*MELCHIOR
The Tides of the Planet Earth, 2nd edition

PARSONS *et al*
Biological Oceanographic Processes, 3rd edition

PARSONS *et al*
A Manual of Chemical and Biological Methods for Seawater Analysis

PICKARD & EMERY
Descriptive Physical Oceanography, 4th edition

TCHERNIA
Descriptive Regional Oceanography

Journals

Acta Oceanologica Sinica
Continental Shelf Research
Deep-Sea Research
Marine Pollution Bulletin
Ocean Engineering
Progress in Oceanography

Full details of all Pergamon publications/free specimen copy of any Pergamon journal available on request from your nearest Pergamon office.

*Not available under the terms of the Pergamon textbook inspection copy service.

Introductory Dynamical Oceanography

SECOND EDITION

by

STEPHEN POND, B.Sc., Ph.D.
Professor

and

GEORGE L. PICKARD, M.A., D.Phil., F.R.S.C.
Emeritus Professor and former Director, Department of Oceanography,
University of British Columbia, Vancouver, Canada

PERGAMON PRESS

OXFORD · NEW YORK · TORONTO · SYDNEY · FRANKFURT

U.K.	Pergamon Press Ltd., Headington Hill Hall, Oxford OX3 0BW, England
U.S.A.	Pergamon Press Inc., Maxwell House, Fairview Park, Elmsford, New York 10523, U.S.A.
CANADA	Pergamon Press Canada Ltd., Suite 104, 150 Consumers Road, Willowdale, Ontario M2J 1P9, Canada
AUSTRALIA	Pergamon Press (Aust.) Pty. Ltd., P.O. Box 544, Potts Point, N.S.W. 2011, Australia
FEDERAL REPUBLIC OF GERMANY	Pergamon Press GmbH, Hammerweg 6, D-6242 Kronberg, Federal Republic of Germany
JAPAN	Pergamon Press Ltd., 8th Floor, Matsuoka Central Building, 1-7-1 Nishishinjuku, Shinjuku-ku, Tokyo 160, Japan
BRAZIL	Pergamon Editora Ltda., Rua Eça de Queiros, 346, CEP 04011, São Paulo, Brazil
PEOPLE'S REPUBLIC OF CHINA	Pergamon Press, Qianmen Hotel, Beijing, People's Republic of China

First edition 1978
Second edition 1983
Reprinted (with corrections) 1986

Library of Congress Cataloging in Publication Data
Pond, Stephen.
Introductory dynamical oceanography.
(Pergamon international library of science,
technology, engineering, and social studies)
Rev. ed. of: Introductory dynamic oceanography.
1978.
Bibliography: p.
1. Oceanography. 2. Hydrodynamics. I. Pickard,
George L. II. Title. III. Series
GC201.2.P66   1983   551.47   83-2375

British Library Cataloguing in Publication Data
Pond, Stephen
Introductory dynamical oceanography.—2nd ed.
—(Pergamon international library)
1. Ocean currents   2. Ocean waves
I. Title
551.47   GC201

ISBN 0-08-028729-8 (Hard cover)
ISBN 0-08-028728-X (Flexi cover)

Printed in Great Britain by
Redwood Burn Ltd Trowbridge, Wiltshire

Preface

THE PURPOSE of this book is to present an introduction to dynamical physical oceanography at a level suitable for senior-year undergraduate students in the sciences and for graduate students entering oceanography.

Our aims are to introduce the basic objectives and procedures and to state some of the present limitations of dynamical oceanography and its relations to the material of descriptive (synoptic) oceanography. We hope that the presentation will serve to introduce the field to physicists intending to specialize in physical oceanography, to help oceanographers in other disciplines to learn enough about the physics of the ocean circulation to discuss with the physical oceanographer the aspects which they need to understand for their own work, and to give those in allied fields an appreciation of what the dynamical oceanographer is trying to do in contributing to our overall knowledge of the oceans.

The presentation involves the use of mathematics, as the essence of the dynamical approach is to deduce quantitative information about the movements of the ocean from mathematical statements of the basic principles of physics as they apply to the ocean waters. The level is such that undergraduates who have taken a course in calculus should be able to follow the essentials of the mathematical arguments, while students in the physical sciences should have no difficulty at all. Non-physical science students should not be disheartened by the mathematics because a course with much of this material has been taken for many years by biological science students, among others, at the University of British Columbia to complement a course in descriptive physical oceanography. For students with little calculus background we emphasize the physical assumptions made in setting up and in solving the equations, so that the limitations inherent will be clear, and then we stress the interpretation of the solutions obtained. Some intermediate mathematical steps are provided for those interested in following them. The student with limited mathematical background should concentrate on the verbal physical interpretations and not worry about the details of the mathematics. At the same time, Appendix 1 provides a brief review of most of the mathematical procedures and symbolism used in the text and of some aspects of fluid mechanics which relate to dynamical oceanography. The non-physicist may find parts of Chapter 7 somewhat difficult to follow at first. If so, re-reading this chapter after reading Chapters 8, 9 and, perhaps, 10 would probably be worthwhile.

We have tried to make the text self-contained within its field but we consider it essential that students interested in dynamical oceanography should first acquaint themselves with the observational aspects of physical oceanography in order to be aware of the characteristics of the oceans which the dynamical oceanographer is endeavouring to understand and explain. A text such as *Descriptive Physical Oceanography* by Pickard and Emery (1982) or other introductions to this aspect listed in the Bibliography would provide the necessary background.

In assembling the text for the First Edition we added significantly to the original course material on which it was based so that it is unlikely that all of the present material could be covered in a course of twenty-five or so lectures as we had done in the past. However, we assume that, when using it as a course text, an instructor will select what is appropriate for his class and will leave the remaining material for later reference or will consider presenting the material in a longer course.

In preparing the Second Edition we have clarified a number of minor points and have introduced the Practical Salinity Scale 1978 as the basic definition of this quantity and the International Equation of State 1980 as the most up-to-date statement of the relations between salinity, temperature, pressure and density for sea water. Information on the earlier and present equations of state has been brought together in the new Appendix 3. A section on the beta-spiral (a method for determining absolute velocities from the density field) has been added, the discussion of mixed-layer models has been updated (Chapter 10), eddy resolving numerical models described briefly (Chapter 11) and Chapters 12 and 13 on Waves and on Tides have been substantially revised and enlarged. (Note that the latter two are each extensive fields and the present treatments, which are more descriptive than analytical, are mainly intended to acquaint the non-specialist reader with some of the essential features. References to more detailed treatments for the physical oceanographer will be found in the Bibliography.)

We believe that the material covered in this text will make it complete enough for the non-physicist and also usable as an introduction for physical science students. For graduate students in physical oceanography the book should serve as a basic introduction, to be supplemented either in lectures or with references to the literature, e.g. further discussion (with more complete mathematical theory) of such topics as turbulence, vorticity, equatorial circulation, boundary layers, thermocline and thermohaline circulation theories, as well as the previously mentioned waves and tides. For the convenience of the reader wishing to obtain fuller information we have included in the text references to material in books or journal articles. These are listed in the Bibliography (Sections B.2 and B.4 respectively.)

If the physicists find that some concepts are introduced in a rather elementary fashion we ask them to bear with us as these are ones which, in our experience, have given trouble to non-physicists. The physicist might even find the more extensive verbal explanations a pleasant relief from the multitudes of equations with limited physical explanations sometimes encountered.

As a focus for the book we have concentrated on the large-scale average circulation. We are aware that much attention is being paid to studies of variability, e.g. as eddies in the open ocean and on smaller scales in coastal regions, and that, in many situations, short-term variations may be much larger than those of the longer-term mean. We consider that detailed discussions of many of these topics are matters for more advanced study. Estuarine and coastal dynamics have not been presented although there is occasional mention of them. There are already several texts on estuarine oceanography, mentioned in the Bibliography (Section B.2), and the text by Csanady (1982) describes the application of the principles of dynamical oceanography to coastal waters and shallow seas.

Finally, it must be re-stated that this text is intended simply as an *introduction* to dynamical oceanography. More sophisticated treatments are available for many aspects. Good introductions to the earlier mathematical studies are Stommel's *The Gulf Stream* (1965) and Robinson's compilation of papers (1963). Excellent reviews of many aspects of physical oceanography (descriptive, dynamical and instrumental) are available in *Evolution of Physical Oceanography*, edited by Warren and Wunsch (1981), as a tribute to Henry Stommel on his sixtieth birthday. This volume also includes a wide range of references to journal and text articles in the field.

Contents

14. Some Presently Active and Future Work

Appendix 1. Mathematical Review with Some Elementary Fluid Mechanics

Appendix 2. Units Used in Physical Oceanography

List of Main
Symbols used in Text

BOLD type (e.g. $\mathbf{V}$) indicates a vector quantity; ordinary type (e.g. V) indicates the magnitude of that quantity.

ROMAN LETTERS

$\mathbf{a}$, a	Acceleration
A (or δA)	Area; wave amplitude (Chap. 12)
A_x, A_y, A_z, A_H	Kinematic eddy viscosity for x, y, z and horizontal directions. (In the latter part of Chap. 9, A is used for brevity for A_H)
b	Estuary width (Chap. 13)
B	Radius of inertial motion circle (Chap. 8)
C; C	Conductivity (Chap. 2); speed of sound (Chap. 5)
$\mathbf{C}$; C; C_b	Wave phase velocity; speed; speed of bore (Chapt. 13)
C_s, C_l	Phase speeds of short (deep-water) and long (shallow-water) waves
$\mathbf{C}_g$, C_g	Wave group velocity, speed
$\mathbf{C}_i$, C_i	Internal wave phase velocity, speed
$\mathbf{C}_{gi}$, C_{gi}	Internal wave group velocity, speed
C_{sol}	Solitary wave phase speed
Cl	Chlorinity
$\mathbf{C}_H$	Horizontal Coriolis parameter ($= 2\Omega \sin\phi\, \mathbf{V}_H \times \mathbf{k}$)
C_D	Aerodynamic drag coefficient
d	Relative density (Chap. 2); level of interface of two-layer system (Chap. 9); distance to wave generation area (Chap. 12)
D	Depth (Chap. 2); geopotential in mixed units system (Chap. 8); thickness of a layer (Chaps. 9, 10)
D_E	Ekman depth
D_z	Particle orbit diameter in wave motion
E	Stability (Hesselberg); wave energy density (Chap. 12)

E_p; E_p^0	Potential; standard potential energy of a water column
E_x, E_y, E_z, E_H	Ekman numbers
f	Coriolis parameter $= 2\Omega \sin \phi =$ planetary vorticity
F, F	Force; with subscript $=$ a particular force or force component
F	Form ratio for tides (Chap. 13)
g, g	Acceleration due to gravity (taken as 9.80 m s^{-2} in this text)
$\mathbf{g}_f$	Gravitational attraction of earth on unit mass in an inertial coordinate system
G	Gravitational constant
h	Water depth; mixed layer depth (Chap. 10)
H	Scale depth (Chaps. 4, 7); wave height (Chap. 12)
H_s	Significant wave height (average of highest one-third waves)
i	Angle between an isobaric surface and a level (horizontal) surface
i, **j**, **k**; j	Unit vectors in the x, y, z-directions; $j = \sqrt{-1}$ (Chap. 9)
k, k	Vector radian wave number and its magnitude
K; [K]	Degrees Kelvin; temperature as a physical dimension
K_{15}, R_t	Electrical conductivity ratio (App. 2)
K_x, K_y, K_z, K_H	Kinematic eddy diffusivity for x, y, z and horizontal directions
K_T; K	Tangent bulk modulus; secant bulk modulus (Apps. 1, 3)
[L]	Length as a physical dimension
L	Horizontal scale length; basin length (Chap. 13)
L_c	Resonant basin length (Chap. 13)
m, M; [M]	Mass; mass as a physical dimension
M_2, S_2, K_1, etc.	Tide-producing force constituents
M; M_x, M_y	Vector mass transport (per unit width); mass transport per unit width in x, y-directions
n	Integer; normal coordinate (i.e. perpendicular to some surface or line), internal wave mode number
n_H	Normal coordinate in horizontal plane
N	Brunt-Väisälä frequency
p; p'; p_w	Pressure; fluctuating component of pressure; component of pressure due to wave elevation relative to undisturbed water level
q	Stands for (any) quantity or variable (App. 1)
Q; Q_x, Q_y	Total transport; volume transport per unit width in x, y-directions
Q_T	Temperature (heat) source function
r	Distance between centres of two masses

R	Gas constant (Chap. 2); distance from centre of earth (Chaps. 6, 13); ratio of *in situ* conductivity to standard conductivity
Re, Ri, Ro	Reynolds, Richardson, Rossby Numbers
s in δs	Element of surface area
$S; s$	Salinity; salinity as a subscript
t	Time; temperature as a subscript and in some formulae
T	Absolute temperature (Chap. 2)
$[T]$	Time as a physical dimension
T	*In situ* temperature ($^\circ$C); scale time (Chap. 7); period (Chaps. 12, 13)
T_f	One pendulum day (Chap. 8); fundamental period (Chap. 13)
u, v, w	Velocity components in x, y, z-directions. Subscripts used: b = barotropic, c = baroclinic, E = Ekman, g = geostrophic
u', v', w'	Fluctuating components of velocity
U, V, W	Characteristic values for x, y, z velocity components
$\mathbf{V}; \mathbf{V}_H$	Vector velocity $= \mathbf{i}u + \mathbf{j}v + \mathbf{k}w$; vector velocity in horizontal plane $= \mathbf{i}u + \mathbf{j}v$
V_b, V_c	Barotropic, baroclinic parts of V_H
V_1, V_2	Horizontal velocity components normal to a vertical section at levels 1 and 2
V_0	Speed of Ekman flow at the surface
$V, \delta V$	Volume, element of volume
w_E	Vertical velocity component at the bottom of the Ekman layer
W	Work (Chap. 8); wind speed (Chap. 9); width of the western boundary current (Chaps. 9, 11) (W' = non-dimensional form, eastern boundary current, Chap. 9)
x, y, z	Rectangular position coordinates in the east–west, north–south and vertical directions

GREEK LETTERS

α (alpha)	Specific volume
β (beta)	$= \partial f/\partial y =$ variation of Coriolis parameter with latitude; compressibility (App. 2)
$\Gamma; \gamma$ (gamma)	Adiabatic temperature gradient; non-dimensional y-coordinate
δ (delta)	Specific volume anomaly $= \delta_s + \delta_t + \delta_{s,t} + \delta_{t,p} + \delta_{s,p} + \delta_{s,t,p}$
$\Delta_{s,t}$ (delta)	Thermosteric anomaly $= \delta_s + \delta_t + \delta_{s,t}$

$\varepsilon_{s,p}$, $\varepsilon_{t,p}$ (epsilon) Density anomaly terms
ζ (zeta) Relative vorticity
η (eta) Vertical displacement of the water surface; as a subscript it indicates a surface value for the quantity
θ (theta) Potential temperature; angle.
κ_s, κ_t (kappa) Kinematic molecular diffusivity for salt, temperature (heat)
λ (lambda) Rossby radius of deformation (subscripts i or e indicate internal or external Rossby radii)
Λ; Λ_s, Λ_l; Λ_i: (lambda) Wavelength; wavelengths of short (deep-water), long (shallow-water) waves; internal waves
μ (mu) Dynamic molecular viscosity
ν (nu) Kinematic molecular viscosity
ξ (xi) Non-dimensional x-coordinate
π (pi) Ratio of circumference to diameter of a circle
ρ (rho) Density
σ_t (sigma-t) Density at atmospheric pressure—1000 kg m^{-3}
σ_θ (sigma-θ) Equivalent of σ_t using potential temperature (θ)
τ (tau); τ' Frictional stress; non-dimensional form (Chap. 9)
ϕ (phi, l.c.) Geographic latitude
Φ (phi, u.c.); $\Delta\Phi_{std}$; $\Delta\Phi$ Geopotential; standard geopotential "distance"; geopotential anomaly
χ (chi) Potential energy anomaly
ψ (psi); ψ' Stream function; non-dimensional form (Chap. 9)
ω (omega, l.c.) Angular speed of revolution of moon about earth; frequency in radians per second (Chap. 12)
Ω (omega, u.c.) Angular speed of rotation of earth about its axis

ABBREVIATIONS FOR FREQUENTLY USED UNITS

(SI) m metre kg kilogram s second
km kilometre J joule h hour
(other) Sv = sverdrup = 10^6 m^3 s^{-1} (for volume transport)

MATHEMATICAL SYMBOLS

= is equal to **bold type** vector quantity (e.g. **V**)
$\simeq$ is approximately equal to · scalar product (of two vectors)
$\equiv$ is equivalent to
~ is of order of × vector product (of two vectors)
> is greater than
$\gg$ is much greater than ‾ (overbar) average quantity

$<$	is less than	| |	absolute value *or* magnitude of a vector
$\ll$	is much less than		
$\geqslant, \leqslant$	is greater (less) than or equal to	$\therefore$	therefore
		$\because$	because
$\gtrsim, \lesssim$	is greater (less) than or approximately equal to	$f(\ )$	is a function of ( )
		$0(\ )$	is of order of ( )
		$\exp(\ )$	exponential function of ( )
∇ (grad)	gradient operator		
$\nabla\cdot$ (div)	divergence operator	Δx	a finite change of x
$\nabla_H\cdot$	horizontal divergence operator	δx	a small (finite) change of x
∇^2	Laplace operator		
∇^4	biharmonic operator	$\dfrac{dq}{dx}$	derivative of q with respect to x
$\text{curl}_z \tau$	$= \dfrac{\partial \tau_y}{\partial x} - \dfrac{\partial \tau_x}{\partial y}$	$\dfrac{\partial q}{\partial x}$	partial derivative of q with respect to x
$\sum, \displaystyle\sum$	summation sign	$\dfrac{dq}{dt}$	total (or individual) derivative of q (see App. 1)
		$\int, \displaystyle\int$	integral sign

Trigonometric functions:

Circular   :   sin = sine, cos = cosine, tan = tangent,

Hyperbolic: $\sinh x = [\exp(x) - \exp(-x)]/2$, $\tanh x = \sinh x/\cosh x$,
$\cosh x = [\exp(x) + \exp(-x)]/2$, $\coth x = 1/\tanh x$.

Acknowledgements

WE ARE indebted to our colleagues Drs. R. W. Burling, P. Crean, P. H. LeBlond and T. R. Osborn for reading parts of the manuscript and offering many helpful suggestions, and to numerous students whose questions have often provoked us into more careful scrutiny of concepts and deductions than we had given them before. Discussions with Bruce Hamon stimulated the development of the intuitive derivation of the Coriolis terms in Chapter 6 and the clarification of a number of other points.

We gratefully acknowledge permission from authors and publishers to reproduce or adapt the following material:

Fig. 8.8   From K. Wyrtki, 1974. Hawaii Institute of Geophysics.

Figs. 9.10, 9.11   From R. O. Reid, 1948. *Journal of Marine Research.*

Fig. 9.12   From H. Stommel, 1948. *American Geophysical Union.*

Fig. 9.15(a)   From W. H. Munk, 1950. *Journal of Meteorology.*

Fig. 9.15(b)   From W. H. Munk and G. F. Carrier, 1950. *Tellus.*

Fig. 10.1   From H. Stommel, 1958. *Deep-Sea Research.*

Fig. 11.1   From J. J. O'Brien, 1971. *Invest. Pesqu.*

Fig. 11.2   From M. D. Cox, 1970. *Deep-Sea Research.*

Figs. 11.3, 11.4   From W. R. Holland and A. D. Hirschman, 1972. *Journal of Physical Oceanography.*

Fig. 11.5(a), 11.6   From M. D. Cox, 1975. National Academy of Sciences, Washington, D.C.

Fig. 11.5(b)   From K. Bryan and M. D. Cox, 1972. *Journal of Physical Oceanography.*

Fig. 13.7   From G. Godin, 1980. Department of Fisheries and Oceans, Canada.

Fig. 13.8   From Y. Accad and C. L. Perkeris, 1978. Royal Society of London.

Finally we are indebted to the many physicists and oceanographers, past and present, on whose works this book is based.

1

Introduction

Oceanography is the study of the ocean making use of the various basic sciences, physics, chemistry, biology and geology, with mathematics being used as an aid to parts of all of these studies. Particular attention is paid to the ocean as an environment both for the organisms which inhabit it naturally and in relation to man's activities, and also to its interaction with the atmosphere, the environment in which man lives.

The physicist's contribution is to study the distribution of properties such as temperature, salinity, density, transparency, etc., which distinguish one water mass from another, and to study and understand the motions of the ocean in response to the forces acting on it.

Some of the problems which have been recognized and studied are:

> Why are the gross mid-latitude surface circulations in the ocean clockwise in the northern hemisphere but anti-clockwise in the southern hemisphere?
>
> Why are these circulations concentrated and swift at the western sides (Gulf Stream, Kuroshio, etc.) but broad and slow elsewhere?
>
> What is the reason for the eastward circulation of the Southern Ocean around Antarctica?
>
> What is the distribution with depth of the ocean currents?
>
> What is (are) the reason(s) for the complicated equatorial flow patterns?
>
> What are the details of the mechanisms of transfer of momentum and energy between atmosphere and ocean?
>
> What are the characteristics and causes of surface and internal waves?
>
> What are the relations between submarine earthquakes and tsunamis and how do the latter behave in the deep ocean and at the shore?
>
> What are the characteristics and significances of turbulent motions in the oceans?

To some of these questions we have answers, to some we have partial answers, and as our studies progress new problems become apparent.

Physical studies are carried out both by direct observation of the properties and movements and also by applying the basic physical principles of mechanics and thermodynamics to determine the motions. The observational approach is called *descriptive* or *synoptic* oceanography because the physicist tries to reduce his observations to a simple synopsis or summary. The essential feature of the second approach, *dynamical* oceanography, is to use physical laws to endeavour to obtain mathematical relations between the forces acting on the ocean waters and their consequent motions. In either case, the ultimate objective is to learn enough about the structure and motions of the ocean to be able to predict its future state.

In principle, to achieve this objective the dynamical approach is most likely to be successful because it should result in analytical expressions which can be used for prediction into the future, whereas the synoptic approach simply describes what has happened in the past. In practice, it turns out that some characteristics of the oceans do not change much with time, or they repeat themselves with recognized periods, so that a good description of the present state may be applicable for some time into the future. However, some features do change and the amplitudes of cyclic variations may alter, so that a quantitative understanding of the relations between the causative forces and the reaction of the ocean is desirable. Therefore, the preliminary quantitative description of the ocean and its movements prepared by the synoptic ocean-ographer is used by the dynamical oceanographer to suggest what kinds of motion he may expect and what forces may be causing them, so helping him to start his theoretical study. Also, if he meets mathematical difficulties in his analysis, as often happens, the available observations may suggest what mathematical approximations may be made while keeping the investigation physically realistic.

When the dynamical oceanographer has made a preliminary analysis, it will probably suggest the need for more extensive or sophisticated observations; when these have been made he may refine his analysis. Successions of improved observations and analysis will hopefully lead to better and better understanding of the physics of the ocean, and improvement in our ability to predict its future state.

Systematic physical observations of the ocean have been made for a century or so, the rate of accumulation of data having increased enormously during the last 25 years. An introduction to the available information is presented in *Descriptive Physical Oceanography*, 4th Edition, Pickard and Emery, 1982, and in other texts listed in the Suggestions for Further Reading later in this book. The purpose of the present text is to provide a parallel introduction to dynamical oceanography. The systematic dynamical study of the circulation started at the end of the last century when Scandinavian meteorologists, recognizing the similarity between the dynamics of the atmosphere and of the ocean, turned their attention to the latter. Studies of some phenomena started much earlier with Newton's (*circa* 1687) and Laplace's (*circa* 1775) studies of the tides, and Gerstner's (1802) and Stokes' (1874) studies of waves, as examples.

The sequence of topics in this book starts with a description of the properties of sea water relevant to dynamical oceanography, and a summary of the basic physical laws or principles which will be used. The principle of conservation of mass is used in the form of conservation of volume which, for the incompressibility approximation used, places mild restrictions on the possible motions. An example is given of its application. A definition of static stability is presented and discussed, and the possibility of double diffusive instability is mentioned. Then the forces which may be acting in the sea are classified and some simple examples given. Following this chapter, and occupying a large part of the present exposition is a discussion of the application of Newton's Second Law of Motion to relate the forces acting to the resultant motion of the sea on a rotating earth—the field of geophysical fluid dynamics. Because of analytical (mathematical) difficulties in solving some of the forms of the equations of motion, numerical methods for solving the equations are increasingly being applied; the basic principles, successes and limitations of this technique are summarized. Some aspects of the longer period (a month or so) time-varying motions and their interactions with the large-scale flow are discussed. A brief account of ideas related to the thermohaline circulation is presented and short, mainly descriptive, accounts of the characteristics of waves (surface and internal) and tides are given. Finally, some discussion is offered of what appear to be the presently active and future areas for research in dynamical oceanography.

Three appendices are included. Appendix 1 is a brief review of mathematical techniques used and of simple hydrodynamical principles for the reader with a limited background in these fields. Appendix 2 is concerned with units. In the mathematical solution of equations, when all quantities are represented by letter symbols, the question of what physical system of units should be used in measurement does not arise. However, as soon as numerical calculations are to be made, to compare the mathematics with observations made of the real world, it is necessary to select a system of units. Unfortunately, in most of the physical oceanographic literature a mixed system of units has been used. It is basically the CGS system using centimetres, grams, seconds and calories as the fundamental units. However, although density is expressed in grams cm^{-3}, depths are expressed in metres, and horizontal distances often in nautical miles, pressure is expressed in decibars, abbreviated as dbar or db (because the depth in metres and the pressure then have numerically almost the same value), and a quantity called "dynamic height" expressed in "dynamic metres" is introduced although it is dimensionally work per unit mass. Because the International System of Units (SI) is now coming into general use, and is often required for publication in many journals, we have elected to depart from the conventional oceanographic units and introduce the SI units systematically in this text. Appendix 2 then contains a glossary of physical oceanographic terms and conversion factors between the SI units and the old mixed system units. Appendix 3 contains information on sources of values for the density and specific volume of sea water, both as tables based on the older Knudsen-Ekman (pre-1980) equation of state and the more recently established data embodied in the International Equation of State 1980.

For coordinate axes we will generally use a right-handed system with the positive x-axis directed horizontally to the east, the positive y-axis horizontally to the north, and the positive z-axis vertically upward, with the origin normally at mean sea level. Note that the term "depth", the distance below the surface, is taken to be positive as is the usual practice; thus with the origin at the surface, z is the negative of the depth, i.e. for a depth of 100 m then $z = -100$ m.

2

Properties of Sea Water relevant to Physical Oceanography

2.1 Introduction

The physical properties of pure water relevant to fluid dynamics studies are functions of pressure (p) and temperature (T) while those of sea water are functions of pressure, temperature and salinity (S). The salinity of sea water is essentially a measure of the mass of dissolved salts in one kilogram of sea water. An average value for the oceans is about 35 grams per kilogram. It is not practical to measure this quantity directly, and for a long time it was determined indirectly by measuring the halide content (Cl, mostly chloride) by a silver nitrate titration and using, on the basis of previous careful analyses, a linear relation ($S = 1.806\,55 \times Cl$) between total dissolved salts and halide. However, it has been demonstrated that there is a close relationship between the salinity and the electrical conductivity of sea water and, as this latter property can now be measured easily and precisely, it has for some time been used to estimate salinity. The procedure is described in *Descriptive Physical Oceanography* (Pickard and Emery, 1982). A "Practical Salinity Scale, 1978" (PSS 78) has been adopted (Unesco, 1979, 1981) based on conductivity measurements as described in Appendix 2. Note that a salinity of 35.00 grams per kilogram, formerly written as $35.00\,^\circ/_{oo}$, is now written on the PSS 78 simply as 35.00.

The effect of the dissolved salts is to alter the physical properties from those of pure water in degree rather than to develop new properties, e.g. small changes in compressibility, thermal expansion, refractivity, and larger changes in the freezing point, density, temperature of maximum density and electrical conductivity. Although water is a very common substance it has extreme values for many physical properties, e.g. high specific heat so that ocean currents carry much thermal energy, a high latent heat of fusion so that in polar regions where there is ice in the water the temperature is maintained close to the melting point, a high latent heat of evaporation which is important in heat transfer from sea to air, and a high molecular heat conductivity. (This latter property is over-shadowed under most circumstances by "eddy" transfer processes due to the turbulent motion of ocean waters. The eddy or turbulent heat transfer effects are discussed briefly in Chapter 10.)

5

2.2 Density

From the point of view of dynamical oceanography the most important aspect is the quantitative manner in which *density* varies with changes in temperature, salinity and pressure. Density (ρ) decreases as temperature increases and increases as salinity and pressure increase. The variation of density expressed as *sigma-t* [$\sigma_t = (\rho - 1000)$ kg m^{-3}] is shown in Fig. 2.1 for temperatures from -2 to $30°$C and salinities from 20 to 40 which cover the ranges found in the open oceans. About 90 % of the volume of the world ocean has values in the much smaller range from -2 to $10°$C and 34 to 35 in salinity as shown in the figure. This is mostly subsurface water, the remainder of the range of properties in the figure represents limited volumes of surface waters. (In the vicinity of rivers or melting ice, salinity values lower than 20 may be found.) The relation between density and the parameters temperature and salinity is non-linear, more so in temperature than in salinity, and density is less sensitive to temperature changes at low temperatures than at high temperatures. Note that pure water has a density maximum close to $4°$C at atmospheric pressure but as salinity increases this temperature decreases to about $-1.4°$C at salinity $= 25$ (where the freezing point has the same value). A useful rule-of-thumb is that density increases by approximately 1 part in 1000 (i.e. by 1 kg m^{-3}) for a change of temperature of $-5°$C, for a change of salinity of $+1$ or for a change

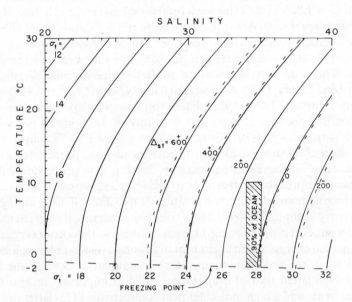

FIG. 2.1 *Values of density (as σ_t) and thermosteric anomaly ($\Delta_{s,t}$) as functions of temperature and salinity over ranges appropriate to most of the oceans. (90 % of the world ocean has temperature and salinity values within the shaded rectangle.)*

of pressure of $+2000$ kPa (kilopascals) $= +200$ dbar (equivalent to a depth increase of about 200 m). For comparison, values of $\Delta_{s,t}$ $(10^{-8}\,\mathrm{m^3\,kg^{-1}})$ defined in Section 2.23 are also shown in Fig. 2.1.

2.21 Measurement of density, temperature and salinity

For many calculations, the physical oceanographer needs to know the distribution of density both horizontally and vertically in the sea but at present there is no method available for measuring it accurately *in situ*. It can be measured directly in the laboratory but the standard methods are slow. In practice, temperature, salinity and pressure are measured *in situ* and the density is deduced from tables or polynomial expressions which have been prepared from laboratory determinations of this property.

In situ temperature (T) is measured either with a specially designed mercury-in-glass thermometer (reversing thermometer) which records the temperature at the moment when a sample of sea water is taken at depth or with an electrical resistance thermometer.

Because sea water is slightly compressible, a sample brought from depth to the surface will expand and therefore tend to cool. The temperature of a sample brought adiabatically to the surface (i.e. without thermal contact with the surrounding water) will therefore be cooler than *in situ*. The temperature which it would have at the surface in these circumstances is called the *potential* temperature (θ). This value is used when comparing water samples at significantly different depths or when considering vertical motions over considerable depth ranges (to eliminate the compressibility effect).

In the laboratory, salinity is determined from measurements of the electrical conductivity and temperature (because at constant pressure conductivity $= f(S, T)$, a compact way of stating that the conductivity "is a function of", i.e. varies with, S and T). At sea, pressure is determined from the depth of sampling and the density of the water column. Alternatively, conductivity, temperature and pressure sensors may be mounted together in an underwater unit which is lowered through the depth range of interest and the instrument (a CTD) either records the data internally or transmits them as electrical signals to instruments on deck where a continuous record of temperature and conductivity (or salinity) against depth is obtained. The relations between salinity, conductivity, temperature and pressure have been redetermined in recent years (for references see Unesco (1979, 1981) and Lewis (1980)); absolute redeterminations of conductivity and density are still to be made.

2.22 Relative density, sigma-t and specific volume

In their discussions, physical oceanographers sometimes use density (ρ), sometimes relative density (d), sometimes specific volume ($\alpha = 1/\rho$) and sometimes a quantity called "sigma-t" (σ_t).

In discussing the dynamics of the oceans, the density of sea water enters into many of the equations. It should be noted that the quantity which the oceanographer derives from tables or polynomial expressions using values of temperature, salinity and pressure is strictly speaking the relative density because the laboratory determinations on which the tables, etc., are based have consisted of comparisons of sea water with pure water, not absolute measurements of density which are much more difficult to make. The relative density can be given with an accuracy of about 3 in 10^6 but the density proper is only accurate to about 10 in 10^6. Fortunately it is density differences which are important in most cases and for these the greater uncertainty in the absolute values is not significant. We will follow the common practice and talk about ρ for sea water as "density" with physical dimensions $[ML^{-3}]$.

It should also be noted that in oceanographic practice, when specifying the conditions for a water sample the quantity p (sea pressure) refers to the hydrostatic pressure, i.e. the pressure due only to the column of water above a point in the sea, so that $p = 0$ implies that the sample is at atmospheric pressure.

Sigma-t, defined as $\sigma_t = (\rho(s, t, o) - 1000)\,\mathrm{kg\,m^{-3}}$, was introduced for brevity. The density of sea water at atmospheric pressure varies from about $1000\,\mathrm{kg\,m^{-3}}$ $(1.000\,\mathrm{g\,cm^{-3}})$ for almost fresh water to about $1028\,\mathrm{kg\,m^{-3}}$ for the densest ocean water. As the variation is entirely in the last two figures (four if density is expressed to the second decimal place) it is convenient for descriptive purposes simply to use these two (or four) figures, e.g. for sea water of $T = 10.00°C, S = 35.00$ and $p = 0$, then $\rho = 1026.95\,\mathrm{kg\,m^{-3}}$ and $\sigma_t = 26.95$. (Although σ_t has units of $\mathrm{kg\,m^{-3}}$ it is usual to omit them when quoting values.)

(Note that when symbols are needed as subscripts or in functional expressions, we will follow the practice of using lower-case italic letters as temperature (t), salinity (s) and pressure (p), e.g. σ_t, $\rho(s, t, p)$, $\Delta_{s,t}$, etc.)

It should also be noted that σ_t is the special case of the more general *in situ* quantity $\sigma(s, t, p) = [\rho(s, t, p) - 1000]$ which includes the effect of pressure but is less often used than σ_t. Sigma-t is more often used because it allows a much better estimate of what the density difference between the two water types would be if they were at the same level and hence it is a better indicator of static stability (which will be discussed in Chapter 5). When one is considering the motion of a water parcel over a considerable depth range, it may be desirable to eliminate the effect of adiabatic heating or cooling by using the potential temperature θ rather than the *in situ* temperature T and calculating the potential density $\rho(s, \theta, 0)$ or $\sigma_\theta = [\rho(s, \theta, 0) - 1000]$.

For orientation and for comparison, some values for density for fresh water and for sea water will be given. For the open ocean, temperatures range from $-2°C$ to $+30°C$, salinity from 30 to 38 and sea pressure from 0 to about $10^5\,\mathrm{kPa}$ $(= 10^4\,\mathrm{dbar})$ corresponding to the maximum ocean depth of about 10 000 m. Near rivers or melting ice, the salinity may fall to 0 in the surface layer, while values of 42 or more occur in the Red Sea. Some sample values for density are given in Table 2.1. The density of fresh water at standard atmospheric pressure (i.e. zero hydrostatic pressure) has its maximum value of 999.95 kg m^{-3} at 3.98°C. Values in brackets in Table 2.1 correspond to conditions which do not occur in lakes or oceans.

TABLE 2.1   *Values of density* in situ *for fresh and sea water* $(kg\,m^{-3})$
(*International Equation of State* 1980)

Sea pressure	Approx. depth	Fresh water S = 0 temperature		Average sea water S = 35 temperature		Red Sea (winter) S = 40 temperature
10²kPa	m	0°	30°C	0°	30°C	18°C
0	0	999.8	995.7	1028.1	1021.7	1029.1
10	100	1000.4	996.1	1028.6	1022.2	1029.6
100	1000	1004.9	(1000.0)	1032.8	(1026.0)	1033.4
400	4000	(1019.3)	(1012.7)	1046.4	(1038.1)	1045.8
1000	10000	(1045.3)	(1035.9)	1071.0	(1060.4)	(1068.7)

Specific volume (α) is the reciprocal of density ($\alpha = 1/\rho$) and has the units of m^3 kg^{-1}. Two other density related quantities are used in oceanography. These are *specific volume anomaly* (δ) and *thermosteric anomaly* $[\Delta_{s,t}]$ which will be defined in the next section. The relations between the units for density and specific volume, etc., used here and those used in earlier texts are discussed in Appendices 2 and 3 for reference.

2.23   *Density and specific volume as functions of temperature, salinity and pressure*

For a one-component fluid, the density $\rho = \rho(T, p)$ where T is the absolute (thermodynamic) temperature and the specific volume $\alpha = \alpha(T, p)$, i.e. they are functions of temperature and pressure only. For a perfect gas, the relation (the "equation of state") has the simple form $\alpha = RT/p$ where R is the Gas Constant. For fresh water, the relationship is more complicated; for sea water, which is a multi-component fluid, the dissolved salts add further complication and $\alpha = \alpha(s, T, p)$. (The situation is similar in the atmosphere where water vapour plays an analogous role to salinity in sea water and changes the relatively simple equation of state for dry air to a more complicated form for moist air.) For sea water, it is usual to use the Celsius temperature T, rather than the absolute temperature T, and the relationship $\alpha = \alpha(s, t, p)$ can be presented either as tables entered with the observed values of S, T and p or as a polynomial in these parameters.

From the early measurements of the variation of density with temperature, salinity and pressure (Knudsen, 1901, for the variation with T and S at $p = 0$, and Ekman, 1908, for the effect of pressure) it was found that the most convenient way to express the results was in terms of the specific volume as follows:

$$\alpha(s,\, t,\, p) = \alpha(35,\, 0,\, p) + \delta_s + \delta_t + \delta_{s,t} + \delta_{s,p} + \delta_{t,p} + \delta_{s,t,p} \qquad (2.1)$$

or as

$$\alpha(s,\, t,\, p) - \alpha(35,\, 0,\, p) = \delta = \Delta_{s,t} + \delta_{s,p} + \delta_{t,p} + \delta_{s,t,p}.$$

In these expressions, $\alpha(s,\, t,\, p)$ is the specific volume of a sample of water of salinity S, temperature T and sea pressure p. Then $\alpha(35,\, 0,\, p)$ is the specific volume of an arbitrary standard sea water of $S = 35, T = 0°C$ and pressure p at the depth of the sample. This term expresses most of the effect of pressure on specific volume. The term δ, the *specific volume anomaly*, represents the sum of the six anomaly terms in equation (2.1). The quantity $\Delta_{s,t} = \delta_s + \delta_t + \delta_{s,t}$ accounts for most of the effect of salinity and temperature, disregarding pressure, and is called the *thermosteric anomaly*. The terms $\delta_{s,p}$ and $\delta_{t,p}$ account respectively for most of the combined effects of salinity and pressure and of temperature and pressure. The last anomaly term, $\delta_{s,t,p}$, is so small that it is always neglected with the present accuracy of determination of the parameters S, T and p. In water of depth less than about 1000 m, the thermosteric anomaly, $\Delta_{s,t}$, is the major component of δ and the pressure terms $\delta_{s,p}$ and $\delta_{t,p}$ may often be neglected. $\Delta_{s,t}$ has, in recent years, to some extent replaced σ_t as a parameter for describing density characteristics in the upper layer of the ocean because it can be used more directly than σ_t in first-order dynamic calculations.

From $\alpha(s,\, t,\, 0) = 1/\rho(s,\, t,\, 0) = 1/(1000 + \sigma_t)$ and $\alpha(s,\, t,\, 0) = \alpha(35,\, 0,\, 0) + \Delta_{s,t}$ and noting that $\alpha(35,\, 0,\, 0) = 0.972\,66 \times 10^{-3}\,\text{m}^3\,\text{kg}^{-1}$ it is easy to show that

$$\Delta_{s,t} = \left(\frac{1000}{1000 + \sigma_t} - 0.972\,66\right) 10^{-3}\,\text{m}^3\,\text{kg}^{-1}.$$

A few values are as follows:

$\sigma_t =$	23.00	24.00	25.00	26.00	27.00	28.00 kg m^{-3}
$\Delta_{s,t} =$	485.7	390.3	295.0	199.9	105.0	$10.3 \times 10^{-8}\,\text{m}^3\,\text{kg}^{-1}$.

Recently the equation of state for sea water has been redetermined and an International Equation of State, 1980 (IES 80), has been accepted for use after 1 January 1982. It is in the form of a polynomial in S, T and p, and it is necessary to use Practical Salinity (PSS 78) to make use of the increased precision of the IES 80 over the previous (Knudsen-Ekman) tables. References to sources of tables of $\alpha(35,\, 0,\, p)$ and the specific volume anomaly (δ) terms needed for using the older method to determine $\alpha(s,\, t,\, p)$, and to details of the IES 80, are given in Appendix 3. In addition, Appendix 3 includes tables indicating the sizes of the various anomaly terms for selected values of S, T and p, and values for the current accuracy of measurement of these parameters and the related accuracies of ρ and α to be expected.

Systematic differences between the older (Knudsen-Ekman) equation of state and the IES 80 are about 9×10^{-3} kg m^{-3} at $p = 0$, 33×10^{-3} kg m^3 at $p = 500$ bar ($= 5 \times 10^4$ kPa $\simeq 5000$ m depth) and 89×10^{-3} kg m^{-3} at $p = 1000$ bar ($\simeq 10000$ m depth), the new values being the higher. (Note that the IES 80 uses the bar for pressure although this is not an SI unit.) These values are equivalent to about 0.01, 0.03 and 0.09 kg m^{-3} respectively in σ_t.

3

The Basic Physical Laws used in Oceanography and Classifications of Forces and Motions in the Sea

3.1 Basic laws

The following basic laws of physics are taken as axiomatic in developing the study of the dynamics of the ocean:

(1) Conservation of mass.
(2) Conservation of energy.
(3) Newtons's First Law of Motion that if there is no resultant force acting on a body there will be no change of motion of the body.
(4) Newton's Second Law of Motion that the rate of change of motion of a body is directly proportional to the resultant force acting upon it and is in the direction of that force.
(5) Newton's Third Law of Motion that for any force acting upon a body there is an equal and opposite force acting on some other body.
(6) Conservation of angular momentum.
(7) Newton's Law of Gravitation.

Strictly speaking the first two are related but in dynamical oceanography we are not concerned with the Mass $\rightleftharpoons$ Energy conversion and therefore it is convenient to keep them separate. Conservation of mass is fundamental but in oceanography it is usually used in the form called the "equation of continuity" which actually expresses conservation of either mass/unit volume (density) or of volume. Laws 3, 4 and 5 are aspects of conservation of linear momentum.

The two types of energy whose conservation is important in oceanography are heat and mechanical. Conservation of heat, or the heat budget, is most important when discussing the distribution of temperature as a property of the ocean waters in descriptive oceanography and an account of this subject may be found in *Descriptive Physical Oceanography* (Pickard and Emery, 1982). Conservation of mechanical energy will be considered when treating waves, while the conversion of mechanical to heat energy will be taken for granted as a loss process for the former but will not be discussed in any detail in this text but

12

left for more advanced treatments. (This source of heat is negligible in the heat budget.)

Dynamical oceanography is concerned with the forces acting on the ocean waters and with the motions which ensue. In some cases the motions occur under a system of forces which are in balance so that no resultant force acts—this is the case covered by Newton's First Law of Motion. In other cases there is a resultant force and acceleration occurs, the relations between force and acceleration being determined by Newton's Second Law. In this text, except in the case of waves and tides, we will be concerned almost entirely with unaccelerated motion, i.e. with applications of the First Law, in interpreting the dynamical behaviour of the oceans.

We will not use angular momentum directly in this text but rather a related quantity called vorticity. It should be noted that both linear and angular momentum may not be conserved when we measure them relative to the earth because the latter is itself rotating in space and the effects of rotation must be taken into account in the development of the equations of motion of the ocean waters.

The Law of Gravitation, as such, is principally applied in discussing the dynamics of the astronomical tides of the ocean, although it is also important in determining the hydrostatic pressure distribution and in causing motion when density changes occur.

3.2 Classification of forces and motions

The important forces can be divided into two classes, *primary* which cause motion, and *secondary* which result from motion. The primary forces are (1) *gravitation*, both terrestrial, giving rise to pressure forces, and that due to the sun and moon, (2) *wind stress* which may be tangential (friction) and normal (pressure) to the sea surface, (3) *atmospheric pressure*, and (4) *seismic* (from sea bottom movements). Gravitation is a body force in that it acts on the total mass of the water, while the others are boundary forces, acting first at a water surface, although their effects may penetrate beyond the surface into the body of the water.

The secondary forces which come into being when water starts to move are (5) *Coriolis force*, an apparent force on a moving body when its motion is observed relative to the rotating earth, and (6) *friction* acting at the boundary of the fluid and tending to oppose its motion or acting within the fluid and tending to make the motion more uniform. Friction also tends to dissipate the mechanical energy of the motion, converting the kinetic energy of the fluid into heat energy. Again, Coriolis force is a body force whereas boundary friction is initially a surface acting force whose effects penetrate in diminishing degree into the body of the fluid.

A common classification of motion is as (1) *thermohaline* motion which results when the density of water changes in a limited region so that the differential action of gravity causes relative motion. The density changes will be due to changes of temperature and/or salinity, hence "thermohaline" motions. ("Thermosaline" might appear to be the more logical term but "thermohaline" is the more common one and is etymologically correct. It dates back to the earlier days of oceanography when the salinity was determined by a silver nitrate titration procedure which determined the total *halogen* content of the sea-water sample, and the salinity was then determined from this quantity.) Evaporation, cooling and freezing (which raises the salinity of the unfrozen water) all increase the density of sea water and may cause it to sink vertically, with subsequent horizontal movement at its own density level or along the bottom, (2) *wind-driven* motions such as the major ocean circulations of the upper layers, surface waves and upwelling, (3) *tidal* currents which are essentially horizontal and *internal waves* of tidal period, (4) *tsunami* or *seismic* sea waves resulting from movements of the ocean bottom during undersea earthquakes, (5) *turbulent* motions resulting from velocity shear (change of velocity with respect to one or more spatial coordinates), usually at the water boundaries, and (6) various motions such as *internal waves, gravitational–gyroscopic waves, Rossby* or *planetary waves*, etc.

It would also be possible to classify motions by their scale (from the smallest turbulent eddies of millimetre scale to the dimensions of the major ocean circulations), by speed, or by the method of determination, but the classification above in terms chiefly of the causal forces is the most used.

A few comments may be made on the various forces acting on the sea, before we proceed to discuss the relations between forces and motions. Terrestrial gravity gives rise to the property possessed by mass near the earth which we call weight, and there are two consequences. In the first place, weight gives rise to the phenomenon of hydrostatic pressure in a fluid. In the second place, because a fluid has weight there will be a component force down slope if a fluid surface is not horizontal and, as a fluid cannot withstand shear strain, it will tend to flow down slope (see Appendix 1).

The astronomical gravitational forces due to the moon and sun fluctuate periodically as the earth rotates and as these two bodies circulate around the earth, and give rise to the periodic motions which we call tidal currents and tides. (The gravitational forces due to the other bodies in the solar system are negligible compared with those of the earth's moon and of the sun.)

The wind, in blowing relative to the water surface, transfers momentum and energy to the upper layer of water giving rise both to the fluctuating motions of waves and to the steadier ocean currents. The generation and development of waves involves fluctuations of the normal stress (pressure) of short period of the order of seconds, and perhaps fluctuations of the tangential stress on the water surface, but the ocean currents express an integrated (averaged) effect of the stress.

Differences in atmospheric pressure can cause differences in water levels (called the inverted barometer effect) and hence currents. For instance, at the centre of a cyclonic storm the air pressure is low and the water level tends to rise; there is an inflow of water to raise the level by as much as 30 cm for a drop in atmospheric pressure of 3 kPa ($\equiv$ 30 mbar). Then as the storm moves, these patterns tend to follow and may cause inundations of low-lying coastal areas.

Sudden motions of the ocean bottom, such as occur in a dip-slip earthquake when there is a component of motion perpendicular to the bottom, cause corresponding movements of the sea surface above. If, for instance, a hump is formed on the sea surface the water immediately begins to flow outward from the hump and a train of waves develops and moves outward from the earthquake location. These are tsunamis or seismic sea waves.

The "Coriolis force" is an apparent force which acts perpendicular to the velocity vector of a moving body on the surface of the earth. It is the name given to a term which appears when the (vector) equation of motion is transformed from a fixed frame of reference (relative to the "fixed" stars) to a frame of reference fixed in the earth which is itself rotating about its axis. This force will be discussed in Chapter 6.

The secondary frictional forces are not physically of a different nature from the primary ones but are introduced as secondary forces because they do not arise until motion has been generated, and they then usually tend to oppose the motion rather than maintain it (as does wind stress). Frictional effects arise because of the molecular nature of the fluid and may be much enhanced if the flow is turbulent.

4

The Equation
of Continuity of Volume

4.1 The concept of continuity of volume

It was stated in the previous chapter that the law of conservation of mass is used in oceanography in the form of an equation of continuity of volume or of density. Before deriving this equation we will consider the physical significance of volume continuity.

For a stationary fluid the idea of continuity of volume is trivial, so let us consider a moving fluid which is assumed to be incompressible (i.e. the volume is not affected by pressure, see Appendix 1) and uniform in kind. (Sea water is not exactly incompressible but this assumption is a good approximation for many applications because the volume changes are small. For instance, a change of pressure corresponding to a change of depth of 1000 m would change the volume of a sample of average sea water by less than 0.5 %. There are few situations in ocean currents where depth (and therefore pressure) changes as large as this occur along the flow path.)

Suppose that we had a hand-basin with two taps, a waste pipe at the bottom and an overflow pipe near the top. With the taps full on (and a good flow of water from them) it is possible that the water level would rise in the basin. It may reach a final stage when the basin fills completely and water starts to drip over the edge—then we will have continuity of volume of water in the basin, which could be expressed as

$$\frac{\text{Hot}}{\text{inflow}} + \frac{\text{Cold}}{\text{inflow}} = \frac{\text{Waste}}{\text{outflow}} + \frac{\text{Overflow pipe}}{\text{outflow}} + \frac{\text{Drip out}}{\text{over edge}}$$

or as

$$+ \text{Hot} + \text{Cold} - \text{Waste} - \text{Overflow} - \text{Drip} = 0.$$

This is rather a domestic example. Consider now the following. An oceanographer is studying a long, narrow coastal inlet which has a river at the inland end. He notes that the upper layer is flowing to seaward and that the thickness of this layer is constant (within observational error of 10–20%) from river to sea. He also notes that the seaward flow is slow at the river end of the inlet but becomes substantially faster toward the seaward end. He wonders why, because this behaviour implies that for a section of the inlet between AB and A'B' (Fig. 4.1(a)), where the water speed increases from u_3 to u_4, there must be more volume of water flowing out across the section A'B'C'D' than in across ABCD (Fig. 4.1(b)). If we consider the horizontal flow only, there appears to be a lack of continuity of volume. However, the inlet is observed not to be emptying and to produce a balance there must be an upward flow (w) from the lower to the upper layer across CDD'C' (Fig. 4.1(b)) so that we have

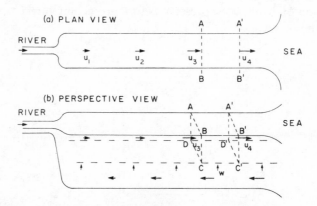

FIG. 4.1   *Continuity of volume for an inlet: (a) plan view of upper layer, outflow (u) increasing to seaward; (b) perspective view—outflow in the upper layer, inflow in the lower layer and upward flow (w) from lower to upper layers.*

$$u_3 \times \text{area ABCD} + w \times \text{area CDD'C'} = u_4 \times \text{area A'B'C'D'}$$

or $$u_3 \times \text{area ABCD} + w \times \text{area CDD'C'} - u_4 \times \text{area A'B'C'D'} = 0,$$

expressing continuity of volume for the upper layer. (We have tacitly assumed no rainfall on the water surface.) The water which leaves the lower layer must be replaced. Because there can be no flow through the bottom of the inlet, an inward horizontal flow from the sea must develop as shown. This is an example of "estuarine flow". Note that this is an idealization of the average behaviour in an inlet. Real inlets are often more complicated, e.g. the inflow might be laterally adjacent to the outflow, there might be more than one region of outflow and/or inflow in a cross-section, and wind-driven and tidal motions might be large enough to make observations of the average flow very difficult. However, the principle remains valid that there must be replacement of the deeper water which the river flow picks up and carries out of the inlet.

4.2 The derivation of the equation of continuity of volume

We will now consider the conservation of mass in order to derive a general equation for applying continuity of volume. In Fig. 4.2 is represented a rectangular volume fixed in space with sides of lengths δx, δy and δz in a moving fluid. Consider first the flow parallel to the x-axis. At the left face the velocity is u and the fluid density ρ, while at the right face the velocity is $u + \delta u$ and the density $\rho + \delta \rho$. These last two expressions may be written approximately as $u + (\partial u/\partial x)\delta x$ and $\rho + (\partial \rho/\partial x)\delta x$ where terms in $(\delta x)^2$, etc., have been neglected because they will vanish when we take the limit as $\delta x \to 0$. (A reader unfamiliar with the notation being used should read Appendix 1 up to the section on hydrostatic pressure before proceeding further in the text.) Then, in the x-direction

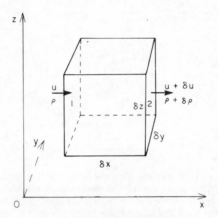

FIG. 4.2 *Continuity of volume—components of flow in the x-direction.*

the mass flow into the volume $= \rho\, u\, \delta y\, \delta z$ (mass/unit time)

and the mass flow out of the volume $= \left(\rho + \dfrac{\partial \rho}{\partial x}\delta x \right)\left(u + \dfrac{\partial u}{\partial x}\delta x \ \right)\delta y\, \delta z$

so that the net flow out of the volume in the x-direction is the difference

$$\left[u\frac{\partial \rho}{\partial x} + \rho\frac{\partial u}{\partial x} + \frac{\partial \rho}{\partial x}\frac{\partial u}{\partial x}\delta x \right]\delta x\, \delta y\, \delta z = \left[\frac{\partial(\rho u)}{\partial x} + 0\,(\delta x) \right]\delta x\, \delta y\, \delta z.$$

Here we have combined the first two terms using the rule for derivatives of products (Appendix 1) and the expression "$0\,(\delta x)$" has been written for the last term in the brackets of the previous equation and indicates that the term is "of the order of" (i.e. size) δx times some finite number. By taking δx sufficiently small (mathematically we take the limit as $\delta x \to 0$) this term must become negligible compared with the other term in the square bracket provided that the multiplier $(\partial \rho / \partial x)(\partial u / \partial x)$ is finite as we expect for a physical system. (The reader may be concerned about treating a medium actually made up of molecules as continuous and about whether taking the limit as $\delta x \to 0$ is realistic. In fact, there are no practical problems in this limiting process. For further discussion, which is beyond the scope of this book, the reader may consult a text on continuum fluid mechanics, such as that by Batchelor (1967) in the Further Reading list, Bibliography, Section B.2.1.)

Then, taking into account the mass flow in all three component directions and neglecting the terms which vanish in the limit as δx, δy, $\delta z \to 0$, the

$$\text{total flow out} = \left[\frac{\partial(\rho u)}{\partial x} + \frac{\partial(\rho v)}{\partial y} + \frac{\partial(\rho w)}{\partial z} \right] \delta x \, \delta y \, \delta z$$

where v and w are the velocity components in the y and z directions respectively.

The mass remaining in the small volume $\delta x \, \delta y \, \delta z$ changes by $(\partial \rho / \partial t) \, \delta x \, \delta y \, \delta z$ per unit time. If mass is to be conserved, the sum of the effects must be zero, i.e.

$$\frac{\partial \rho}{\partial t} + \frac{\partial(\rho u)}{\partial x} + \frac{\partial(\rho v)}{\partial y} + \frac{\partial(\rho w)}{\partial z} = 0. \tag{4.1}$$

Now the rate of change of density with the moving fluid (the individual derivative, see Appendix 1) is

$$\frac{d\rho}{dt} = \frac{\partial \rho}{\partial t} + u \frac{\partial \rho}{\partial x} + v \frac{\partial \rho}{\partial y} + w \frac{\partial \rho}{\partial z}. \tag{4.2}$$

Combining equations (4.1) and (4.2) we have

$$\frac{1}{\rho} \frac{d\rho}{dt} + \left[\frac{\partial u}{\partial x} + \frac{\partial v}{\partial y} + \frac{\partial w}{\partial z} \right] = 0. \tag{4.3}$$

This is called the equation of continuity (of volume). The first term is the fractional rate of change (change/unit time) of density for a small piece or parcel of the moving fluid (a "fluid element"); the second term is the fractional rate of change of volume for the element as we shall show in a moment. The equation expresses conservation (continuity) of volume, i.e. the relation between volume and density changes. Notice that we do *not* assume that the fluid has the same density everywhere (a homogeneous fluid), which is important in application to the ocean whose water is not homogeneous. The effects of pressure and heat change are included in equations (4.1) and (4.3) since they do not affect the mass of a fluid element appreciably. Salt-exchange effects are not included; if we assume that an element exchanges an equal mass of salt and water, equation (4.3) gives the correct relation between density and volume changes. However, in the more likely case that about the same number of molecules of water are replaced by salt ions the mass increases and the volume will not decrease as much as equation (4.3) predicts. In the ocean the effect is small enough to ignore when considering mass or volume conservation. In fact all of these effects are quite small; here they may be ignored and the volume of a fluid element assumed to be constant, i.e. the fluid may be treated as effectively incompressible.

If a fluid is incompressible, as may be taken to be the case for sea water in most circumstances, then $(1/\rho)(\mathrm{d}\rho/\mathrm{d}t) = 0$, as shown in Appendix 1, and the equation of continuity becomes

$$\frac{\partial u}{\partial x} + \frac{\partial v}{\partial y} + \frac{\partial w}{\partial z} = 0. \tag{4.4}$$

We note that in deriving the equation of continuity, instead of considering the rate of change of mass in a volume $(\delta x\ \delta y\ \delta z)$ fixed in space we could consider the rate of change of volume of a fluid element. If we consider the volume $\delta V = \delta x\ \delta y\ \delta z$ of Fig. 4.2 to be moving with the fluid, then in time δt (eventually we take the limit as $\delta t \to 0$, so we neglect terms proportional to $(\delta t)^2$ and higher powers) side 1 moves $(u\ \delta t)$ while side 2 moves $[u + (\partial u/\partial x)\delta x]\delta t$. The change in volume is $(\partial u/\partial x)\delta x\ \delta y\ \delta z\ \delta t$.

Consideration of movements in the y and z directions as well gives the rate of change of the volume $\delta V = \delta x\ \delta y\ \delta z$ as $\mathrm{d}(\delta V)/\mathrm{d}t = \delta V(\partial u/\partial x + \partial v/\partial y + \partial w/\partial z)$. This rate of change of volume must be balanced by a corresponding rate of change of density, because the mass, δm, of the fluid element must be constant (assuming salt exchange effects are negligible or balance) although its shape, density and volume may change. The signs are opposite as an increase in volume decreases ρ and vice versa. Now $\delta m = \rho\ \delta V$ and with the mass constant

$$\frac{\mathrm{d}}{\mathrm{d}t}(\delta m) = \frac{\mathrm{d}\rho}{\mathrm{d}t}(\delta V) + \rho\frac{\mathrm{d}}{\mathrm{d}t}(\delta V) = 0$$

and

$$\frac{\mathrm{d}\rho}{\mathrm{d}t} = -\frac{\rho}{\delta V}\frac{\mathrm{d}(\delta V)}{\mathrm{d}t} = -\rho\left[\frac{\partial u}{\partial x} + \frac{\partial v}{\partial y} + \frac{\partial w}{\partial z}\right] \quad \text{as before.}$$

It may be noted that terms like $\partial u/\partial x$ or $\delta u/\delta x$ may be positive or negative. In Fig. 4.3, case (a), fluid is flowing from A where its velocity component is u to B where it is $(u + \delta u)$. Then along the flow direction, δu is +, δx is + and therefore $\delta u/\delta x$ is +. In case (b), the velocity decreases from C to D and δu is −, δx is + and therefore $\delta u/\delta x$ is −. In case (c), the flow is to the left from F to E, δu is − (since it makes the velocity at E more negative than at F), δx is −, and therefore $\delta u/\delta x$ is +.

4.3   An application of the equation of continuity

As an example of the application of the equation of continuity, we consider the estimation of vertical velocities in the open ocean. These are difficult to

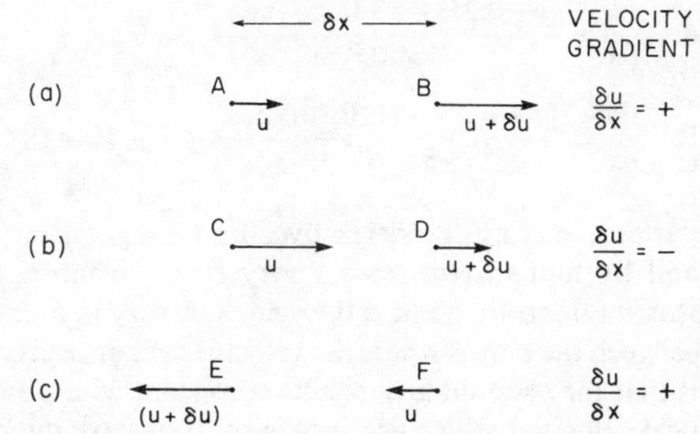

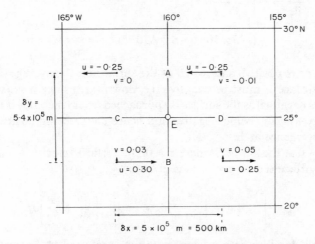

FIG. 4.3 *Signs of the velocity gradient.*

measure directly because they are very small in magnitude, but some information about them may be deduced with the aid of the equation of continuity from a knowledge of horizontal velocities which are larger and more easily measured. From continuity we have

$$\frac{\partial w}{\partial z} = -\left[\frac{\partial u}{\partial x} + \frac{\partial v}{\partial y}\right]$$

and from data about u and v (horizontal components) we may learn something about w (vertical), near a point in the sea such as E in Fig. 4.4, where some surface current data are presented from Pilot Charts for a tropic region.

FIG. 4.4 *Example of horizontal flows for calculation of vertical flow from continuity of volume (speeds in $m\,s^{-1}$).*

At point A: $\dfrac{\partial u}{\partial x} \simeq \dfrac{[(-0.25)-(-0.25)]\,\text{m s}^{-1}}{5 \times 10^5 \text{ m}} = 0$

and at point B: $\dfrac{\partial u}{\partial x} \simeq \dfrac{[(+0.25)-(+0.30)]\,\text{m s}^{-1}}{5 \times 10^5 \text{ m}} = -10 \times 10^{-8}\,\text{s}^{-1}.$

Here we are approximating the derivatives by taking velocity differences over a finite (and by human size scale a very large) distance. Such "finite difference" approximations are good if the velocities vary in a smooth, nearly linear manner between the points where the velocities are measured. The test of such methods is that they should give results consistent with observations; in the case of vertical velocities which are very hard to observe directly, because they are so small, such tests must be indirect. We should also point out that in this particular example the available velocity information only consists of averages over approximately 5° squares at best, so finer scale estimates are not possible.

Therefore at the centre point E of the area, taking the mean of the values at A and B,

$$\frac{\partial u}{\partial x} = -5 \times 10^{-8}\,\text{s}^{-1}.$$

Similarly, calculating $\partial v/\partial y$ at C and D and taking the mean for point E we get

$$\frac{\partial v}{\partial y} = -8.3 \times 10^{-8}\,\text{s}^{-1}.$$

Then from $\dfrac{\partial w}{\partial z} = -\left[\dfrac{\partial u}{\partial x} + \dfrac{\partial v}{\partial y} \right]$

we have $\dfrac{\partial w}{\partial z} = -(-5 \times 10^{-8} - 8.3 \times 10^{-8})\text{s}^{-1} = +13.3 \times 10^{-8}\,\text{s}^{-1}.$

Now $\partial w/\partial z$ is positive and $w = 0$ at the surface on the average. Therefore below the surface w must be negative (i.e. downward) since it must increase (become less negative) as the surface is approached where the value is to be zero. Thus $(\partial u/\partial x + \partial v/\partial y)$ being negative and hence $\partial w/\partial z$ being positive implies surface convergence at E.

Since $w = 0$ at the surface, where $z = 0$, the vertical velocity, w_h, at depth h below the surface (i.e. at $z = -h$) is given by

$$w_h = \int_0^{-h} \text{d}w = \int_0^{-h} \frac{\partial w}{\partial z}\text{d}z = -\int_0^{-h} \left[\frac{\partial u}{\partial x} + \frac{\partial v}{\partial y} \right]\text{d}z.$$

(A reader not familiar with integral notation is referred to Appendix 1.)

In the numerical example above, if the convergence were constant from the sea surface to the bottom of the homogeneous layer, taken to be at 50 m depth, the vertical velocity at 50 m depth would be

$$w_{50} = \int_0^{-50} (13.3 \times 10^{-8})dz = -6.7 \times 10^{-6} \text{ m s}^{-1} = 0.58 \text{ m day}^{-1} \text{ down.}$$

The fact that the layer is homogeneous and therefore presumably subject to considerable mixing (otherwise it would probably be stratified) provides some evidence that the convergence is uniform. With sufficient stirring, the horizontal velocities u and v should be independent of depth in the homogeneous layer. Then $[\partial u/\partial x + \partial v/\partial y]$ should also be independent of depth.

As the vertical velocities at shallower depths are correspondingly less, the time for a particle of water to sink from near the surface to 50 m depth would be quite large. As $w = 13.3 \times 10^{-8}z$ and the time (δt) to sink δz is $\delta t = \delta z/w$,

$$t_1^2 = \int_{z_1}^{z_2} \frac{1}{w}dz = 7.5 \times 10^6 \int_{z_1}^{z_2} \frac{1}{z}dz = 7.5 \times 10^6 \ln(z_2/z_1) \text{ seconds}$$
$$= 87 \ln(z_2/z_1) \text{ days}$$

where ln = natural logarithm.

For example, the time to sink from 1 m depth to 50 m depth would be $(87 \ln 50) = 340$ days or almost one year!

It should be noted that this numerical example has been presented principally to illustrate the *use* of the equation of continuity and is for an open ocean situation. For regions of active upwelling, usually found along the east sides of the oceans, the vertical velocities may be greater—recent measurements indicate values of the order of 10^{-4} m s$^{-1} \simeq 10$ m day^{-1} off the west coast of North America. Nevertheless these speeds are much less than the typical horizontal ones.

Continuity also shows why $w \ll u$ or v. The ocean is very thin, the depth to width ratio being similar to that of a sheet of very thin paper! For uniform convergence, the vertical velocity component at $z = -H$ is $w_H \simeq HU/L$ where U is the change of horizontal velocity component over a distance L. Since $(H/L) = 0(10^{-3})$ or less for the whole ocean, then $w_H = 0(10^{-3}U)$ or less, i.e. of the order of one one-thousandth of the horizontal velocities or less. (This rough calculation is an example of "scaling", used to estimate the approximate magnitude of a quantity or of a term in an equation to judge its possible significance or size relative to other quantities.)

Finally, the effects neglected in deriving the equation of continuity (4.4) are very small compared with $(\partial u/\partial x + \partial v/\partial y)$. Thus the assumption of constant volume allows us to estimate $\partial w/\partial z$ and hence w; in fact, the errors in the estimation are much larger than the neglected effects.

5

Stability and Double Diffusion

5.1 Static stability

Here we consider whether or not the variation of density with depth in the ocean is likely to cause the water to move vertically. If there is light fluid on top of heavy fluid then there will be no tendency for motion to occur. However, if there is heavy fluid above light fluid there will be a tendency for the heavy fluid to sink and the light to rise—the density distribution is unstable. Thus we must examine the vertical density gradient to determine whether the fluid is stable, i.e. resists vertical motion, is neutral, i.e. offers no resistance to vertical motion, or is unstable, i.e. tends to move vertically of its own accord. If $\partial \rho / \partial z < 0$ (density increases with depth) we might expect the fluid to be statically stable so that if no motion is occurring the density distribution will not cause motion to occur. If $\partial \rho / \partial z > 0$, we expect the fluid to be unstable.

When considering the density distribution in relation to stability we cannot ignore compressibility, i.e. the variation of density with pressure, which means with depth. In the case of neutral stability, if a fluid parcel is moved up or down adiabatically (that is, with no heat exchange with its surroundings) and without salt exchange with its surroundings and then brought to a stop it will not tend to move further because wherever it is moved to it will have the same density as the surrounding fluid. Density must increase with depth because a parcel which is moved down will be compressed but must then have the same density as its surroundings. Thus in the neutral stability case, $\partial \rho / \partial z < 0$ and the fluid appears to be quite stable if the pressure effect is overlooked. At the same time, compression causes heating which decreases density, although not nearly enough to overcome the increase of density due to increased pressure and so, in the neutral stability case, temperature increases with increased depth, assuming that salinity effects can be ignored. If we neglect pressure effects we might think that the fluid is slightly unstable because we have colder fluid above warmer fluid.

To take the compressibility into account one might try to consider the potential density or its anomaly, σ_θ, defined in Chapter 2. It is the density of the fluid when taken adiabatically to a reference pressure with the adiabatic temperature change taken into account. This procedure can be used in the

24

atmosphere (provided that no condensation or evaporation occurs). In fact, an apparent or virtual potential temperature is used; this is the potential temperature which dry air would have if it had the same potential density as the moist air. Unfortunately, because of the complicated and non-linear equation of state for sea water, using potential density to determine the static stability does not always work in the ocean. For example, North Atlantic Deep Water has a slightly larger potential density than Antarctic Bottom Water. However, the former is found *above*, not below the latter. The temperature and salinity differences between these two water masses are sufficient that the variation of compressibility with these parameters leads to the *in situ* density (at the same depth) of the Antarctic Water being slightly greater than that of the North Atlantic Water, so that the Antarctic Water flows under the North Atlantic Water.

The apparent instability in this case is caused by the fact that the reference pressure for σ_θ is taken at the surface ($p = 0$). If a reference pressure close to the *in situ* pressure is used then zero vertical variation of this potential density will indicate neutral stability. However, to consider the stability over the whole water column no single reference pressure is satisfactory in all cases and it is necessary to calculate a local value of stability as a function of depth as described below.

5.11   *Criterion for static stability (E)*

Suppose that the density of a stationary water mass changes with depth in some arbitrary manner and that at level 1 (Fig. 5.1), depth $= -z$, pressure $= p$, the *in situ* water properties are (ρ, S, T). Then a parcel of water is moved a short distance vertically from level 1 to level 2 without exchanging heat or salt with its surroundings. At level 2, the depth $= -(z + \delta z)$ and pressure $= (p + \delta p)$, and

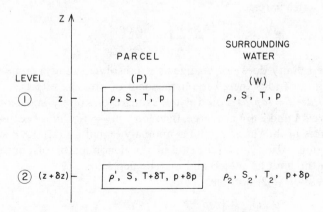

FIG. 5.1   *Water properties for calculations of stability.*

the surrounding water properties are (ρ_2, S_2, T_2). The water properties of the parcel at level 2 will be $(\rho', S, T + \delta T)$ and its pressure $= p + \delta p$. Here δT is the adiabatic change of temperature due to change of pressure, i.e. $\delta T = (dT/dp)_{ad}$ δp (where the subscript 'ad' = adiabatic). As $\delta p = -\rho g \, \delta z$ (refer to Appendix 1), $\delta T = -(dT/dp)_{ad} \rho g \, \delta z = -\Gamma \delta z$ where Γ stands for the adiabatic temperature gradient. It is the change of temperature with depth caused by pressure change and is positive, i.e. compression causes the temperature to increase. At level 2 the restoring force on the parcel of volume δV_2 will be F = (buoyant upthrust − weight). By Archimedes' Principle, buoyant upthrust = weight of surrounding fluid displaced. Hence

$$F = \delta V_2 \rho_2 g - \delta V_2 \rho' g$$
$$= \delta V_2 g(\rho_2 - \rho') \tag{5.1}$$

and its acceleration if released will be

$$a_z = \frac{F}{M} = \frac{\delta V_2 g(\rho_2 - \rho')}{\delta V_2 \rho'}$$

$$= \frac{g\left[\rho + \left(\dfrac{\partial \rho}{\partial z}\delta z\right)_W - \rho - \left(\dfrac{\partial \rho}{\partial z}\delta z\right)_P\right]}{\rho\left[1 + \left(\dfrac{1}{\rho}\dfrac{\partial \rho}{\partial z}\delta z\right)_P\right]} \tag{5.2}$$

where the subscript W refers to the surrounding water and the subscript P to the parcel.

In equation (5.2) the change of density of the surrounding water

$$\left[\frac{\partial \rho}{\partial z}\delta z\right]_W = \left[\frac{\partial \rho}{\partial S}\frac{\partial S}{\partial z} + \frac{\partial \rho}{\partial T}\frac{\partial T}{\partial z} + \frac{\partial \rho}{\partial p}\frac{\partial p}{\partial z}\right]_W \delta z$$

and of the water parcel

$$\left[\frac{\partial \rho}{\partial z}\delta z\right]_P = \left[-\frac{\partial \rho}{\partial T}\Gamma + \frac{\partial \rho}{\partial p}\frac{\partial p}{\partial z}\right]_P \delta z$$

because the salinity does not change as it (absolute salinity) is measured in grams/kilogram and is therefore independent of pressure effects.

Now $(dp/dz)_W = (dp/dz)_P$, and if the changes of salinity and temperature between levels 1 and 2 are not large, then $(\partial \rho/\partial p)_W = (\partial \rho/\partial p)_P$ because the $\delta_{s,p}$ and $\delta_{t,p}$ terms in the specific volume anomaly (equation (2.1)) are small and slowly varying. Also, $(1/\rho)(\partial \rho/\partial z)\delta z$ in the denominator of equation (5.2) vanishes in the limit as $\delta z \to 0$ and may be neglected.

Then equation (5.2) becomes

$$\frac{a_z}{g} = \frac{1}{\rho}\left[\frac{\partial \rho}{\partial S}\frac{\partial S}{\partial z} + \frac{\partial \rho}{\partial T}\left(\frac{\partial T}{\partial z} + \Gamma\right)\right]\delta z \tag{5.3}$$

which is the ratio of the restoring acceleration of the displaced parcel to the acceleration due to gravity. Hesselberg defined the *stability E* of the water column as

$$E = \left(-\frac{a_z}{g} \right) \text{ for } \delta z = \text{unit length,}$$

i.e.
$$E = -\frac{1}{\rho} \left[\frac{\partial \rho}{\partial S} \frac{\partial S}{\partial z} + \frac{\partial \rho}{\partial T} \left(\frac{\partial T}{\partial z} + \Gamma \right) \right] \text{m}^{-1}. \tag{5.4}$$

If $E > 0$, i.e. positive, the water is stable and a parcel displaced a short distance vertically will tend to return to its original position. Because it has inertia it will tend to overshoot its original position and then to oscillate about it, hence the stability of the water may be related to the occurrence of internal waves (Chapter 12). If $E = 0$, the water is neutrally stable and a displaced parcel will tend to remain in its displaced position. If $E < 0$, i.e. negative, the water will be unstable and a parcel which is displaced will tend to continue its displacement, i.e. overturn of the water should occur.

5.12 Numerical values for stability

In the open ocean, values of E in the upper 1000 m are of the order of $100 \times 10^{-8} \text{ m}^{-1}$ to $1000 \times 10^{-8} \text{ m}^{-1}$, the largest values generally occurring in the upper few hundred metres. Below 1000 m depth, values decrease to less than $100 \times 10^{-8} \text{ m}^{-1}$ and in deep trenches values close to $1 \times 10^{-8} \text{ m}^{-1}$ are found. In these latter cases, $\partial S/\partial z$ is generally very small so that its effect on stability is negligible. Then as $E \to 0$ this means that $\partial T/\partial z \to -\Gamma$, i.e. the temperature change with depth *in situ* is close to the adiabatic rate due to change of pressure. The adiabatic rate increases from about 0.14°C/1000 m at 5000 m to 0.19°C/1000 m at 9000 m depth, the temperature changes being positive for increase of depth, i.e. the *in situ* temperature increases with depth in deep trenches.

Note that in equation (5.4), $\partial \rho/\partial S$ and $\partial \rho/\partial T$ are taken holding the other variables fixed (T, p and S, p respectively) at the local *in situ* values. This formula is not computationally very convenient because tables for density are not commonly available—it is the specific volume which is normally tabulated. To use such tables we use $\alpha = 1/\rho$ and hence $(1/\alpha)(\partial \alpha/\partial S) = -(1/\rho)(\partial \rho/\partial S)$ and $(1/\alpha)(\partial \alpha/\partial T) = -(1/\rho)(\partial \rho/\partial T)$. Making use of the expansion of α of equation (2.1) (omitting the $\delta(s, t, p)$ term which is negligible) equation (5.4) becomes

$$E = \frac{1}{\alpha} \left[\frac{\partial \Delta_{s,t}}{\partial S} \frac{\partial S}{\partial z} + \frac{\partial \Delta_{s,t}}{\partial T} \frac{\partial T}{\partial z} + \frac{\partial \delta_{s,p}}{\partial S} \frac{\partial S}{\partial z} + \frac{\partial \delta_{t,p}}{\partial T} \frac{\partial T}{\partial z} + \Gamma \left(\frac{\partial \Delta_{s,t}}{\partial T} + \frac{\partial \delta_{t,p}}{\partial T} \right) \right]. \tag{5.5}$$

The first two terms usually dominate and may be recognized as an expansion of $(\partial \Delta_{s,t}/\partial z)$. The term involving Γ is generally quite small and may be ignored except in deep water where E is small. If $E = 0$, the neutral stability case, omitting the Γ term would give an apparent E of about $-2 \times 10^{-8}\,\mathrm{m}^{-1}$ near the surface and about $-4 \times 10^{-8}\,\mathrm{m}^{-1}$ at great depth, so the water would appear slightly unstable if the adiabatic temperature change with pressure (i.e. depth) were neglected as noted before. The importance of the other terms can be estimated by comparing them with the first two. The first and third terms have $(1/\alpha)(\partial S/\partial z)$ as a common factor so we need only compare coefficients of this common factor. $(\partial \delta_{s,p}/\partial S)$ is of opposite sign to $(\partial \Delta_{s,t}/\partial S)$; its magnitude is much smaller near the surface but increases to about 10% of $(\partial \Delta_{s,t}/\partial S)$ at 5000 m depth and about 15% at 10 000 m depth. $(\partial \delta_{t,p}/\partial T)$ has the same sign as $(\partial \Delta_{s,t}/\partial T)$; it also is relatively small near the surface but becomes comparable at depths greater than about 2000 m and may dominate at great depths.

It is not easy to give a general rule about the relative importance of the temperature and salinity terms. As a first approximation, the first two terms of equation (5.5) may be used

$$E = \frac{1}{\alpha}\left[\frac{\partial \Delta_{s,t}}{\partial S}\frac{\partial S}{\partial z} + \frac{\partial \Delta_{s,t}}{\partial T}\frac{\partial T}{\partial z} \right] = \frac{1}{\alpha}\frac{\partial \Delta_{s,t}}{\partial z}. \tag{5.6}$$

If the calculated values of E are less than $50 \times 10^{-8}\,\mathrm{m}^{-1}$, then the other terms should be included.

The thermosteric anomaly $\Delta_{s,t}$ is normally calculated from σ_t since they are directly related, as shown in Chapter 2, and σ_t is usually calculated and tabulated along with S and T values during the first stage of data processing. Thus it is convenient to have an approximate formula for E in terms of σ_t. Suppose that we expand the *in situ* density in a manner similar to that used for α (equation (2.1) as

$$\rho = 1000 + \sigma_t + \varepsilon_{s,p} + \varepsilon_{t,p} \tag{5.7}$$

where a term of the form $\varepsilon_{s,t,p}$ has been omitted because it will be negligible. Substituting this expansion into equation (5.4) and using

$$\frac{\partial \sigma_t}{\partial S}\frac{\partial S}{\partial z} + \frac{\partial \sigma_t}{\partial T}\frac{\partial T}{\partial z} = \frac{\partial \sigma_t}{\partial z}$$

gives
$$E = -\frac{1}{\rho}\left[\frac{\partial \sigma_t}{\partial z} + \frac{\partial \varepsilon_{s,p}}{\partial S}\frac{\partial S}{\partial z} + \frac{\partial \varepsilon_{t,p}}{\partial T}\frac{\partial T}{\partial z} + \frac{\partial \rho}{\partial T}\Gamma \right]. \tag{5.8}$$

The equivalent approximation to equation (5.6) is

$$E = -\frac{1}{\rho}\frac{\partial \sigma_t}{\partial z}. \tag{5.9}$$

From equations (5.6) and (5.9) we see that a first approximation to stability is that $\Delta_{s,t}$ shall decrease with depth or that σ_t shall increase with depth. Thus one can get an estimate of the sign of E just by looking at the tabulated values of these quantities. This is one of the reasons why σ_t (or $\Delta_{s,t}$) is used rather than *in situ* values. (Another reason is that flow along constant σ_t surfaces is easy since it is not restricted by static stability when equation (5.9) is a good approximation.) If one included the Γ term of equation (5.8) in equation (5.9) then it would essentially be equivalent to $E = -(1/\rho)(\partial\sigma_\theta/\partial z)$. However, as we go a long way from the reference pressure ($p = 0$) the terms, other than the Γ term, not in the approximate equations (5.6) and (5.9) become more important. Neglect of them leads to the apparent instability between the North Atlantic Deep Water and the Antarctic Bottom Water mentioned earlier.

Much of the effect of the pressure on the density cancelled out in deriving equation (5.4). Note also that the part of the pressure effect which cancelled is quite large. Suppose that we had just considered the gradient of *in situ* density. If the stability of the water were neutral, the *in situ* gradient must be the same as that for the water parcel. Now

$$\frac{1}{\rho}\left(\frac{\partial\rho}{\partial z}\right)_P = \frac{1}{\rho}\left(\frac{\partial\rho}{\partial p}\right)_{ad}\frac{\partial p}{\partial z} = -g\left(\frac{\partial\rho}{\partial p}\right)_{ad},$$

but

$$\left(\frac{\partial\rho}{\partial p}\right)_{ad} = \frac{1}{C^2} \text{ where } C \text{ is the speed of sound,}$$

so

$$-\frac{1}{\rho}\left(\frac{\partial\rho}{\partial z}\right)_w = \frac{g}{C^2} \simeq 400 \times 10^{-8}\,\text{m}^{-1}$$

and as stated earlier using the *in situ* density gives a false impression of quite stable conditions when the stability is actually neutral! If one wishes to use *in situ* density, $\rho(s, t, p)$, then to correct for compressibility the stability is given by

$$E = -\frac{1}{\rho}\frac{\partial\rho}{\partial z} - \frac{g}{C^2}. \tag{5.10}$$

To use this formulation for E one would use the equation of state (IES 80) directly to determine ρ and a suitable equation for C which is also a function of S, T and p.

Although the water should be unstable and expected to turn over whenever E is negative, in practice it is not uncommon to find values of $E = -25$ to -50 $\times 10^{-8}\,\text{m}^{-1}$ in the upper 50 m of the sea with indications that the stratification is stable. As already shown, the neglect of the adiabatic temperature gradient and the other terms which are in equations (5.5) and (5.8) but not in (5.6) and (5.9) cannot account for such observations. It may be that some of these cases are in fact associated with weak convection but the observations are not detailed enough to detect it. Such apparent unstable situations may also be due to observational errors. In practice, E is calculated using finite differences with observations from discrete levels. The error in a σ_t observation may easily be 5×10^{-3} (see Appendix 3) and the error, $\Delta\sigma_t$, in the difference between two levels could easily be 10^{-2}. With a depth difference Δz of 20 m, the error in $E \simeq \pm (1/\rho)(\Delta\sigma_t/\Delta z)$ may be $\Delta E \simeq \pm 50 \times 10^{-8} \, \mathrm{m}^{-1}$. At greater depths where Δz is larger (because the difference between observation levels is usually greater) the errors will be smaller, e.g. for $\Delta z = 500$ m and an error in $\Delta\sigma_t$ of 10^{-2}, the error in E is only $2 \times 10^{-8} \, \mathrm{m}^{-1}$.

Tables of values for $\partial\rho/\partial S$, $\partial\rho/\partial T$ and Γ (as $\partial\theta/\partial z$) for the calculation of E using equation (5.4) are given in Neumann and Pierson (1966) or one may use tables of $\Delta_{s,t}$, $\delta_{s,p}$, $\delta_{t,p}$ in equation (5.5).

5.13   Buoyancy frequency (N)

The *Brunt–Väisälä* (or *buoyancy*) *frequency* N is given by

$$N^2 = (gE) = g\left[-\frac{1}{\rho}\frac{\partial\rho(s,t,p)}{\partial z} - \frac{g}{C^2} \right] \simeq g\left[-\frac{1}{\rho}\frac{\partial\sigma_t}{\partial z} \right] (\text{radians s}^{-1})^2.$$

$$(5.11)$$

The frequency in cycles s^{-1} (hertz) is $N/2\pi = (gE)^{1/2}/2\pi$. It can be shown that this is the maximum frequency of internal waves in water of stability E. High values of N are usually found in the main pycnocline zone, i.e. where the vertical density gradient is greatest. This is usually in the thermocline in oceanic waters (where density variations are determined chiefly by temperature variations) or in the halocline in coastal waters (where density variations may be determined chiefly by salinity variations).

5.2   Double diffusion

Even though the water column may be statically stable at a particular time, instability may develop because sea water is a multi-component fluid and the rates at which heat and salt diffuse molecularly are different. A result is that if two water masses of the same density but different combinations of temperature and salinity are in contact, one above the other, the differential ("double") diffusion of these two properties may give rise to density changes which render the layers unstable. This is an active area of research and reviews of the subject may be found in Turner (1973, 1981). The details are beyond the scope of the present book but the general ideas are interesting and double diffusion may play a significant role in small-scale mixing in the oceans and in the formation of "fine" structure, the small scale (one to a few metres) vertical variations in temperature and salinity which have been found in the oceans as observations have improved with the use of continuously recording STD or CTD instruments (Salinity or Conductivity, Temperature, Depth).

We consider the stability of cases with positive static stability but with no motion initially, because if there is motion, particularly turbulent motion generated by velocity shear or strong static instability, turbulent diffusion will dominate and probably prevent double-diffusion effects from becoming important. However, it seems that the ocean is sufficiently statically stable in some parts that shear generated turbulence is suppressed and double-diffusive effects may be important.

Suppose that there is a layer of warmer, saltier water above cooler, fresher water, such that the upper layer is of the same density as or is less dense than the lower layer. Then the saltier water at the interface will lose heat to the cooler water below faster than it will lose salt because the rate of molecular diffusion of heat is about 100 times that for salt. If the density difference between the layers is small, the saltier water above may become heavier than the cooler, fresher layer below and sink downward into this layer. Likewise the cold, fresh water below the interface gains heat faster than salt and may become light enough to rise into the upper layer. The situation is referred to as one of "double-diffusive instability". The falling and rising motion occurs (in laboratory experiments later verified by field observations, Williams, 1975) in the form of thin columns and the phenomenon is called "salt fingering". There is evidence for its occurrence in the ocean at the lower surface of the outflow of warm, saline Mediterranean water from the Strait of Gibraltar into the cooler, less saline Atlantic water.

If a layer of colder, fresher water is above a layer of warmer, saltier water, the water just above the interface becomes lighter than that above it and tends to rise while water below gets heavier and tends to sink. This phenomenon is called "layering" and may lead to fairly homogeneous layers separated by thinner regions of high gradients of temperature and salinity. There is evidence for its occurrence in the Arctic Ocean among other locations.

Both of these processes could lead to the vertical transports of heat and salt being greater than the molecular diffusion rates, and to greater mixing than would occur if these processes were not possible. Of course, once the motion begins it may become dynamically unstable and break down into smaller scale turbulent motions and become very complicated. Dynamic instability is discussed briefly in the next section and also in Chapter 7.

The final possibility of a warmer, fresher layer above a cooler, saltier layer does not allow a double-diffusive instability. The fresher water cools so that it does not tend to rise but it cannot get colder than the saltier water below and therefore does not tend to sink. Similarly, the saltier water does not tend to move up or down. For double diffusion to occur, the gradients of temperature and salinity across the interface must have the same sign; then, since they affect density oppositely, double diffusion may occur.

5.3 Dynamic stability

Even if the water is statically stable and double diffusion is not permitted by the temperature and salinity distributions, if motion is initiated it may be dynamically unstable and it may break down into smaller-sized irregular turbulent motions. This possibility will be discussed further after we have examined the equations of motion.

Turbulent flows are familiar to everyone although they may not normally be labelled as such. Examples are the flow in most rivers, the gusty wind and the flow of water out of a tap, among many others. All these flows are very irregular both as a function of time at a fixed point and from point to point at a given time. The strong mixing caused by turbulent flow is often used, e.g. in stirring milk and sugar into coffee. After the stirring is stopped, the flow will gradually become more regular providing an example of non-turbulent flow, a type of flow which is less familiar in everyday experience.

6

The Equation of Motion in Oceanography

6.1 The form of the equation of motion

Here we consider how Newton's Second Law of Motion ($\mathbf{F} = m\mathbf{a}$) can be written in a form which can be applied in oceanography. (Bold face type, e.g. $\mathbf{F}$, indicates that the quantity is a vector. Further discussion is given in Appendix 1.)

This relation says that if a resultant force $\mathbf{F}$ acts on a body of mass m, the body will acquire an acceleration or rate of change of velocity, $\mathbf{a}$. Notice the adjective "resultant", which means that more than one force may be acting simultaneously and we must first find the resultant of these, i.e. the net force, by appropriate vector addition; the acceleration will then be in the direction of this resultant force. In practice, vector equations such as $\mathbf{F} = m\mathbf{a}$ are usually broken down into three component equations so that the sum of the x-components of the forces equals the product of the mass times the x-component of the acceleration, etc.

The relation implies that if $\mathbf{F} = 0$, then $\mathbf{a} = 0$, i.e. there will be no *change* of motion but there may be persistent motion. This situation is governed by Newton's First Law of Motion, a special case of the Second Law.

Also, if we *observe* that $\mathbf{a} = 0$, we can conclude that the resultant $\mathbf{F} = 0$. In principle, this conclusion could mean that no forces at all were acting but in practice on earth this situation never occurs. For instance, there is always weight acting and often a reacting force balancing it, and if there is motion there is generally friction acting. When unaccelerated motion occurs, if we have information about some of the forces acting we can often learn something about the other forces. Notice also that $\mathbf{a} = 0$ implies motion in a straight line; whenever we observe that the motion is curved there is a centripetal acceleration (see Appendix 1) and therefore a resultant (centripetal) force must be acting.

It is convenient to write $\mathbf{a} = \mathbf{F}/m$ and think of the Law as stating that the observed acceleration is due to the resultant force acting *per unit mass*. In words we write acceleration = (pressure + gravity + frictional + tidal) forces/unit mass.

To make physical-mathematical deductions from this statement of a physical law we must first write mathematical statements of the forms of the forces; then we can try to "obtain solutions" to the equations as explained below.

In vector form the equation of motion is

$$\frac{d\mathbf{V}}{dt} = -\alpha \nabla p - 2\mathbf{\Omega} \times \mathbf{V} + \mathbf{g} + \mathbf{F}. \tag{6.1}$$

Pressure + Coriolis + Gravity + Other forces
(all per unit mass)

(We will explain shortly how we arrived at this equation and the meaning of the symbols ∇ and $\times$. $\mathbf{V}$ is the total (vector) velocity.)

This equation can be written as three component equations with the coordinates x, y and z and their respective velocity components u, v and w being positive in the east, north and upward directions respectively and the origin of coordinates being at the sea surface as

		Pressure	Coriolis	Gravity	Other forces per unit mass

$$(x) \quad \frac{du}{dt} = -\alpha \frac{\partial p}{\partial x} + 2\Omega \sin\phi\, v - 2\Omega \cos\phi\, w \qquad\quad + F_x$$

$$(y) \quad \frac{dv}{dt} = -\alpha \frac{\partial p}{\partial y} - 2\Omega \sin\phi\, u \qquad\qquad\qquad\quad\ + F_y \tag{6.2}$$

$$(z) \quad \frac{dw}{dt} = -\alpha \frac{\partial p}{\partial z} + 2\Omega \cos\phi\, u \qquad - g \qquad + F_z.$$

These equations (6.1) or (6.2) are called the equation(s) of motion; they are also referred to as the equations of conservation of linear momentum. Similar equations may be written for conservation of angular momentum as well but we will use a related quantity called "vorticity" in this book, as is customary in fluid mechanics, and use of the word "momentum" will imply linear momentum unless otherwise stated. (It will be shown in Section 7.3 that the two terms containing $\cos\phi$ are negligible in the real ocean for most dynamics.)

6.2 Obtaining solutions to the equations, including boundary conditions

In these equations, the quantities u, v and w are the components of the velocity of the water and they describe the "motion of the ocean"—they are what the physical oceanographer, particularly the dynamical oceanographer, wants to learn about. Together with the pressure p, they form the four unknowns in the equations. If we add the equation of continuity (4.4) we have four equations and four unknowns. (When we have as many equations as unknowns, then it is possible, at least in principle, to find solutions.) The other quantities assumed to be known are: x, y, z for position, $\alpha = $ specific volume (from the observed distributions of temperature and salinity with pressure (or depth)), $\Omega = $ the magnitude of the angular velocity of rotation of the earth, ϕ = geographic latitude (being positive in the northern hemisphere and negative in the southern). $\mathbf{F}$ and its components F_x, etc., represent frictional and tidal forces which we will introduce later. We shall consider the more complicated situation when S, T and α are also taken to be unknown in Chapter 10 when discussing thermohaline effects.

Then "obtaining solutions" to the equations of motion means finding (or guessing) values for u, v and w, in terms of the known quantities, which "satisfy the equations". This statement means that if one substitutes these values for u, v and w in the equations, these will balance, i.e. the numerical value of the time differential of u (i.e. du/dt) on the left of the first of equations (6.2) must be identical with that of the right-hand side when numerical values for the quantities are substituted there. And similarly for the other two equations. Note that the same expressions for u, etc., must be used in all equations where they occur and the solutions must be valid for all points and times being considered.

It must also be noted that the expressions for u, v and w must simultaneously satisfy two other conditions:

(1) the equation of continuity $\dfrac{\partial u}{\partial x} + \dfrac{\partial v}{\partial y} + \dfrac{\partial w}{\partial z} = 0$

and (2) the boundary conditions.

With regard to item (1), the assumption of incompressibility leading to this simple form of continuity eliminates acoustic (sound) waves from the possible solutions to equations (6.2) because such waves depend for their existence on the medium being compressible. We shall not discuss sound waves explicitly in this book although many of the properties of waves given in Chapter 12 are also applicable to them.

Item (2) above means that the velocity components u, v and w must behave in a reasonable manner at the boundaries of the ocean, i.e. at the bottom, shore and air/sea surface. For instance, if x occurs in the expression (solution) for u, it must be in such a way that u becomes zero at a north–south shore and w must become zero at the ocean bottom if it is level. More generally, there can be no flow through the boundaries, so the component of flow normal (i.e. perpendicular) to the boundary must vanish. Next to solid boundaries the component of flow along the boundary (the "tangential" component) must vanish too, i.e. there must be "no slip" at solid boundaries. This condition is based on the observed behaviour of almost all real fluids and is a consequence of the molecular nature of materials and the surface interactions between the solid and fluid, i.e. the fluid/solid friction at the boundary.

Sometimes it may be possible to relax the "no-slip" condition. The velocity along the boundary may decrease to zero at the boundary in a relatively thin "boundary layer". One may be able to find a solution for the interior region of the flow which does not satisfy the no-slip condition and a "boundary-layer solution" which goes from the interior solution to no-slip in the thin layer. If one is only interested in the interior then one can consider the interior solution alone and use a "free-slip" boundary condition to find it. Caution is required, however, because the assumption that the effects of the boundary do not penetrate into the region of interest may be incorrect and erroneous solutions may result.

One procedure for obtaining a solution is simple (even crude)—invent one and then see if it will satisfy all the conditions; then examine it, using observed information, to see if it describes a likely motion of the water. (This method is practical because typical solutions are already known to many forms of differential equations.) We can ease the task by simplifying the equations by ignoring the **F** terms and by ignoring the acceleration terms. A number of useful solutions for this case are known and will be discussed in Chapter 8.

The procedure becomes more difficult when expressions for fluid friction are introduced as **F** terms, because we are still uncertain about the physical details of turbulent fluid friction except in a few special cases. Finding solutions is even more difficult when the acceleration terms are included. The equations then become non-linear and very difficult to deal with mathematically if we need analytic solutions, i.e. algebraic expressions for the velocities. (The alternative procedure of solving the equations numerically will be discussed in Chapter 11.)

"*Non-linear*" means that the unknowns occur in combination in the equations, e.g. as $v(\partial u/\partial y)$, and non-linear equations seem to be impossible to solve in general. (One may be able to show that unique solutions should exist but have no general way of finding them.) The non-linearity of the equations and the presence of turbulence are not unrelated as we shall discuss in the next chapter.

6.3   The derivation of the terms in the equation of motion

6.31   *The pressure term*

Imagine a rectangular volume, in a fluid, of sides δx, δy and δz (Fig. 6.1). Then the force in the x-direction on this volume due to the hydrostatic pressure will be $+ p\delta y\delta z$ on the left face and $- (p + \delta p)\delta y\delta z$ on the right face, where the minus sign indicates that it acts in the negative x-direction. The net "pressure force" in the x-direction is the sum of these two or $-\mathbf{i} \ \delta p\delta y\delta z =$

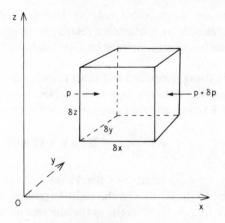

FIG. 6.1 *For the derivation of the pressure term in the equation of motion.*

$-\mathbf{i}\,(\partial p/\partial x)\delta x\delta y\delta z$, where the unit vector $\mathbf{i}$ denotes the x-direction. (Here, as usual, we have omitted terms of higher order in δx which vanish in the limit as δx becomes small. From now on, this approach will be used where appropriate without stating so explicitly every time.) The force per unit volume is $-\mathbf{i}(\partial p/\partial x)$, and the force per unit mass $= -\mathbf{i}(\partial p/\partial x)(1/\rho) = -\mathbf{i}\alpha(\partial p/\partial x)$. Then considering all directions, the total pressure force/unit mass will be

$$-\alpha\left(\mathbf{i}\frac{\partial p}{\partial x} + \mathbf{j}\frac{\partial p}{\partial y} + \mathbf{k}\frac{\partial p}{\partial z}\right) = -\alpha\nabla p.$$

Here, ∇ is short for $(\mathbf{i}\partial/\partial x + \mathbf{j}\partial/\partial y + \mathbf{k}\partial/\partial z)$, called the "gradient operator", and $\mathbf{j}$ and $\mathbf{k}$ are unit vectors denoting the y and z directions respectively. The gradient of a property (e.g. ∇p) is always a vector. The minus sign indicates that if p increases to the right, then the pressure force acts to the left. (Derivatives of quantities, in this case p, in a particular direction, e.g. $\partial p/\partial x$, are often called gradients too, as a convenient term for the component in a particular direction of the total gradient.)

6.32 *Transforming from axes fixed in space to axes fixed in the rotating earth*

The Coriolis term arises because we normally make observations relative to axes fixed to the earth which is itself rotating about its axis. The equation of motion $\mathbf{F} = m\mathbf{a}$, however, applies only when $\mathbf{a}$ is measured relative to axes "fixed in space", i.e. in what is called an inertial coordinate system which is one whose origin is not accelerating. For practical purposes this is a system *fixed relative to the distant stars.* Obviously it is more convenient for the oceanographer to make his measurements relative to points and directions on earth and so the equation of motion must be adjusted to suit this rotating frame of reference.

A mathematically straightforward and exact transformation from ideal axes fixed in space to practical rotating axes (e.g. see Neumann and Pierson, 1966; Lacombe, 1965; or Batchelor, 1967) gives in vector form

$$\mathbf{a}_f = \left(\frac{d\mathbf{V}'}{dt}\right)_f = \left(\frac{d\mathbf{V}}{dt}\right)_e + 2\mathbf{\Omega} \times \mathbf{V} + \mathbf{\Omega} \times (\mathbf{\Omega} \times \mathbf{R}) \tag{6.3}$$

where the subscript f means relative to fixed axes and the subscript e means relative to axes fixed to the earth. On the right-hand side, the first term is the acceleration relative to axes fixed to the earth, the second term is the Coriolis acceleration and the third term is the centripetal acceleration required to make an object on the earth's surface rotate with the earth. The other symbols are: $\mathbf{V}'$ = velocity relative to fixed axes, $\mathbf{V}$ = velocity relative to the earth, $\mathbf{R}$ = the vector distance of the body from the centre of the earth, and $\mathbf{\Omega}$ = angular velocity of rotation of the earth. Its magnitude is 2π radians in one sidereal day or 7.29×10^{-5} rad s^{-1}. (One sidereal day = 23 h 56 min 4 s = 86 164 s is the time required for the earth to rotate once about its axis, relative to the fixed stars. Since the earth revolves about the sun it must turn a little further to point back to the sun and complete one solar day—hence the solar day is a little longer than the sidereal day.) The "$\times$" in a term such as $2\mathbf{\Omega} \times \mathbf{V}$ represents what is called a vector product. The reader unfamiliar with this vector operation need not be concerned because we will write down and use the components of this operation in the component equations. (An intuitive derivation of the Coriolis term will be given in Section 6.35.)

The equation of motion relative to fixed axes is

$$\left(\frac{d\mathbf{V}}{dt}\right)_f = -\alpha \nabla p + \mathbf{g}_f + \mathbf{F}. \tag{6.4}$$

When transformed to earth axes using equation (6.3) we get

$$\left(\frac{d\mathbf{V}}{dt}\right)_e = -\alpha \nabla p - 2\mathbf{\Omega} \times \mathbf{V} + \mathbf{g}_f - \mathbf{\Omega} \times (\mathbf{\Omega} \times \mathbf{R}) + \mathbf{F}. \tag{6.5}$$

In this equation, the term on the left is the acceleration relative to the earth and the terms on the right are the forces per unit mass acting, i.e. the accelerations due to these forces. The transformation simply adds two *apparent* forces/unit mass, i.e. $-2\mathbf{\Omega} \times \mathbf{V}$, termed the *Coriolis acceleration*, and $-\mathbf{\Omega} \times (\mathbf{\Omega} \times \mathbf{R})$, the negative of the centripetal acceleration (which is sometimes called the centrifugal acceleration (force/unit mass), see Appendix 1). The true forces in equation (6.4) are unchanged.

6.33   *Gravitation and gravity*

Gravitation is the name given to the attractive force between masses, recognized first by Newton. Its magnitude is expressed by $F_g = G(M_1 M_2)/r^2$ where M_1 and M_2 are the sizes of the two masses and r is the distance between their centres. It is an attractive force acting along the line connecting the centres of the masses. (This expression is only true for two masses whose sizes are small compared with r or for two spheres whose density distribution is radially symmetrical. These conditions are sufficiently well satisfied in the case of the earth and a small object on it, and for the earth and moon when we consider tidal theory.) G is the Gravitational Constant. The gravitational force provides the $\mathbf{g}_f$ in the absolute equation of motion (6.4). In the relative equation (6.5), the term $\mathbf{\Omega} \times (\mathbf{\Omega} \times \mathbf{R})$ is the centripetal acceleration required to make a body at a distance $\mathbf{R}$ from the centre of the earth circulate about the earth's axis with angular velocity $\mathbf{\Omega}$. As usual for bodies in contact with the earth it is provided by a portion of the gravitational acceleration $\mathbf{g}_f$, as shown in Fig. 6.2. (The maximum value of the magnitude of the centripetal acceleration is only about 0.3 % of the gravitational acceleration.) The difference, $[\mathbf{g}_f - \mathbf{\Omega} \times (\mathbf{\Omega} \times \mathbf{R})]$, is referred to as the *acceleration due to gravity*, i.e. it is the familiar acceleration $\mathbf{g}$ of a body falling freely near the earth (in the absence of friction). In future we combine $[\mathbf{g}_f - \mathbf{\Omega} \times (\mathbf{\Omega} \times \mathbf{R})]$ as $\mathbf{g}$. At the surface of the earth it depends only on geographical position. It is a maximum at the poles (where the needed centripetal acceleration vanishes and $\mathbf{g}_f$ is also a maximum because the polar radius is slightly less than the equatorial radius) and is a minimum at the equator (where the needed centripetal acceleration is a maximum and $\mathbf{g}_f$ is a

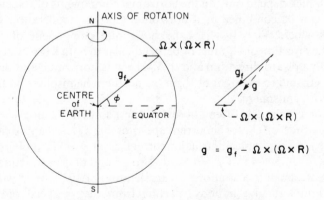

FIG. 6.2   *Showing how the gravitational acceleration* ($\mathbf{g}_f$) *is reduced to the acceleration due to gravity* ($\mathbf{g}$) *in providing the centripetal acceleration* ($\mathbf{\Omega} \times (\mathbf{\Omega} \times \mathbf{R})$) *required. Note that the size of the centripetal acceleration is exaggerated. Its magnitude is* ($\Omega^2 R \cos \phi$) *and it acts perpendicular to the axis of rotation.*

minimum). However, as the variation of **g** from pole to equator is only about 0.5 % we will neglect it, and will also neglect the very small variation with depth below the ocean surface, and will take the value of g as constant at $9.80 \, \mathrm{m \, s^{-2}}$.

Notice that in the component equations (6.2) the acceleration due to gravity occurs only in the z-component equation because the z-axis is, by definition, taken parallel to the local direction of this acceleration. It appears with a minus sign because the acceleration is down while the positive direction for z is taken as up.

6.34   *The Coriolis terms*

The terms containing $\mathbf{\Omega}$ in equation (6.1) and Ω in equation (6.2) and the $(\mathbf{\Omega} \times \mathbf{V})$ term in equation (6.5) are called the *Coriolis acceleration* terms (named after G. Coriolis, 1835, although they had been recognized by others before him). As will be seen from equations (6.2) there are four terms. Of these, the component $2\Omega \cos \phi \, \omega$ in the x-equation is very small, and this component of the Coriolis acceleration will be neglected. In addition, the Coriolis term $2\Omega \cos \phi$ u in the z-equation is small compared with the pressure term and with g, but is not necessarily small compared with their difference which is itself usually small in the sea. However, this z-component Coriolis term is usually neglected in dynamical oceanography. (Note that this component will affect gravity measurements taken on a ship moving east–west and it is also of significance in tidal studies.)

Only two Coriolis terms are left and these are in the x- and y-component equations, which depend only on the horizontal components of velocity. These two terms can be combined as a horizontal Coriolis acceleration $\mathbf{C}_H = 2\Omega$ $\sin \phi \mathbf{V}_H \times \mathbf{k}$ where $\mathbf{V}_H = \mathbf{i}u + \mathbf{j}v =$ the horizontal component of the total velocity. The direction of $\mathbf{C}_H$ must be perpendicular to both $\mathbf{k}$ (the unit vector in the vertically upward direction) and to $\mathbf{V}_H$, i.e. it is horizontal and directed at right angles to and to the right of $\mathbf{V}_H$ in the northern hemisphere, to the left in the southern hemisphere.

The factor $2\Omega \sin \phi$ is often abbreviated to f so that $2\Omega \sin \phi \, u = fu$, etc.

The magnitude of C_H for a current speed of $1 \, \mathrm{m \, s^{-1}}$ (or approximately 2 knots) which is fairly typical for major currents is: at $\phi = 90°$ (pole), $C_H = 1.5$ $\times 10^{-4} \, \mathrm{m \, s^{-2}}$; at $\phi = 45°$, $C_H = 1 \times 10^{-4} \, \mathrm{m \, s^{-2}}$ and at $\phi = 0°$ (equator), C_H $= 0$. These are small accelerations; an acceleration of $10^{-4} \, \mathrm{m \, s^{-2}}$ would take about 40 hours to give a body, starting from rest, a speed of $14 \, \mathrm{m \, s^{-1}}$ ($\simeq 50 \, \mathrm{km \, h^{-1}}$ or 30 miles per hour)! Another way to put it is to say that a body lying on a frictionless slope of $10^{-4}/9.8 \simeq 1$ in 10^5 or 1 cm drop in 1 km horizontally would experience the acceleration of $10^{-4} \, \mathrm{m \, s^{-2}}$. This slope is of the same magnitude as is calculated for the mean slopes of the sea (neglecting wave slopes).

6.35   *The Coriolis terms—an intuitive derivation*

To obtain a physical picture of the apparent effect of the rotation of the earth on objects moving near to it, consider the following hypothetical situation. A long-range gun mounted at the north pole is aimed along a meridian directly at a target fixed on earth and some distance to the south. In plan view, a projectile fired from the gun will travel in a plane fixed relative to the "fixed stars" but the target will be carried to the east by the rotating earth during the flight of the projectile. From the point of view of the gunner, also rotating with the earth, the projectile will appear to curve to the west, i.e. to the right, of its aimed direction. (From the south pole, a projectile would appear to curve to the left.) Since the change of direction represents acceleration, the earthbound observer interprets the apparent curved path of the projectile as due to some force acting across its direction of motion. This is the so-called "Coriolis force". We say "so-called" because it is not a real force but simply a way of interpreting the observations made by an observer in a rotating frame of reference. It is intuitively obvious that a projectile fired from the equator in the equatorial plane will not appear to deviate left or right of its original direction of flight. Using a similar argument, one concludes that a projectile fired due north or south from any point on the earth's surface, not only from the poles, will appear to curve right or left when it is in the northern or southern hemisphere respectively. When the gun is not at either pole one must remember that the projectile will have the eastward velocity associated with the earth's rotation at the point of firing.

It is less easy to see why an object projected east or west from a latitude between pole and equator will also deviate right or left, depending on which hemisphere it is in. One way to see this is to imagine a plane tangential to the earth's surface at some latitude ϕ in the northern hemisphere. Then the angular velocity of the earth, Ω, a quasi-vector, may be resolved into a vertical component, $\Omega \sin \phi$, and a horizontal one, $\Omega \cos \phi$. Considering the vertical component first, the tangential plane is rotating in space at angular velocity $\Omega \sin \phi$, anticlockwise as seen from above in the northern hemisphere (clockwise in the southern hemisphere), and a projectile fired in any direction will appear, to an observer fixed on the plane, to curve to the right of its original direction (or to the left in the southern hemisphere). It is clear that the apparent tendency to curve will be maximal at the poles ($\sin \phi = \pm 1$), and that the angular velocity $\Omega \sin \phi$ of the tangent plane leads to the $u \sin \phi$ and $v \sin \phi$ Coriolis terms in the component equations of motion. It is left as a mental exercise for the student to identify how the other two Coriolis terms in the equations of motion arise from the horizontal angular velocity component of the tangential plane.

In the above exposition we used a projectile as a moving body as it is obviously decoupled from the solid earth while in flight. Any moving body will have a Coriolis "force" acting on it but other forces, e.g. road friction on a car's tires, may overcome it and prevent the body from following a curved path in coordinates fixed to the earth. Indeed the Coriolis "force" is too small to be directly experienced in everyday life (unless one plays catch on a merry-go-round). However, for the ocean and atmosphere the horizontal forces are small and the Coriolis "force" is important.

The origin of the factor 2 in the Coriolis terms in the equations of motion can be seen from the following. In the plan view of Fig. 6.3, if AB is the initial direction to the target B and the flight path of the projectile in space, then in time t the projectile will travel $AB = Vt$ and the target will be displaced $BB' = Vt(\Omega \sin \phi)t$. The apparent acceleration will then be the second differential of the apparent displacement, i.e.

$$\frac{\mathrm{d}^2}{\mathrm{d}t^2}(\Omega \sin \phi)Vt^2 = 2\Omega \sin \phi V.$$

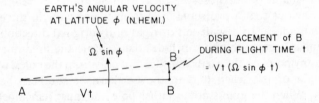

FIG. 6.3   *For explanation of the origin of the factor 2 in the Coriolis terms (e.g. $2\Omega \sin \phi \, v$).*

6.36   *Other accelerations*

The final term **F** (force/unit mass, i.e. acceleration in the equations of motion in this chapter) reminds us that there are other forces to be taken into account, such as the gravitational attraction of the moon and sun, friction between wind and water, friction at solid boundaries, friction within the water mass, etc. In Chapter 8, we will ignore these complicating factors and just examine some solutions to the equations which have been simplified by assuming that **F** = 0; friction will be introduced in Chapter 9 and the gravitational forces of the moon and the sun in Chapter 13.

Before looking at the **F** = 0 solutions we will examine, in the next chapter, some of the characteristics and magnitudes of the terms in the equations of motion after briefly discussing, in the next section, the coordinates to be used.

6.4   Coordinate systems

In writing the vector equation of motion (6.1) in component form (6.2) we used *rectangular* or *Cartesian* coordinates because the equations then have a fairly simple expression. The vector form of the equation (6.1) is valid in any coordinate system, an advantage in using vector notation. (Indeed, one may derive an equation in component form in one system, e.g. our derivation of the pressure force in rectangular coordinates, transform it to vector form and expect it to be valid in any coordinate system when properly transformed.)

If we are considering the motion of the ocean over the whole earth, then rectangular coordinates are clearly not appropriate. Spherical coordinates must be used. The earth is not exactly spherical but is approximately elliptical in cross-section from north to south pole with an ellipticity of about 1/300. (This shape or "figure" of the earth is a consequence of the fact that the gravitational acceleration is partially used to provide the necessary centripetal acceleration.) However, the error involved in using the equations in spherical form is only about 0.5 % and can be neglected (Veronis, 1973).

In this book, we shall write the component equations in rectangular form. The equations are simplest in this form so that it is easiest to illustrate the principles with them. Also, for many phenomena they are a consistent approximation. If the horizontal area being considered is not too large then we can work on a plane tangent to the sphere and use a rectangular system with negligible errors. For phenomena of relatively small scale, e.g. 100 km or so, this tangent plane is called the *f-plane* because for such small north–south distances the Coriolis parameter, $f = 2\Omega \sin \phi$, may be taken to be constant at its value at the centre of the area. (In the Arctic Ocean, where f is near to its maximum and varying only slowly, this approximation of using rectangular (or cylindrical) coordinates with $f = $ constant may be used for many phenomena over rather larger regions.) For relatively larger areas, with ϕ varying over a few tens of degrees, between mid-latitudes and the equator, the tangent plane approximation is called the *β(beta)-plane*. Here, while a rectangular coordinate approximation is used, the variation of f with latitude is taken as $f = (f_0 + \beta y)$ (the y-axis being north–south) where f_0 is the value of f at the mid-latitude of the region and $\beta = \partial f / \partial y$ is given the value at the mid-latitude of the area. The quantity β is termed the variation of the Coriolis parameter with latitude.

7

The Role of the Non-linear Terms and the Magnitudes of Terms in the Equations of Motion

BEFORE discussing some special cases of the application of the equation of motion, we will examine in this chapter the role of the non-linear terms in the equation and will make some estimates of their quantitative significance.

7.1 The non-linear terms in the equation of motion

7.11 *The friction term for the instantaneous velocity*

Consider the equation of motion per unit mass (equation (6.2)) for the x-component,

$$\frac{du}{dt} = -\alpha\frac{\partial p}{\partial x} + 2\Omega \sin \phi v - 2\Omega \cos \phi w + friction + tidal \ forces. \quad (7.1)$$

The tidal force terms can be written down using Newton's Law of Gravitation and can be taken to be known, although when this equation was being examined in the early days of fluid mechanics attention was directed chiefly to laboratory flows where such terms are not important. We shall omit the tidal terms until we consider their effects in Chapter 13. In order to get a system which can be solved, an expression for the friction is needed. One can consider the forces on a small element of fluid associated with the molecular nature of the fluid and differences in velocity within the fluid. Based on observational evidence, Newton hypothesized that the (tangential) frictional stress was related to the velocity shear as (for example) $\tau_x = \mu \partial u/\partial y$ where μ is a molecular viscosity coefficient which is a property of the particular fluid. This was later verified using non-turbulent flows. Then it can be shown that the net friction force per unit mass in the x-direction on a small mass of fluid is given by

$$F_x = v\left(\frac{\partial^2 u}{\partial x^2} + \frac{\partial^2 u}{\partial y^2} + \frac{\partial^2 u}{\partial z^2}\right)$$

and this term represents molecular friction in equation (7.1). The quantity $v = \mu/\rho$ is the kinematic molecular viscosity, $v = v(s, t, p)$. A typical value for water is $10^{-6} \, \mathrm{m^2 \, s^{-1}}$ with a range of 0.8 to 1.8 times this value. Thus the frictional effects could be expressed in terms of the velocity, and a closed system of equations could be obtained, i.e. as many equations as unknowns and therefore, in principle, a solvable system.

In the derivation of the above expression for F_x it has been assumed that the fluid is incompressible and terms of the form $(\partial v/\partial x)(\partial u/\partial x)$ have been neglected because they are small compared with those retained in realistic oceanographic cases. The derivation of the friction term in this form was done by Navier and Stokes and the equations of motion including it are called the "Navier–Stokes" equations. (Details of the derivation which is mathematically straightforward, although the algebra may be complicated depending on the notation used, may be found in more advanced texts, e.g. *Introduction to Fluid Mechanics*, Batchelor, 1967.) As an example we shall carry out a derivation of the $v(\partial^2 u/\partial z^2)$ term later (Section 9.3) when discussing the wind-driven circulation in Chapter 9. A term representing the resistance, due to molecular viscosity, to compression has been omitted. It is not important for the solutions which we shall consider but viscosity does lead to damping of sound waves and may need to be retained when studying acoustics in the sea.

Equation (7.1) applies to the instantaneous velocity of the fluid and is an excellent approximation to describe the behaviour of fluids such as water, i.e. it agrees with all the experimental results within measurement error. However, real oceanic and atmospheric fluid motions may be turbulent (very irregular in space and time) so that it may not be practical to solve the equations exactly for them. (For example, the details of boundary conditions and the initial state of the fluid are never well enough known.)

7.12   *What is the source of the difficulty?*

The term du/dt applies to the acceleration of a piece of the fluid. The terms of the right hand side are written in Eulerian form (see Appendix 1). To use the equation we must write du/dt in Eulerian terms. (We could attempt to write the right-hand side in Lagrangian terms but it is more difficult to do so and the same problem arises as in the Eulerian approach.) In writing du/dt in Eulerian terms (velocities at fixed points as functions of time) we must take account of the fact that a particle of fluid, when it moves to another point, must arrive at that point at a later time with the velocity appropriate to the new point and time.

The form which the acceleration term takes is then the *total* or *individual* derivative discussed in Appendix 1 for precisely this purpose and equation (7.1) becomes

$$\underset{\substack{\text{local rate} \\ \text{of change} \\ \text{with time}}}{\frac{du}{dt} = \frac{\partial u}{\partial t}} + \underset{\substack{\text{advective rates of} \\ \text{change due to motion}}}{u\frac{\partial u}{\partial x} + v\frac{\partial u}{\partial y} + w\frac{\partial u}{\partial z}} = \text{(terms on the right} \atop \text{as in equation (7.1))} \qquad (7.2)$$

The advective terms are called "non-linear" because the velocities occur as squares (e.g. $u(\partial u/\partial x) = (1/2)[\partial(u^2)/\partial x]$) or as products between velocity components and derivatives of other components (e.g. $v(\partial u/\partial y)$). Because of these non-linear terms a small perturbation (variation) may grow into a large fluctuation—these terms can cause instability and lie behind the presence of the turbulence which occurs whenever they are sufficiently large compared with the frictional terms which tend to remove velocity differences.

7.13 *Scaling and the Reynolds Number*

To estimate what is meant by "large" in this case, let us consider the ratio $(u\partial u/\partial x)/(v\partial^2 u/\partial x^2)$ of one of the non-linear terms to one of the molecular friction terms. If we take both u and ∂u to be of order U (a typical velocity magnitude) and ∂x of order L (a typical distance over which the velocity varies by U), then the ratio above is of the order $(U^2/L)/(vU/L^2) = UL/v = Re$ which is called the *Reynolds Number* (*Re*) for a fluid flow. It is a measure of the ratio of the non-linear, also called inertial, terms to frictional terms in the equation of motion.

This process of *scaling* or *ordering* terms is very often used in fluid mechanics because we cannot solve the full equations. By doing this scaling we may find that some terms may be neglected and thus simplify the analysis. We shall look at all of the terms presently but here we wish to use the Reynolds Number example to explain in detail the assumptions involved and the limitations of the approach. As a specific example consider flow in a pipe, the problem on which Osborne Reynolds worked. Here the radius of the pipe provides a length scale, L, for variations of velocity which goes from zero at the pipe wall to a maximum at the centre. The flow rate at the centre, U, provides a velocity magnitude. Now, to estimate the size of the non-linear terms (velocity times velocity gradient) we use U^2/L. The non-linear terms will not be of exactly this value and may vary in size from one part of the flow to another but they should be proportional to this quantity with a constant of proportionality of order 1 (i.e. between 0.1 and 10) if we have chosen our scales properly. Likewise, $v(\partial^2 u/\partial x^2)$, etc., should be proportional to vU/L^2. Hence the ratio, the Reynolds Number, gives an estimate of the relative importance of the two terms. This is another example of a finite difference approximation—in this case a rather crude one.

This dimensionless number, *Re*, is a very important quantity in determining the character of the flow. Indeed it can be shown that for a uniform density fluid with no rotation the solution to the equations is completely determined by the geometry of the boundaries and the Reynolds Number, although we may not be able to find the solution in a particular case. (If a free surface, i.e. a non-solid boundary, is present, the dimensionless parameter U^2/gL, called the *Froude Number*, must be considered too.) Flows which have the same geometry and Reynolds Number are said to be "dynamically similar", that is when scaled properly, the flows are identical. As a specific example, if we double the pipe diameter and adjust the total volume flow per unit time so that the velocities are half as large, and UL/v remains the same, the two flows will look the same. Thus dynamical similarity can be used to organize one's experimental efforts. For example, if we are trying to determine under what conditions the flow in the pipe will change from smooth, laminar flow to irregular, turbulent flow, the principle of dynamical similarity tells us that the transition will, for a given geometry, occur at the same particular value of Re. Thus to investigate the change we need to vary the flow rate only and not the pipe size or viscosity which are more difficult. The flow in the pipe will not be turbulent if $Re < 1000$. If the entrance to the pipe is smoothly flared and care is taken to reduce velocity fluctuations before the fluid enters the pipe it is possible to maintain laminar flow for Re up to about 100 000. This variation in the value of Re at transition is caused by changes in geometry, in this case the entrance conditions. The fact that transition to turbulence does not occur until $Re \simeq 1000$ shows that the non-linear terms do not become of dominant importance until this value is reached. In the pipe flow they are zero until transition to turbulence occurs but in many flows (e.g. around objects in the fluid) they usually begin to modify the solution (from the one assuming that they are zero) when $Re \simeq 1$, although turbulence will not be produced until Re becomes much larger. Once Re becomes larger than 10^5 to 10^6, depending on geometry, turbulence is very likely to occur unless there is some stabilizing influence such as density stratification as we shall discuss later in the chapter.

If we use the Gulf Stream as an oceanographic example, $U \sim 1$ m s^{-1} and $L \sim 100$ km $= 10^5$ m and $v \sim 10^{-6}$ m^2 s^{-1}, so that $Re \sim 10^{11}$ and the flow will definitely be turbulent.

We conclude from this example that the non-linear effects are very strong compared with the molecular friction effects. We can, in fact, ignore molecular friction in the open sea; it only becomes important very close to solid boundaries and in removing energy from turbulent flow at small scales to prevent it from growing without limit, i.e. molecular friction is important only for low values of Re which occur at low values of U and/or of L.

7.14   *Reynolds stresses*

Although molecular friction may be neglected in most aspects of the dynamics of ocean motions, it must not be assumed that there are no forces opposing the motions or giving rise to redistribution of energy and other properties. When the motion is turbulent, so that it includes rapidly fluctuating components in addition to any mean flow, then the non-linear terms give rise to terms in the equations of motion which have the physical character of friction and they, and similar terms in the heat- and salt-conservation equations (discussed in Chapter 10), give rise to more rapid distribution of momentum, heat and salt than would occur with purely molecular processes. These are the so-called *Reynolds stresses* (forces/unit area) and *fluxes* (transports/unit area) which appear in the equations for the mean or average motion of a turbulent fluid (the Reynolds equations, named after Osborne Reynolds who first derived them) and the equations for heat and salt conservation.

7.2   Equations for the mean or average flow

Because of the nature of turbulent flow it does not appear to be practical to solve for the detailed velocities so let us examine the possibility of writing equations for the average motion. The average used will be taken to be a *time average* over a suitable period (which might be a few minutes up to several months, depending on the phenomenon). Following Osborne Reynolds, who first suggested the approach, the variables u, v, w and p are split into a *mean* and a *fluctuating* part, e.g. $u = \bar{u} + u'$ where the overbar denotes an average and $\bar{u}' = 0$ by definition; $\bar{u} + u'$, etc., are then substituted into equation (7.2) and the average is taken.

Consider first the average of $\partial u/\partial t$,

$$\frac{\overline{\partial u}}{\partial t} = \frac{1}{T} \int_0^T \frac{\partial u}{\partial t} \, dt = \frac{[u(T) - u(0)]}{T} \quad \text{where } T \text{ is the averaging period.}$$

Now u must have some upper limit because the available energy sources are limited and frictional losses always occur (and usually increase as u increases). Thus as T gets large, this term will become negligible.

In practice we might wish to consider separation into a time-varying mean and fluctuations about it, if we were interested in variations due to tides or in seasonal changes. The sketch of Fig. 7.1 illustrates how we might do so. (Note that the separation is not always as straightforward as shown in the figure but such problems do not affect the general results of this section; a more detailed discussion is left for more advanced texts.) Thus we retain $\partial \bar{u}/\partial t$ in the average of equation (7.2) so that later we can consider how important this term is in the equation for the mean flow. For the other terms the order of averaging and differentiating may be interchanged if required. First let us consider the terms on the right-hand side of equation (7.2).

The pressure term becomes

$$\overline{(\bar{\alpha} + \alpha') \frac{\partial(\bar{p} + p')}{\partial x}} = \bar{\alpha} \frac{\partial \bar{p}}{\partial x} + \bar{\alpha} \overline{\frac{\partial p'}{\partial x}} + \overline{\alpha' \frac{\partial \bar{p}}{\partial x}} + \overline{\alpha' \frac{\partial p'}{\partial x}}$$

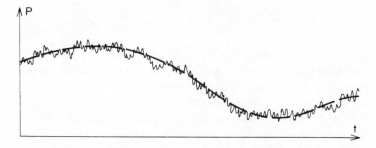

FIG. 7.1 *Time variation of some property P at a point. The solid curve is the total quantity P(t) as a function of time t, the dashed curve is a possible mean $\bar{P}$ while the difference between the two at any instant would be taken as the fluctuating part P'.*

In the second term on the right, α is already an average and does not change in the averaging process, so this term is $\bar{\alpha}\overline{(\partial p'/\partial x)} = 0$ because $\bar{p}'\ (= 0)$ is independent of x. Likewise the third term vanishes. In general, any average of a term which contains a single fluctuating quantity will vanish. The term $\overline{\alpha'(\partial p'/\partial x)}$ may not vanish if fluctuations in α' and p' are related (e.g. α' might be either of the same or of the opposite sign to $(\partial p'/\partial x)$ on average (although not always)). However, the α variations in the ocean are very small compared with $\bar{\alpha}$, and $\overline{\partial p'/\partial x}$ will be of the order $\partial \bar{p}/\partial x$ or less, so $\overline{\alpha'(\partial p'/\partial x)}$ is negligible compared with $\bar{\alpha}(\partial \bar{p}/\partial x)$.

The first Coriolis term $\overline{2\Omega \sin \phi (\bar{v} + v')} = 2\Omega \sin \phi (\bar{v} + \bar{v}')$ because $\Omega \sin \phi$ is a constant at a particular location and the average of a sum is the sum of the averages. Now $\overline{v'} = 0$, so this Coriolis term becomes $2\Omega \sin \phi\ \bar{v}$; likewise the other Coriolis term becomes $2\Omega \cos \phi\ \bar{w}$ when averaged.

In the frictional term, for example,

$$\nu \frac{\partial^2 \overline{(\bar{u} + u')}}{\partial x^2} = \nu \left(\frac{\partial^2 \bar{u}}{\partial x^2} + \frac{\partial^2 \overline{u'}}{\partial x^2} \right) = \nu \frac{\partial^2 \bar{u}}{\partial x^2}$$

because $\overline{u'} = 0$ everywhere and therefore its spatial derivative must vanish.

Thus all the terms on the right-hand side of equation (7.2) are of the same form for both the total flow and for the mean flow, which is also true for the corresponding terms in the y and z equations.

Now let us examine the advective acceleration terms on the left-hand side of equation (7.2) which become

$$(\bar{u} + u')\frac{\partial(\bar{u} + u')}{\partial x} + (\bar{v} + v')\frac{\partial(\bar{u} + u')}{\partial y} + (\bar{w} + w')\frac{\partial(\bar{u} + u')}{\partial z}.$$

When we average them any term which contains a single fluctuating quantity will vanish on average, as we showed above. Hence the average becomes

$$\left(\bar{u}\frac{\partial \bar{u}}{\partial x} + \bar{v}\frac{\partial \bar{u}}{\partial y} + \bar{w}\frac{\partial \bar{u}}{\partial z}\right) + \left(\overline{u'\frac{\partial u'}{\partial x}} + \overline{v'\frac{\partial u'}{\partial y}} + \overline{w'\frac{\partial u'}{\partial z}}\right).$$

The first three terms combined with $\partial \bar{u}/\partial t$ give $d\bar{u}/dt$, the total or individual derivative using the mean rather than the total velocity, leaving the other three terms involving the fluctuating components.

Hence, if we collect everything together the *Reynolds equation* for $\bar{u}$ has the same form as (7.2), with mean quantities replacing total quantities, plus the three new terms above involving velocity fluctuations. These new terms must represent the effect of the velocity fluctuations or "turbulence" on the mean motion. Note that they arise from the non-linear terms in the Navier–Stokes equation; the non-linear nature of the equations and the possible existence of turbulence and its possible frictional effects on the mean flow are not unrelated, as noted above. Note too that equations (6.1) and (6.2), because we did not write a specific form for the friction, may be taken to be applicable to either the total or the mean velocities.

We are still faced with the problem of writing down specific expressions for the friction in order to produce a closed (i.e. complete) set of equations. The approach of Navier and Stokes produces a set which is closed but which cannot be applied to high Reynolds Number turbulent oceanic flows in practice. Reynolds' approach shows how the non-linear terms give rise to turbulence effects on the mean flow and gives explicit expressions for these effects in terms of the velocity fluctuations. However, the system is still not closed because we have now added three more unknowns, u', v' and w'. In principle one might attempt to observe these turbulence terms. Such observations can be made, with great difficulty, at a single location but to do so in the detail required for a large region is simply not practical. To improve our understanding of turbulence effects and to be able to "parameterize" these effects in terms of mean flow quantities and their gradients (i.e. space derivatives) many more observations will be needed. (By *parameterize* we mean write down expressions for the turbulence quantities in terms of quantities (parameters) which we can observe more easily or calculate from our equations, in this case the mean velocities and their gradients and the density distribution or perhaps the static stability based on it.) This problem of "closure"—completing the equation set—remains as a fundamental problem in studying turbulent flows and remains unsolved in any general way. It is possible to work out from the Navier–Stokes equation an equation describing the behaviour of such terms as $\overline{u'u'}$, $\overline{u'v'}$, $\overline{u'w'}$, etc., from which the turbulence terms above can be calculated. However, these equations contain terms involving averages of triple products of fluctuating quantities which also come from the non-linear terms. An equation for these can be obtained but it involves quadruple products and so on *ad infinitum*. There are always more unknowns than equations. One must use observational knowledge and physical intuition to provide the necessary additional equations. The test of any closure scheme is that predictions using it agree with observations. In the following we shall outline the simplest closure scheme—an analogy with molecular friction effects.

7.21 *Reynolds stresses and eddy viscosity*

The analogy which will be discussed, of introducing an "eddy" or "turbulent" viscosity of much greater magnitude than the molecular one, does not produce very exact results except in special cases. However, if we can show by using this analogy that turbulent friction effects are small we can perhaps "solve" the equations ignoring friction and expect realistic results. Note also that in the mean flow or Reynolds equation the non-linear terms are not likely to be dominant. The "breakdown" into turbulence will mix momentum and reduce the spatial derivatives of the mean flow to the point where the non-linear terms based on the mean flow do not dominate, i.e. are not the largest terms. Thus the mean flow equations are likely to be, at worst, weakly non-linear. Provided that we can overcome the closure problem in a reasonable way we have a good chance of solving the equations, by numerical calculations on a computer if necessary (as described in Chapter 11), although many analytical methods (i.e. writing down appropriate mathematical formulae for the solutions) are available for weakly non-linear equations.

First we shall rewrite the turbulence terms using the equation of continuity of volume for an incompressible fluid, $\partial u/\partial x + \partial v/\partial y + \partial w/\partial z = 0$. If we average this equation then $\partial \bar{u}/\partial x + \partial \bar{v}/\partial y + \partial \bar{w}/\partial z = 0$; subtracting this average equation from the original equation then $\partial u'/\partial x + \partial v'/\partial y + \partial w'/\partial z = 0$ or $\mathbf{V}\cdot\mathbf{V}' = 0$ in mathematical shorthand. Thus the total velocity, the mean velocity and the fluctuating velocity all satisfy continuity of volume. To the turbulence terms of the Reynolds equation we add $u'(\mathbf{V}\cdot\mathbf{V}')$ which is zero and so does not change the value, only the mathematical form. For the x-component

$$u'\frac{\partial u'}{\partial x} + v'\frac{\partial u'}{\partial y} + w'\frac{\partial u'}{\partial z} + u'\left(\frac{\partial u'}{\partial x} + \frac{\partial v'}{\partial y} + \frac{\partial w'}{\partial z}\right) = \frac{\partial}{\partial x}\overline{u'u'} + \frac{\partial}{\partial y}\overline{u'v'} + \frac{\partial}{\partial z}\overline{u'w'}$$

and the equation for $\bar{u}$ is

$$\frac{d\bar{u}}{dt} = -\bar{\alpha}\frac{\partial \bar{p}}{\partial x} + 2\Omega \sin\phi \; \bar{v} - 2\Omega \cos\phi \; \bar{w} + \nu\left(\frac{\partial^2 \bar{u}}{\partial x^2} + \frac{\partial^2 \bar{u}}{\partial y^2} + \frac{\partial^2 \bar{u}}{\partial z^2}\right)$$

$$-\frac{\partial}{\partial x}\overline{u'u'} - \frac{\partial}{\partial y}\overline{u'v'} - \frac{\partial}{\partial z}\overline{u'w'}. \tag{7.3}$$

This is the Reynolds equation for the x-component of velocity.

Now a term such as $\nu(\partial^2 \bar{u}/\partial x^2)$ can be written as $\partial(\nu\partial\bar{u}/\partial x)/\partial x$ (the form which this term had before it was assumed that v is essentially uniform as shown later in Section 9.3), and $\rho v(\partial \bar{u}/\partial x)$ is the stress (force/unit area) in the x-direction due to molecular effects and a gradient of $\bar{u}$ in the x-direction. We can therefore identify $-\rho \overline{u'u'}$ as a stress due to the turbulence. It is the derivatives of these stresses which produce net forces on a small volume of fluid (just as does the derivative of the pressure as shown in Section 6.31). The stress mechanisms are qualitatively similar—the molecular effect is produced by molecules bouncing back and forth and exchanging momentum, the turbulent stress is produced by "chunks" of fluid moving back and forth and exchanging momentum with the surrounding fluid. The latter is more effective because the distance moved and the mass involved are *much* larger. Stresses such as $-\rho \overline{u'u'}$, $-\rho \overline{u'v'}$, $-\rho \overline{u'w'}$ (and other averaged quadratic products of u', v' and w') are termed Reynolds stresses, again after Osborne Reynolds who first derived them. By analogy with the molecular case we might suppose that these stresses are related to the mean velocity gradients by some sort of "viscosity" (an *eddy* or *turbulent* viscosity) as, for example,

$$-\overline{u'u'} = A_x \frac{\partial \bar{u}}{\partial x}; \quad -\overline{u'v'} = A_y \frac{\partial \bar{u}}{\partial y}; \quad -\overline{u'w'} = A_z \frac{\partial \bar{u}}{\partial z}. \tag{7.4}$$

Unlike the molecular case we use different values of eddy viscosity (A_x, etc.) for different directions, since they may be different (particularly between the vertical and horizontal directions because of static stability).

The above is the simplest way to define eddy viscosity. However, this definition does not preserve the symmetry of the Reynolds stresses, e.g. $-\overline{v'u'}$ $= A_x(\partial \bar{v}/\partial x)$ is not necessarily the same as $-\overline{u'v'} = A_y(\partial \bar{u}/\partial y)$ although it should be. Usually either $\partial \bar{v}/\partial x$ or $\partial \bar{u}/\partial y$ will dominate and one can pick from the flow which is appropriate. It is possible to get round this problem but equation (7.4) (and (7.5) following) are sufficient for our purposes and we shall leave further discussion of this point for more advanced texts.

Then a term such as $-\partial(\overline{u'u'})/\partial x$ becomes $\partial[A_x(\partial \bar{u}/\partial x)]/\partial x$. It is quite common to take A_x outside the derivative, either based on the argument (again in analogy with molecular viscosity) that terms such as $(\partial A_x/\partial x)(\partial \bar{u}/\partial x)$ are less important, or that the analogy is crude (which it is) and this further assumption is no worse than the initial one (which may or may not be true, depending on the case). With this final neglect of space variations of the A's relative to the other terms, the turbulent friction terms become, in the x-direction,

$$A_x \frac{\partial^2 \bar{u}}{\partial x^2} + A_y \frac{\partial^2 \bar{u}}{\partial y^2} + A_z \frac{\partial^2 \bar{u}}{\partial z^2} \tag{7.5}$$

where the A's are called "eddy" viscosities. Note that they are, like v, kinematic (dimensions $[L^2 T^{-1}]$ with units $m^2\,s^{-1}$) and the terms in expression (7.5) have dimensions of force/unit mass, i.e. acceleration. One must multiply the A's by ρ to get dynamic viscosity which, when multiplied by $\partial^2 \bar{u}/\partial x^2$, gives a force acting on unit volume. In the CGS system of units, when $\rho \sim 1 \text{ g cm}^{-3}$, the dynamic and kinematic viscosities have about the same numerical value but in SI with $\rho \sim 1000 \text{ kg m}^{-3}$ they do not. (In the literature the symbol A (or other symbol for viscosity) has sometimes been used for kinematic and sometimes for dynamic viscosity, so some care is required when extracting numerical values. In this book, kinematic viscosity will be used throughout.) Unlike coefficients of molecular viscosity, the eddy-viscosity coefficients are not constant for a particular fluid and temperature, salinity and pressure but vary with the particular motion involved. They are *not properties of the fluid but of the flow*! Values are up to 10^{11} times those for kinematic molecular viscosity. Many attempts have been made to express A_x, etc., in terms of the mean velocities and their derivatives but no generally applicable results have been obtained. We must therefore remember that the eddy viscosity terms in the above form are just an interim measure to represent one of the effects of turbulence until we understand this feature of fluid motion well enough to represent it more exactly.

The eddy viscosity approach does give good results in some cases, e.g. in the atmospheric surface layer, the first few tens of metres above the earth's surface. In this layer, A_z varies linearly with z and the solution of the equations (which are the same as for the ocean) using the eddy-viscosity form for the friction term agrees very well with observations. Presumably the flow near the ocean bottom could be treated in the same way but observations of the flow in this part of the ocean are quite limited.

When we introduce the eddy viscosity (including the molecular viscosity in it), the equations of motion for the x- and y-components are

$$\frac{du}{dt} = \frac{\partial u}{\partial t} + u\frac{\partial u}{\partial x} + v\frac{\partial u}{\partial y} + w\frac{\partial u}{\partial z} =$$

$$-\alpha\frac{\partial p}{\partial x} + fv - 2\Omega\cos\phi\,w + A_x\frac{\partial^2 u}{\partial x^2} + A_y\frac{\partial^2 u}{\partial y^2} + A_z\frac{\partial^2 u}{\partial z^2} \qquad (7.6x)$$

$$\frac{dv}{dt} = \frac{\partial v}{\partial t} + u\frac{\partial v}{\partial x} + v\frac{\partial v}{\partial y} + w\frac{\partial v}{\partial z} =$$

$$-\alpha\frac{\partial p}{\partial y} - fu \qquad\qquad + A_x\frac{\partial^2 v}{\partial x^2} + A_y\frac{\partial^2 v}{\partial y^2} + A_z\frac{\partial^2 v}{\partial z^2} \qquad (7.6y)$$

where u, v, w, α and p are average quantities the overbar having been omitted for simplicity. (Unless otherwise stated we assume that we are discussing the mean motion equations from now on.)

7.3   Scaling the equations of motion; Rossby number, Ekman number

The equations of motion in the form of equations (7.6) and the corresponding z-component equation (given shortly) are complicated and non-linear (although usually only weakly) and are generally not solvable explicitly. Before giving up in mathematical despair, let us examine the various terms to make rough estimates of their sizes—it may be possible initially to neglect some of them but still leave equations which refer to physical reality in the ocean and describe actual motions, even if only approximately. Later, we can reintroduce some of the neglected terms and obtain more exact mathematical descriptions of the motion.

What we will do is refer to the data bank of descriptive oceanography to find out what may be the sizes of the various terms so that we can decide which are the most important in particular situations.

First let us consider the main body of the oceans away from strong currents (such as the Gulf Stream or Kuroshio) and away from the sea surface where the frictional influence of the wind is important. We can return to these regions later.

The Pacific Ocean is roughly 12 000 km across and the Atlantic 6000 km, so let us take a horizontal length scale (L) of 1000 km = 10^6 m as typical of large-scale features of the ocean circulation. Typical horizontal speeds (U) are of the order of 0.1 m s^{-1}. We will take a vertical scale length (H) of 10^3 m, a reasonable fraction of the total depth (world ocean average depth $\simeq$ 4000 m).

First we will estimate a typical vertical speed (W) using the equation of continuity

$$\frac{\partial w}{\partial z} = -\left(\frac{\partial u}{\partial x} + \frac{\partial v}{\partial y}\right)$$

for which the order of magnitude of the terms is

$$\frac{W}{H} \simeq \frac{U}{L}, \text{ i.e. } W \simeq \frac{UH}{L} \simeq \frac{10^{-1} \times 10^3}{10^6} \simeq 10^{-4} \, m\,s^{-1}.$$

For a typical time scale (T) we take 10 days $\simeq 10^6$ s, considering shorter periods to be turbulent components for the moment. For the Coriolis acceleration, at latitude $\phi = 45°$ then $2\Omega \sin 45° = 2\Omega \cos 45° \simeq 2 \times 7.3 \times 10^{-5} \times 0.71 \simeq 10^{-4}$ s^{-1}, while $g \simeq 10$ m s^{-2}. For the pressure term, $\alpha \simeq 10^{-3}$ m^3 kg^{-1} and $p \simeq 10^4$ kPa = 10^7 Pa for $z = -10^3$ m from the hydrostatic equation.

Values estimated for A_x and A_y vary from 10 to 10^5 m^2 s^{-1} and we will use 10^5 m^2 s^{-1}. For A_z estimates range from 10^{-5} to 10^{-1} m^2 s^{-1} and we will use 10^{-1} m^2 s^{-1}. By using maximum values for A_x, A_y and A_z we should get upper limits for the sizes of the friction terms.

We see that these estimates for eddy viscosity values vary widely. Part of this variation is due to the fact that they are properties of the flow, not of the fluid, and part is due to the way in which they are obtained. For example, by measuring or estimating the other important terms in the equations one may obtain the friction terms by difference and then calculate the eddy viscosity from $A_x = \text{friction}/(\partial^2 u/\partial x^2)$. Alternatively, one may adjust the eddy viscosities in a solution (either analytical or numerical) to make it fit the observations as well as possible. A simple rough approach (probably good within a factor of 100) is to use the fact that the non-linear terms are about the same size as the turbulent friction terms which we pointed out in the previous section. Then

$$\frac{U^2}{L} \simeq A_x \frac{U}{L^2} \simeq A_y \frac{U}{L^2} \simeq A_z \frac{U}{H^2} \quad \text{or} \quad A_x \simeq UL \quad \text{and} \quad A_z \simeq \frac{H^2}{L^2} A_x.$$

Thus with $H/L \simeq 10^{-3}$, $A_z \simeq 10^{-6} A_x$ as for the estimates given. The fact that $A_z \ll A_x$ or A_y is partly due to the static stability caused by stratification which both inhibits vertical turbulent transfers and helps to make the flow nearly horizontal. The facts that $H/L \simeq 10^{-3}$, that the forcing has large horizontal scales and that the rotational effects (the Coriolis terms) are important enter too. Note that $A_x \simeq UL$ is equivalent to saying that a Reynolds Number based on eddy viscosity is of order 1. Using $U = 0.1 \text{ m s}^{-1}$ and $L = 10^6$ m gives A_x and $A_y \simeq 10^5 \text{ m}^2 \text{ s}^{-1}$, at the upper end of the range of estimated values. Lower values may occur because the flows on which they are based are of smaller scale (smaller L or U) or have an eddy viscosity Reynolds Number > 1.

The vertical component equation of motion corresponding to equations (7.6) is

$$\frac{dw}{dt} = \frac{\partial w}{\partial t} + u\frac{\partial w}{\partial x} + v\frac{\partial w}{\partial y} + w\frac{\partial w}{\partial z} =$$

$$-\alpha\frac{\partial p}{\partial z} + 2\Omega\cos\phi\,u - g + A_x\frac{\partial^2 w}{\partial x^2} + A_y\frac{\partial^2 w}{\partial y^2} + A_z\frac{\partial^2 w}{\partial z^2} \quad (7.7)$$

and will have scale sizes as follows:

$$\frac{W}{T} + \frac{UW}{L} + \frac{VW}{L} + \frac{W^2}{H} =$$

$$\alpha\frac{10^7}{H} + 2\Omega\cos\phi\,U - g + 10^5\frac{W}{L^2} + 10^5\frac{W}{L^2} + 10^{-1}\frac{W}{H^2},$$

i.e. $\quad 10^{-10} + 10^{-11} + 10^{-11} + 10^{-11} =$

$$+ 10 + 10^{-5} - 10 + 10^{-11} + 10^{-11} + 10^{-11}.$$

In this equation, all the terms are very much smaller than the pressure term and g and so we can ignore all except these two and will be left with the *hydrostatic equation* (derived in Appendix 1), i.e.

$$\alpha\frac{\partial p}{\partial z} = -g \quad \text{or} \quad dp = -\rho g \, dz \tag{7.8}$$

correct to about 1 part in 10^6, even when the water is moving with typical ocean speeds and even though we have chosen values for the eddy viscosities at the high end of their observed range for the open ocean. (It is left as an exercise for the reader to show that the hydrostatic equation still applies even in faster currents such as the Gulf Stream where the maximum speed is about 3 m s^{-1} and the stream width is ~ 100 km).

Note that the non-linear terms are all of the same size as a result of our estimating the vertical speed W from the horizontal speeds using the equation of continuity. Also the friction terms are all of the same (small) size as a result of our choice of the H, L, A_x and A_z values. This result will hold also for the other component equations and therefore when examining them we will need to look at only one non-linear and one friction term to estimate their sizes.

Looking now at one of the horizontal component equations

$$\frac{\partial u}{\partial t} + u\frac{\partial u}{\partial x} + \cdots = -\alpha\frac{\partial p}{\partial x} + fv - 2\Omega\cos\phi\, w + A_x\frac{\partial^2 u}{\partial x^2} + \cdots$$

The order of magnitude of the terms is

$$\frac{U}{T} + \frac{U^2}{L} + \cdots = -\alpha\frac{\partial p}{\partial x} + fU - 10^{-4}\, W + 10^5\frac{U}{L^2} + \cdots$$

or $10^{-7} + 10^{-8} + \cdots = \quad ? \quad + 10^{-5} - 10^{-8} + 10^{-8} + \cdots$

or relatively

$$10^{-2} + 10^{-3} + \cdots = \quad ? \quad + 1 - 10^3 + 10^{-3} + \cdots$$

The pressure term has been represented by a query here because we do not have direct measurements of $\partial p/\partial x$. We see, however, that it must be of the same size as the Coriolis term (fv) in order to balance the equation. Of the remaining terms the local acceleration term $\partial u/\partial t$ is the largest but even it is only about 1 % of the Coriolis term for typical times of the order of 10 days and will be smaller for longer times. The second Coriolis term $(2\Omega\cos\phi\, w)$ is small because of the typically small values of w. The non-linear terms for the mean motion are negligibly small and so are the friction terms in the interior of the water mass. Therefore, to an order of accuracy of 1 % we have

$$0 = -\alpha\frac{\partial p}{\partial x} + fv$$

$$0 = -\alpha\frac{\partial p}{\partial y} - fu \qquad \text{for the interior of the ocean a few degrees or more away from the equator.} \tag{7.9}$$

$$0 = -\alpha\frac{\partial p}{\partial z} - g$$

These equations describe the relationships between the horizontal pressure distributions and the horizontal velocity components in the ocean, and the distribution of pressure as a function of depth and density distribution ($\alpha = 1/\rho$) which is a function of the distribution of salinity, temperature and pressure. In principle, if we observe the distribution of salinity and temperature as a function of depth in the ocean we can calculate p from the z equation (7.9) and use it to find u and v from the x and y equations. Alternatively, for theoretical studies we could regard the temperature and salinity distributions as unknowns, introduce the equation of state $\alpha = \alpha(s, t, p)$ (from laboratory studies of the properties of sea water as described previously) and the heat- and salt-conservation equations, and solve the set of simultaneous equations (seven in all), an approach which we shall discuss in Chapter 10.

If we take $\partial/\partial y$ of the first of equations (7.9) and $-\partial/\partial x$ of the second equation and add them we get

$$-\alpha\frac{\partial^2 p}{\partial y \partial x} + f\frac{\partial v}{\partial y} + v\frac{\partial f}{\partial y} + \alpha\frac{\partial^2 p}{\partial x \partial y} + f\frac{\partial u}{\partial x} = 0.$$

But

$$\frac{\partial u}{\partial x} + \frac{\partial v}{\partial y} = -\frac{\partial w}{\partial z} \quad \text{and} \quad \frac{\partial^2 p}{\partial y \partial x} = \frac{\partial^2 p}{\partial x \partial y}$$

for a reasonable physical variable such as p. Thus

$$v\frac{\partial f}{\partial y} = f\frac{\partial w}{\partial z}.$$

Now $dy = R\,d\phi$ where R is the radius of the earth ($\simeq 6000$ km), so

$$\frac{\partial f}{\partial y} = \frac{\partial(2\Omega \sin \phi)}{R\,\partial \phi} = \frac{2\Omega \cos \phi}{R}.$$

For $\phi \simeq 45°$, $\sin \phi \simeq \cos \phi$ and then $W = UH/R$ and taking H to be the total depth, $W \simeq 10^{-3}U$. For $\phi = 5°$ and $H = 5000$ m, this scaling gives $W \simeq 10^{-2}U$. Thus flows for which the Coriolis terms are large are essentially horizontal ($\because W \ll U$) even if the horizontal scale is much smaller than the $L = 1000$ km which we chose earlier.

It appears, therefore, that the interior region of the ocean is described by a simple set of equations which can be solved, because non-linear effects are negligible. However, these simple equations do not give us a complete description because the boundary conditions for the interior of the ocean depend on the surface layers where wind friction acts, and on the lateral boundary layers (e.g. the Gulf Stream) where the dynamics are more complicated. A complete solution for the interior requires solutions for the outer regions, so that the problem is not fully solved. We can, however, ignore the boundary-condition problem for a while and make use of the simple equations to find out quite a lot about the motion in the interior.

In the next chapter we shall look at a simple case where there are no true forces—just acceleration provided by the Coriolis effect. In this case we will be looking at a phenomenon of smaller linear scale because if

$$\text{Acceleration term} = \text{Coriolis acceleration,}$$

then $U^2/L \simeq fU$ or $L \simeq U/f$ and for $U = 0.1 \ \mathrm{m \, s^{-1}}$, $f \simeq 10^{-4}$,

then $L \simeq 10^3$ m or 1 km. In this example we have used a possible balance between terms to determine what the length scale must be, another way to use the scaling approach.

In the scaling of the large-scale flow in the interior we found that both non-linear and friction effects were very small. In other regions they may be more important. We found the Coriolis term to dominate—it turns out to be important for almost all large-scale flow phenomena.

To help to classify flow types in other regions it is useful to consider the ratios between non-linear and Coriolis terms and between friction and Coriolis terms

$$\frac{\text{Non-linear term}}{\text{Coriolis term}} = \frac{U^2}{L} \frac{1}{f_0 U} = \frac{U}{f_0 L} = Ro.$$

Here f_0 is a typical value for f for the region being considered and the ratio Ro is called the *Rossby Number*. The second non-dimensional ratio is

$$\frac{\text{Friction term}}{\text{Coriolis term}} = A_x \frac{U}{L^2} \frac{1}{f_0 U} = \frac{A_x}{f_0 L^2} = E_x,$$

$$\text{or} = \frac{A_y}{f_0 L^2} = E_y, \quad \text{or} \quad = \frac{A_z}{f_0 H^2} = E_z.$$

These E's are called *Ekman Numbers*, e.g. E_z is the vertical Ekman Number because it depends on the friction term involving spatial derivatives with respect to the vertical coordinate, often termed for brevity "vertical friction". Likewise E_x and E_y are horizontal Ekman numbers. In the interior, $E_x \sim E_y$ and the symbol E_H is often used. For the interior, $Ro \leqslant 10^{-3}$, $E_z \sim E_H \leqslant 10^{-3}$. In other regions they may not be so small but for the large-scale circulation, values of the order of 1 are an upper limit.

7.4   Dynamic stability

What determines when a flow will become unstable so that it may break down into irregular small-scale motions leading to friction effects which are much larger than those due to the molecular nature of the fluid? As we have already noted, such effects seem to occur in the ocean because the apparent friction effects, as quantified by the eddy viscosities, are much larger than molecular ones. The horizontal eddy viscosity values are 10^7 to 10^{11} times molecular values while the vertical values are 10 to 10^5 times molecular values.

First consider a fluid which is not rotating so that the Coriolis terms can be ignored. Also take the fluid to have constant and uniform density throughout, so that derivatives of density with respect to space coordinates vanish everywhere. This is an idealized example requiring a truly incompressible fluid and is often used without further explanation. It is quite easy, however, to construct a realistic example with the same properties. Take the salinity and potential temperature to be constant throughout. Then the static stability is neutral—there is no buoyant resistance to vertical motion because a displaced parcel always has the same density as the surrounding water. Alternatively, one can say that there are now no buoyancy effects—displaced parcels are never lighter or heavier than their surroundings. The salinity is uniform throughout but temperature and density both increase with depth and are not uniform. However, this real fluid case will behave exactly as the ideal constant density, truly incompressible, case so the results of the ideal case are of practical value.

Now for this simple situation it is the ratio of the non-linear terms to the molecular friction term, i.e. the Reynolds Number, which determines the *dynamic stability*. If $Re > 10^6$, then turbulent flow is likely. Suppose $U = 0.01 \text{ m s}^{-1}$, a rather small speed, then taking $v \simeq 10^{-6} \text{ m}^2 \text{ s}^{-1}$, the characteristic length to make $Re = 10^6$ is $L = 100$ m. As this length is rather small compared with the size of the ocean basins it would seem that turbulent flow is likely to occur everywhere. However, even in this simple case a large value of Re is not sufficient for turbulence to occur. In order that small velocity variations (also called "perturbations") can grow they must have an energy source. It turns out that there is no energy source unless there are gradients in the flow. Thus if the flow is very uniform in velocity there is no energy source and molecular viscosity will smooth out the perturbations. Of course, the ocean is finite and near solid boundaries the velocity vanishes, leading to gradients if there is any flow at all, and therefore turbulence will probably be present there. At the surface the wind acts, leading to velocity gradients and turbulence.

Another possibility is that although the Reynolds Number is large and velocity gradients are present, for a particular type of flow the non-linear terms remain small and the breakdown to turbulence does not occur. Surface waves provide such a case. Although Re may easily be 10^7 or more they are weakly non-linear and not turbulent until wave breaking occurs. While the non-linear effects are small for surface waves they are not entirely negligible, as we shall discuss briefly in Chapter 12.

7.41   *The effect of density variations on dynamic stability*

When density variations occur in a fluid they may enhance or diminish the mechanical effects. The static stability gives a measure of the effect. If it is negative (unstable) the vertical component of velocity fluctuations is enhanced. If it is positive (stable) the vertical component is diminished. If the turbulence persists it will tend to mix the fluid, that is, make the density more uniform in the vertical. In doing so light fluid is mixed down and heavy fluid up, raising the centre of gravity and increasing the gravitational potential energy. This increase in potential energy comes from the kinetic energy of the turbulence which in turn is usually derived from the kinetic energy of the mean flow. The turbulent fluid also loses some energy to heat (internal energy) through molecular viscous effects. If the rate of turbulent energy loss exceeds the rate of gain, the turbulence will die out. Indeed, if the static stability is sufficient, turbulence involving fluctuations of the vertical component will not be possible.

How can we establish a criterion for the relative importance of static stability and the tendency for instability due to the effects of the non-linear terms? As mentioned earlier, generation of turbulence requires a velocity gradient. First consider the case where $v = w = 0$ and u varies with z but not with x or y. Then the only velocity gradient possible is $\partial u/\partial z$ and it needs to be compared with the static stability. The possible generation of turbulence does not depend on the sign of $\partial u/\partial z$; dynamic instability may occur if u is either increasing or decreasing from one level to another—only a change is required, so we consider $(\partial u/\partial z)^2$ as an indicator of the strength of mechanical generation. A measure of the static stability is the *Brunt–Väisälä frequency* (N) given by $N^2 = gE$ (equation (5.11)). Then a measure of the relative importance of mechanical and density effects is the dimensionless *Richardson Number* $Ri = N^2/(\partial u/\partial z)^2$, named after the person who introduced it. (This is sometimes called the "gradient" Richardson Number because it is based on gradients of mean quantities; it is possible to define a slightly different Richardson Number based on the turbulence itself, but this extension is beyond the scope of this book.) If $\partial v/\partial z \neq 0$, $\partial V_H/\partial z$ would replace $\partial u/\partial z$ in the Richardson Number.

If $Ri < 0$, density variations enhance the turbulence; if $Ri > 0$ they tend to reduce it. If only vertical variations of V_H occur and Ri becomes sufficiently large, turbulence is not possible—the stabilizing effect of the density distribution overcomes the potential instability due to the non-linear terms. Assuming small perturbations (fluctuations) about the mean, Miles (1961) showed that a stratified shear flow is stable if $Ri > 1/4$ *everywhere* in the flow. A proof of this theorem and references to laboratory tests of it may be found in *Waves in the Ocean*, LeBlond and Mysak, 1978. This "critical" Richardson Number may not be valid if large perturbations occur. Establishing the value of a "critical" Richardson Number in the ocean is experimentally very difficult because even Ri is not easy to measure and it is necessary to decide just when the fluid becomes barely turbulent and to account for effects of horizontal gradients (derivatives with respect to x and y) of velocity which are impossible to eliminate entirely. Empirically (i.e. experimentally) it seems that when Ri is larger than about 1/4, turbulence cannot be generated by vertical gradients of velocity (i.e. $\partial u/\partial z$ or $\partial v/\partial z$). Of course, if horizontal gradients of velocity are present fluctuations of essentially horizontal velocity may develop even when Ri is much larger than the critical value for damping vertical component fluctuations. An example is the meandering of the Gulf Stream which is known to occur even though Ri is probably considerably larger than its critical value.

Note that the effect of the density variation, principally in the vertical direction, on the mean flow is indirect. It acts on the turbulence modifying the vertical eddy viscosity (and similar coefficients for heat- and salt-turbulent transports). The reason is that the density variations are small (both the fluctuations and in the mean values). Indeed there are no obvious effects of density variations on the mean flow—there are no terms involving derivatives of mean density (or specific volume) in the equations nor are there terms involving specific volume fluctuations. Because these fluctuations are small, we neglect the terms involving them such as $\alpha'(\partial p'/\partial x)$ in deriving the Reynolds equations for the mean motion, e.g. equation (7.3) for the x-component. This approximation is consistent with what is termed the *Boussinesq approximation*. Boussinesq said that, if the density variations are fairly small, to a first approximation we can neglect their effect on the *mass* (i.e. inertia) of the fluid but we must retain their effect on the *weight*. That is, we must include the buoyancy effects but can neglect the variations in horizontal accelerations for a given force due to the mass variations with density (which are at most 3 % if we use an average over the whole ocean for ρ or α). Thus in the horizontal momentum equations (x- and y-directions) we can use an average density over the region being considered but in the z-equation, which reduces to the hydrostatic equation, we must use the actual *in situ* values when calculating the pressure field.

7.5 Effects of rotation

In a non-rotating fluid the Reynolds and the Richardson Numbers must be considered. Once Re is sufficiently large the flow will be turbulent, at least in the horizontal, even though a large enough Ri will restrict fluctuations of the vertical velocity component. Once Re is large enough for turbulence to occur its value is not important for the mean flow directly. The value of Re then determines the scale size at which molecular viscous effects become important for the turbulence itself and prevent fluctuations of smaller scales from becoming large.

The rotation provides another possibility, that the Coriolis terms may affect the flow. For the ocean and atmosphere these appear to be important, often dominant, compared with non-linear and friction terms. The Rossby and Ekman Numbers then determine the relative importance of non-linear effects (for the mean flow) and frictional (usual turbulent) effects compared with the Coriolis effects. While the Reynolds Number based on molecular friction is not important, the ratio of the non-linear terms (for the mean) and the turbulent friction, measured by the Reynolds Number based on eddy viscosity, remains an important parameter. Also, the dominance of the Coriolis terms forces the mean flow to be nearly horizontal, just as does strong stability, making any turbulence associated with it mainly horizontal (or mainly two-dimensional rather than three-dimensional as in a homogeneous non-rotating fluid).

Having discussed the significance of the non-linear, frictional and rotational terms in the equations of motion, we will next examine the simple approximate equations of motion (equations (7.9)) where both the Rossby and Ekman Numbers are so small that for the mean flow both turbulent friction (non-linear effects in the total flow) and, except for the first example of inertial motion, non-linear terms involving mean velocity components may be ignored.

8

Currents without Friction:
Geostrophic Flow

IN THIS chapter we will discuss some of the characteristics of motion which we can deduce from the equations of motion when it is assumed that the **F** terms in equations (6.1) or (6.2) (i.e. friction, gravitation of the moon and sun, etc.) are zero and that there is a steady state, that is the velocities at any point do not change with time (i.e. $\partial u/\partial t = \partial v/\partial t = \partial w/\partial t = 0$). Except for the example of inertial oscillations we shall also assume that the advective acceleration terms may be neglected. These approximations for the large-scale mean circulation in the ocean's interior were justified in the previous chapter.

8.1 Hydrostatic equilibrium

As a preliminary to discussing moving fluids, let us first look at stationary ones. Let us assume that (1) $u = v = w = 0$, i.e. that the fluid is stationary, (2) $d\mathbf{V}/dt = 0$, i.e. the fluid remains stationary, and (3) all the **F** terms are zero. Then, from equations (6.2) we are left with only

$$\alpha \frac{\partial p}{\partial x} = 0, \quad \alpha \frac{\partial p}{\partial y} = 0, \quad \alpha \frac{\partial p}{\partial z} = -g. \tag{8.1}$$

The first two mean that the isobaric (constant pressure) surfaces are horizontal, i.e. there is no pressure term, in fact no force at all in this case, to cause horizontal motion. The third can be written as

$$dp = -\rho g dz, \tag{8.2}$$

which is the *hydrostatic* (pressure) *equation* in differential form, i.e. it gives the pressure dp due to a thin layer dz of fluid of density ρ. If ρ is constant (independent of depth) it becomes $p = -\rho g z$. This is not a very exciting result—really all that it does is confirm that the equations of motion do give a previously known answer (as shown from first principles in Appendix 1) when the fluid is stationary. As we showed in Section 7.3, this equation remains an excellent approximation even for flows at typical ocean speeds.

The reason for the minus sign on the right is because we take the origin of coordinates at the sea surface with z positive upward. Measurements up into the atmosphere are given as, for example, "the masthead is at $+10$ m", while measurements down into the sea are given as, for example, "the fish is at a depth of 50 m, i.e. at $z = -50$ m". The pressure at this depth is (taking $\rho = 1025$ kg m^{-3})

$$p_{50} = -(1025 \times 9.8 \times -50) = +5.02 \times 10^5 \text{ Pa} = +502 \text{ kPa}.$$

An increase in depth of 1 m yields an increase in pressure of about 10 kPa.

8.2 Inertial motion

We first assume that (1) $\partial p/\partial x = \partial p/\partial y = 0$ (i.e. there is no slope of the sea surface and all the pressure surfaces inside the fluid are also horizontal; we shall look at the situation when these terms are not zero presently), (2) that we can ignore the **F** terms as stated above, and (3) that $w = 0$ (i.e. that there is only horizontal motion). Then the x- and y-equations of motion become

$$\frac{du}{dt} = 2\Omega \sin \phi\, v \quad \text{and} \quad \frac{dv}{dt} = -2\Omega \sin \phi\, u. \tag{8.3}$$

The equations (8.3) have solutions

$$u = V_H \sin (2\Omega \sin \phi\, t),$$
$$v = V_H \cos (2\Omega \sin \phi\, t), \tag{8.4}$$

where $V_H^2 = u^2 + v^2$ and t here stands for time. Now these are the equations of motion for a body in the northern hemisphere travelling clockwise in a horizontal circle at constant linear speed V_H and angular speed $2\Omega \sin \phi$. If the radius of the circle is B, then $V_H^2/B = 2\Omega \sin \phi V_H$, i.e. the centripetal acceleration V_H^2/B is provided by the Coriolis acceleration $2\Omega \sin \phi$ V_H (Fig. 8.1). Physically, such motion might be generated when a wind blows steadily in one direction for a time, causing the water to acquire a speed V_H, and then the wind stops and the motion continues without friction (to a first approximation) as a consequence of its "inertia" (properly its momentum), hence the term "*inertial motion*". Flow variations of inertial period are often present in records from current meters. The amplitudes vary depending on the strength of generating mechanisms and they decay due to friction when the generation stops.

Note that equations (8.3) are non-linear but do have solutions, equations (8.4), so non-linear equations can sometimes be solved explicitly. Note also, however, that if we regard the equations as Lagrangian equations for a fluid parcel, they are linear and the terms which would be non-linear in Lagrangian terms (friction) have been assumed to be small, making solution easy.

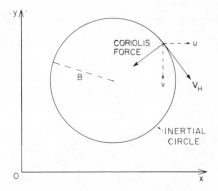

FIG. 8.1   *Relationship of Coriolis force and velocity for inertial motions (northern hemisphere).*

For a speed $V_H = 0.1$ m s^{-1} at latitude $\phi = 45°$, then $B \simeq 1$ km. For $V_H = 1$ m s^{-1}, then $B \simeq 10$ km. The period of revolution is

$$\frac{2\pi}{\text{angular speed}} = \frac{2\pi}{2\Omega \sin \phi} = \frac{1}{2} \frac{1 \text{ sidereal day}}{\sin \phi} = \frac{T_f}{2}$$

because $\Omega = 2\pi/1$ sidereal day. The quantity $T_f = (1 \text{ sidereal day}/\sin \phi)$ is called "one pendulum day" because it is the time required for the plane of vibration of a Foucault pendulum to rotate through 2π radians. The value of $0.5\,T_f$ (one-half pendulum day) is 11.97 h at the pole, 16.93 h at 45° latitude and infinity at the equator.

The direction of rotation in the inertial circle is clockwise viewed from above in the northern hemisphere and anticlockwise in the southern hemisphere. If one thinks of observing the motion in the southern hemisphere by looking down through the earth from the northern hemisphere then the motion also appears clockwise. However, the observer in the southern hemisphere is upside-down relative to the observer in the northern hemisphere and he calls the motion anticlockwise. Likewise, he says that the Coriolis force acts to the left of the velocity in the southern hemisphere. It is a matter of point of view. In the terms used by meteorologists, the motion is anticyclonic in both hemispheres. The term *cyclonic* comes from cyclone, a storm with low pressure at its centre about which the winds are anticlockwise in the northern hemisphere and clockwise in the southern hemisphere. An *anticyclonic* system has high pressure at its centre and winds circulate in the opposite direction. The reason for this behaviour will become clear when we discuss geostrophic flow. Equivalent terms *contra solem* and *cum sole* were occasionally used by oceanographers in the older literature meaning, respectively, against and with the direction of motion of the sun as seen by an observer facing the equator. These terms are related in Fig. 8.2.

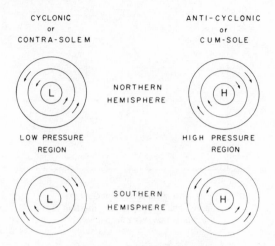

FIG. 8.2 *Directions of rotation around low-*
and high-pressure regions in northern and
southern hemispheres.

8.3 Geopotential

In preparation for the discussion of the geostrophic method for calculating currents we must introduce the concept of geopotential. The quantity $dw = Mgdz$ is the amount of work done (= potential energy gained) in raising a mass M through a vertical distance dz against the force of gravity (ignoring friction). We then define a quantity called *geopotential* (Φ) such that the change of geopotential $d\Phi$ over the vertical distance dz is given by

$$Md\Phi = dw = Mgdz \text{ (joules)},$$

or $d\Phi = gdz \text{ (joules kg}^{-1} = \text{m}^2\text{ s}^{-2}) \text{ (potential energy change/unit mass)}$

$$= -\alpha dp \text{ (from equation (8.2))}.$$

Integrating from z_1 to z_2 we have

$$\int_1^2 d\Phi = \int_1^2 gdz = -\int_1^2 \alpha dp.$$

Now writing $\alpha = \alpha_{35,0,p} + \delta$ from Section 2.23 we get

$$\Phi_2 - \Phi_1 = g(z_2 - z_1) = -\int_1^2 \alpha_{35,0,p}\, dp - \int_1^2 \delta dp$$

$$= -\Delta\Phi_{\text{std}} \qquad -\Delta\Phi. \qquad (8.5)$$

The quantity $(\Phi_2 - \Phi_1)$ is called the *geopotential distance* between the levels z_2 and z_1 where the pressures will be p_2 and p_1. The first quantity on the right of equation (8.5) is called the "*standard* geopotential distance" ($\Delta\Phi_{std}$, a function of p only) while the second is called the "geopotential *anomaly*" ($\Delta\Phi$, a function of S, T and p). In size, the second term is of the order of one-thousandth of the first.

The reader is reminded that although $(\Phi_2 - \Phi_1)$ is called the geopotential "distance" in oceanographic jargon, it really has the units of energy per unit mass ($J\,kg^{-1}$ or $m^2\,s^{-2}$) and for $g = 9.8\,m\,s^{-2}$ and $\delta z = 1\,m$, then $d\Phi = 9.8\,J\,kg^{-1}$. For numerical convenience, oceanographers in the past have used a unit of geopotential called the "dynamic metre" such that $1\,dyn\,m = 10.0\,J\,kg^{-1}$. To indicate that this unit is being used, it is usual to use the symbol D for geopotential. The geopotential distance $(D_2 - D_1)$ is then numerically almost equal to $(z_2 - z_1)$ in metres, e.g. relative to the sea surface:

	SI units	Mixed units
at a geometrical depth in the sea	$= +100\,m$	$+100\,m$
then	$z_2 = -100\,m$	$-100\,m$
the pressure will be about	$p = +1005\,kPa$	$+100.5\,dbar$

and the geopotential distance

relative to the surface $\Phi_2 - \Phi_1 = -980\,J\,kg^{-1}$, $D_2 - D_1 = -98\,dyn\,m$.

It is because of its use in the calculation of geopotential "distance" that tables of α as a function of S, T and p are more common than tables of ρ.

8.31   *Geopotential surfaces and isobaric surfaces*

A surface to which the force of gravity, i.e. the plumb line, is everywhere perpendicular is called a *geopotential surface* because the value of the geopotential must be the same everywhere on the surface. The term "level surface" is taken to mean the same thing. An example of such a surface is the smooth surface of a lake in which there are no currents and where there are no waves, or of a billiard table set up correctly. The reason for specifying "no currents" will be explained in the next section.

An *isobaric surface* is one on which the pressure is everywhere the same. In the above stationary lake the water surface would be the isobaric surface $p = 0$ (atmospheric pressure being assumed constant and ignored). Isobaric surfaces for higher pressure would be deeper in the lake and would be geopotential (level) surfaces as long as the lake was still.

Isobaric surfaces must be level in the stationary state. (Do not confuse this "stationary state" when $u = v = w = 0$ with the "steady state" when u, v and w may be non-zero but do not change with time, i.e. when $\partial u/\partial t = \partial v/\partial t = \partial w/\partial t = 0$, see Appendix 1.) Suppose for the moment that an isobaric surface (dashed line in Fig. 8.3(a)) were inclined to the level surface (full line in Fig. 8.3(a)). The

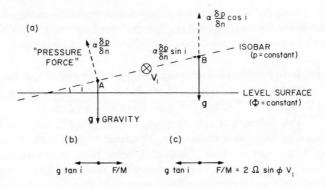

FIG. 8.3 *Pressure terms in relation to isobaric and level surfaces for the northern hemisphere.* V_1 *is into the paper.*

pressure force on a particle of water A of unit mass will be $\alpha \partial p / \partial n$ as shown. ($\partial / \partial n$ means the gradient along a normal, i.e. a perpendicular, to the surface and in the plane of the paper.) In addition, gravity acts on the particle. This is an unstable situation because the two forces cannot balance, as they are not exactly opposed, but must have a resultant to the left. The situation is shown in more detail for particle B where the pressure force has been resolved into

a vertical component $\alpha (\partial p / \partial n) \cos i$ which balances g,   and

a horizontal component $\alpha (\partial p / \partial n) \sin i$ which is unbalanced and would cause accelerated motion to the left, i.e. the situation is not stable.

The component to the left is

$$\alpha \frac{\partial p}{\partial n} \sin i = \left(\alpha \frac{\partial p}{\partial n} \cos i \right) \frac{\sin i}{\cos i} = g \tan i.$$

To stop the acceleration to the left it is necessary to apply to the right a force/unit mass, F/M, equal in size to $g \tan i$ (Fig. 8.3(b)). In Chapter 7 we showed that the Coriolis force was likely to be important, so one way to apply a force to the right would be to generate a Coriolis force by having the water move "into the paper" (in the northern hemisphere) at speed V_1 so that $2\Omega \sin \phi$ $V_1 = F/M = g \tan i$ (Fig. 8.3(c)). In the southern hemisphere, the water would have to come "out of the paper" at speed V_1.

8.4 The geostrophic equation

The Coriolis force is sometimes called the "geostrophic" (= earth turned) force and the equation

$$2\Omega \sin \phi \, V_1 = g \tan i \qquad (8.6)$$

is one version of the *geostrophic equation*, which expresses a balance between the pressure force and the Coriolis force.

In principle, this geostrophic equation should permit us to determine the speed V_1 by measuring the slope i of the isobaric surface. In practice we cannot do this because we cannot determine p directly with the necessary accuracy. Instead we have to determine p from the hydrostatic equation $p = -\int \rho g dz$ after having determined the distribution of density ρ with depth. Even with this method we cannot determine the angle i absolutely. The reason is that we make our measurements from a ship on the surface of the sea and we do not know if the sea surface is level or not (disregarding waves). In fact, if there are currents in the surface waters the sea surface will *not* be level because the geostrophic equation applies there (neglecting wind effects which we shall add in Chapter 9), and motion gives rise to a Coriolis force which requires the water surface to be sloping so that the horizontal component of the pressure gradient can act to balance the Coriolis force. All that we can do is determine the difference between i_1 at level z_1 and i_2 at level z_2 as described shortly. This difference will give us the velocity at level z_1 relative to that at level z_2, and a finite difference estimate of the *velocity shear* $(\partial V/\partial z)$.

The slopes are small, e.g. $2\Omega \sin \phi \simeq 10^{-4}$ at 45° latitude and for V_1 $= 1 \text{ m s}^{-1}$, $\tan i \simeq 10^{-5}$, i.e. the surface rises by 1 m in 100 km, a distance typical of the width of a strong current such as the Gulf Stream.

A technique for determining the absolute slope of the sea surface which is receiving much attention is to use radar or laser altimetry from satellites. Cheney and Marsh (1981a) demonstrated the estimation from *Seasat* satellite radar altimeter observations of the change of sea surface elevation across the Gulf Stream of 140 ± 35 cm or a slope of $(1.2 \pm 0.3) \times 10^{-5}$ for three months in 1978. Substantially improved techniques with smaller standard deviations are anticipated in the future. A review of techniques for satellite altimetry of the sea surface was given by Cheney and Marsh (1981b).

The geostrophic equation applies equally to the atmosphere but the meteorologist is more fortunate than the oceanographer. He can measure air pressure directly at a number of places on the ground or at known levels in the atmosphere and then determine the horizontal pressure-gradient term $(\alpha \, (\partial p/\partial n) \sin i)$ directly and so calculate the geostrophic wind speed. In addition, because the speeds of currents in the ocean are small compared with wind speeds in the atmosphere, the meteorologists can ignore the water slopes and use "mean" sea level as a reference level.

8.41   *Why worry about the geostrophic equation?*

The reason why the oceanographer concerns himself about using the geostrophic equation to determine currents is because direct measurement of ocean currents in sufficient quantity to be useful is technically difficult and expensive.

In shallow water a ship can anchor and hang a current meter over the side to measure the current, or can hang several meters to measure at several depths simultaneously. However, this procedure only gives information about the currents at the one point where the ship is anchored. Also, a ship usually does not remain stationary when anchored but moves about (i.e. it surges and swings) relative to the anchor. Part of this motion will be added to the water motion measured by the current meter and constitutes a source of error for which it is difficult to correct. In the deep ocean, it is much more difficult to anchor and the ship motion error may be much larger than the real water motion.

A more practical method is to use recording current meters which are hung in a "string" from a moored buoy (see *Descriptive Physical Oceanography*, Pickard and Emery (1982) for a description of instruments and techniques or Dobson, Hasse and Davis (1980) for more details). A number of such buoys moored in a pattern in the ocean will provide information about the three-dimensional distribution of currents as a function of time. However, because of the expense, the difficulties of working at sea and the complicated nature of the currents when examined in detail, it is not possible to obtain observations over as much of the ocean as we would like.

Why should we need to measure the currents over a period of time? Why is not one measurement at each place sufficient? Simply because real ocean currents are not steady. They fluctuate in speed and direction and the only way to determine the mean and the variation with time is to make frequent measurements for a sufficient period of time (probably several months at least).

The geostrophic method for calculating the current requires information on the distribution of density in the ocean; it is easier to obtain this information (from measurements of temperature and salinity) than it is to measure currents directly. The method suffers from several disadvantages, but when used intelligently and in parallel with other information it can be very helpful. In fact, most of our knowledge of ocean circulation below the surface has been obtained in this way. The geostrophic method is also useful in strong currents (e.g. the Gulf Stream as we shall show in Section 8.10) in which it is difficult to moor recording current meters.

We should add that currents in the surface layer can be deduced from the navigation records of ships, and most of our surface-layer information has been acquired from this source. The method of using navigation records is to assume that the difference between the intended path, based on the speed and direction of the ship relative to the water, and the one actually followed (determined by astronomical, satellite, etc., navigation) is due to the water currents. Obviously such data are "noisy", that is any one observation may have a large error, but by averaging over many years using all the observations in a particular area (e.g. 5° latitude by 5° longitude) one can obtain the "climatological" or long-term average motion. There are undoubtedly significant variations of the actual motions from these averages; variations of several times the mean seem common according to our limited direct current observation data. There are probably cases of smaller-scale features of the flow than are resolved by such means. For example, the pattern of flow in the equatorial Pacific deduced from observations has become more and more complicated as more detailed observations have been made. (Because of effects of surface wave currents, many current meters do not work well near the surface, so obtaining better observations of surface currents remains a problem. See also Baker, 1981. Surface drifters tracked by satellite are beginning to give direct observations of surface currents in the open ocean but broad coverage by this method is unlikely because of the cost.) Better navigation with the improved electronic facilities now in use will help to improve the quality of the current data derived from ship navigation but it will still be limited to the busy shipping lanes and therefore one does not get good coverage in time and space over most of the ocean.

8.42 *The geostrophic method for calculating relative velocities*

In Fig. 8.4, A and B represent the positions where oceanographic stations have been taken so that the distribution of ρ or α is known along each vertical AA_1A_2 and BB_1B_2. The line AB represents the sea surface which is assumed not to be level but whose slope unfortunately is not known. (In the present state of the art it is not measurable from a ship on the open sea, although the satellite technique mentioned earlier is showing promise and measurements of sea surface height to ± 0.1 m or better, relative to the geoid, are considered likely in the future.) Lines Φ_1 and Φ_2 represent two level surfaces passing through A_1 and A_2 at station A and C_1 and C_2 at station B. The two isobaric surfaces p_1 and

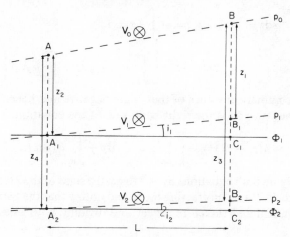

FIG. 8.4 *For the derivation of the geostrophic equation.*

p_2 pass through A_1 and A_2 at station A, and through B_1 and B_2 at station B. The angles of these two isobaric surfaces are i_1 and i_2 relative to geopotential surfaces. If the velocity component, relative to the earth, of the water (into the paper, northern hemisphere) on surface p_1 is V_1 and on p_2 is V_2, then the geostrophic equations for the two levels are

$$2\Omega \sin \phi\, V_1 = g \tan i_1,$$
$$2\Omega \sin \phi\, V_2 = g \tan i_2.$$

Subtracting

$$2\Omega \sin \phi\, (V_1 - V_2) = g\,(\tan i_1 - \tan i_2),$$

i.e. $\quad 2\Omega \sin \phi\,(V_1 - V_2) = g \dfrac{B_1 C_1}{A_1 C_1} - \dfrac{B_2 C_2}{A_2 C_2}$

$$= \frac{g}{L}(B_1 B_2 - C_1 C_2) \quad \text{because } A_1 C_1 = A_2 C_2 = L$$

$$\text{and } B_1 C_1 - B_2 C_2 = B_1 B_2 - C_1 C_2$$

$$= \frac{g}{L}(B_1 B_2 - A_1 A_2) \quad \text{because } C_1 C_2 = A_1 A_2$$

$$= \frac{g}{L}[(z_1 - z_3) - (z_2 - z_4)]. \tag{8.7}$$

Now from the hydrostatic equation

$$g = -\alpha\, dp$$

$$\therefore \int_{B_1}^{B_2} g\, dz = g(z_3 - z_1) = - \int_{p_1}^{p_2} \alpha_B\, dp \tag{8.8}$$

$$= - \left[\int_{p_1}^{p_2} \alpha_{35,0,p}\, dp + \int_{p_1}^{p_2} \delta_B\, dp \right] \text{ from (8.5).}$$

Note that the numerical values of the z's are negative and hence $g(z_3 - z_1)$ is numerically negative as is the right-hand side of the equation.

Similarly $\qquad g(z_4 - z_2) = - \left[\int_{p_1}^{p_2} \alpha_{35,0,p}\, dp + \int_{p_1}^{p_2} \delta_A\, dp \right].$

Then, multiplying both equations by -1 to get the signs of the z terms the same as in equations (8.7) and subtracting them, and noting that the two ($\alpha_{35,0,p}\, dp$) terms are identical and therefore cancel, and dividing both sides by L

$$\frac{g}{L}[(z_1 - z_3) - (z_2 - z_4)] = \frac{1}{L} \left[\int_{p_1}^{p_2} \delta_B\, dp - \int_{p_1}^{p_2} \delta_A\, dp \right]$$

therefore $$(V_1 - V_2) = \frac{1}{L\,2\Omega\sin\phi}\left[\int_{p_1}^{p_2}\delta_B\,\mathrm{d}p - \int_{p_1}^{p_2}\delta_A\,\mathrm{d}p\right]$$

$$= \frac{1}{L\,2\Omega\sin\phi}[\Delta\Phi_B - \Delta\Phi_A]. \qquad (8.9A)$$

(In texts using mixed units this is written

$$(V_1 - V_2) = \frac{10}{L\,2\Omega\sin\phi}[\Delta D_B - \Delta D_A] \text{ where } \Delta D = \int_{p_1}^{p_2}\delta\,\mathrm{d}p; \qquad (8.9B)$$

with L in metres, δ in $\mathrm{cm}^3\,\mathrm{g}^{-1}$ and p in decibars, $(V_1 - V_2)$ will be in $\mathrm{m\,s}^{-1}$.)

Equations (8.9A) and (8.9B) are the practical forms of the geostrophic equation. The measurement of temperature and salinity at a series of depths at each station in principle provides the information needed to calculate the two integrals of the specific volume anomalies δ_A and δ_B (i.e. $\Delta\Phi_A$ and $\Delta\Phi_B$) while L, the distance between the stations, is obtained from navigation. In practice, the variation of specific volume anomaly with depth is never available in a mathematical form suitable for direct integration and it is necessary to determine the values of the integrals (or of ΔD in the mixed system) by numerical summation (Appendix 1, Integrals) as shown in the example in the next sub-section.

In equation (8.9A), if L is expressed in metres, δ in $\mathrm{m}^3\,\mathrm{kg}^{-1}$, p in pascals and $\Omega = 7.29 \times 10^{-5}\,\mathrm{s}^{-1}$, then $(V_1 - V_2)$ will be in metres per second. In practice it is not necessary actually to calculate the pressure from $p = -\int\rho g\,\mathrm{d}z$, it is sufficient to use $p = -10^4 z$ (because the mean $\rho(s, t, p) \simeq 1035\,\mathrm{kg\,m}^{-1}$ to 5000 m depth and therefore $\rho g \simeq 1.014 \times 10^4\,\mathrm{Pa\,m}^{-1}$ or $10^4\,\mathrm{Pa\,m}^{-1}$ within 1.5 %). The reason is simply that over practical distances L, say up to about 100 km, the vertical density structures are generally sufficiently similar that when the two integrals calculated using $p = -10^4 z$ are subtracted, the remaining error will be negligible compared with the observational errors.

The result of the calculation with equations (8.9) is a value for $(V_1 - V_2)$, the difference between the current at pressure p_1 (depth $\simeq 10^{-4}p_1$ metres) from that at pressure p_2 (depth $\simeq 10^{-4}p_2$) *averaged* between the stations A and B. Its direction is perpendicular to the line AB, i.e. it is the component in the direction perpendicular to AB of the actual current difference in the sea. In the northern hemisphere, it would be directed "into the paper" in the Fig. 8.4 or "out of the paper" in the southern hemisphere. With station B to the right of station A, as in Fig. 8.4, a negative value of $(\Delta\Phi_B - \Delta\Phi_A)$ would indicate flow "out of the paper" in the northern hemisphere and "into the paper" in the southern hemisphere. In the figure as drawn, the slopes between the stations are constant but this need not be the case. Equations (8.9) are based on the difference in distance between the isobaric surfaces *at* the two stations and therefore on the difference in the *average* slope, so they give the average value between A and B of the difference

$(V_1 - V_2)$ in the horizontal velocity component (perpendicular to AB). This may be thought of as the *velocity shear* $(V_1 - V_2)/\bar{z}$ where $\bar{z}$ is the mean vertical distance between the isobars p_1 and p_2 between stations A and B. Note also that equations (8.9) are valid no matter what geographical direction the line AB has in the horizontal.

A better way to state the current direction on one pressure surface is to say that if this surface is sloping then, in the northern hemisphere, the current will be along the slope in such a direction that the surface is higher on the right (on the left in the southern hemisphere). This statement is useful when considering surface currents because the sea surface is the isobaric surface $p = 0$. An application will be given in Section 8.5. However, plots of isobars below the surface are not generally made when analysing oceanographic data and a more practical approach is to consider the density distribution which gives rise to the pressure distribution. Since the slopes of p_0, p_1 and p_2 as shown would occur if the water on the right (station B) were lighter (less dense) than on the left (station A), a rule for direction is that, in the northern hemisphere, the current flows *relative to the water just below it* with the "lighter water on its right" (and vice versa in the southern hemisphere). (If the water is less dense at B than at A on the average, it takes a deeper column of water to produce the same pressure change (p_0 to p_1 and p_1 to p_2) as shown in Fig. 8.4.)

As the isopycnals in a section of the real ocean are generally wavy rather than straight and vary irregularly in angle relative to each other over finite depth ranges (e.g. Fig. 8.5), a more practical statement of the relation between density distribution and current direction is that given by Sverdrup *et al.* (*The Oceans*, 1946, p. 449): "In the northern hemisphere, the current at one depth relative to the current at a greater depth flows away from the reader if, on the average, the δ or σ_t curves in a vertical section slope downward from left to right in the

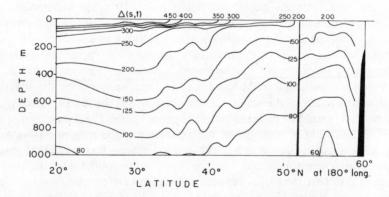

FIG. 8.5   *Vertical section of specific volume (as $\Delta_{s,t}$) for a real ocean region to show the irregular character of isopleths, (Norpac Atlas, 1955, fig. 88).*

interval between two depths, and toward the reader if the curves slope downward from right to left." (Curves of $\Delta_{s,\,t}$ may be substituted for δ curves.) The directions will be opposite in the southern hemisphere. This matter of the relation between current direction and the slopes of isobars and property curves in vertical sections will be discussed further in Sections 8.45 and 8.7.

In Fig. 8.5, if one assumes that the thermosteric anomaly ($\Delta_{s,\,t}$) curves become level in the deep water (well below 1000 m): (a) between 20°N and 30°N, the general flow below about 300 m would be "into the paper", i.e. to the west (the North Equatorial Current), (b) between 30°N and 48°N, the general flow would be to the east (the North Pacific Drift or Subarctic Current), (c) between 48°N and 52°N (the Aleutian Island chain) the general flow would be to the west (the Aleutian Stream which forms the northern side of the cyclonic (anticlockwise) Aleutian Gyre of the northeast Pacific, and (d) between 52°N and the continental slope at about 59°N, the general flow is the cyclonic circulation of the deep basin of the Bering Sea. The "waves" in the $\Delta(s,\,t)$ curves between 33°N and 45°N might be due to eddies within the general flow.

8.43   An example of the calculation of a geostrophic velocity profile

In Tables 8.1 (a and b), the first three columns show the depths and observed temperatures and salinities at two stations, A and B, in the region of the North Atlantic Drift (the extension of the Gulf Stream toward Europe). In successive columns are given the corresponding values at each depth for σ_t, $\Delta_{s,\,t}$, $\delta_{s,\,p}$, $\delta_{t,\,p}$, δ (= the sum of the previous three columns), $\bar{\delta}$ (= the mean value of δ between each successive pair of depths, $\bar{\delta} \times \Delta p$), and finally $\Sigma(\bar{\delta} \times \Delta p) = \Delta\Phi$ which is the sum of the values in the previous column from each successive level to the 1000 m level. Note that $\Delta p = -\Delta z \times 10^4$ Pa $= 10^4 \times$ depth difference in metres has been used in calculating $\Delta\Phi$. As stated previously, this may involve an error up to about 1.5% in Δp, and hence in $\Delta\Phi$, but as the error will be almost the same for both stations, when the difference ($\Delta\Phi_B - \Delta\Phi_A$) is taken later in the calculation it will also have an error of up to 1.5% which is small relative to the errors due to the limited accuracy of the observations.

In Table 8.2, the values calculated for $\Delta\Phi$ for stations B and A are listed and then the difference between them ($\Delta\Phi_B - \Delta\Phi_A$). Finally the relative speed is calculated for each depth (relative to zero speed assumed at 1000 m) from equation (8.9A) using $L = 5 \times 10^4$ m because the stations are 27 minutes of latitude apart (= 27 n ml = 50 km), the stations being at the same longitude, and the mean value for $\sin\phi = 0.655$.

The values of the speed relative to that at 1000 m are plotted against depth in Fig. 8.6(a). The velocity component is directed to the east because relative to 1000 m depth the isopycnals slope down from left to right (Fig. 8.6(b)).

TABLE 8.1.(a)   *Oceanographic data, etc., and calculation of geopotential anomalies* ($\Delta\Phi$)
for station A
(using tables in *Processing Oceanographic Data*, Lafond, 1951)

Station A 41°55′N, 50°09′W				Units of $10^{-8} m^3 kg^{-1}$				Units of $m^3 kg^{-1} Pa = m^2 s^{-2}$		
Depth (m)	T°C	S	σ_t	$\Delta_{s,t}$	$\delta_{s,p}$	$\delta_{t,p}$	δ	$\bar{\delta}$	$\bar{\delta} \times \Delta p$	$\Sigma(\bar{\delta} \times \Delta p)$ $= \Delta\Phi_A$
0	5.99	33.71	26.56	148	0	0	148			6.638
								146	0.365	
25	6.00	33.78	26.61	144	0	0	144			6.273
								135	0.338	
50	10.30	34.86	26.81	125	0	1	126			5.935
								126	0.315	
75	10.30	34.88	26.83	123	0	2	125			5.620
								122	0.305	
100	10.10	34.92	26.89	117	0	2	119			5.315
								112	0.560	
150	10.25	35.17	27.06	101	0	3	104			4.755
								99	0.455	
200	8.85	35.03	27.19	89	0	4	93			4.300
								83	0.830	
300	6.85	34.93	27.41	68	0	5	73			3.470
								65	0.650	
400	5.55	34.93	27.58	52	0	5	57			2.820
								52	1.040	
600	4.55	34.95	27.71	39	0	7	46			1.780
								45	0.900	
800	4.25	34.95	27.74	37	0	8	45			0.880
								44	0.880	
1000	3.90	34.95	27.78	33	0	10	43			0

Note also that $(\Delta\Phi_B - \Delta\Phi_A)/g$ gives an estimate of the height difference (*dynamic topography*) of isobaric surfaces at the two stations, e.g. at the sea surface the difference is 0.13 m, that is the depth of the water column from the sea surface to the pressure level of 10^4 kPa (corresponding to about 1000 m depth) differ by only 0.13 m! If the 10^4 kPa pressure surface is also level, which it will be if the velocity there is zero, then the water surface is 0.13 m higher at station B than at station A. In the mixed units system which used dynamic metres, $\Delta D/0.98$ gave height differences in metres, so the ΔD values were numerically almost the same as the height differences. Also $\Delta\Phi/g$ is the difference in depth over a given pressure difference from the depth the water would have if it were standard water of $S = 35$ and $T = 0°C$.

The above computation has been carried out with the readily available tables of specific volume anomaly terms and the older definition of salinity. If Practical Salinity values are used (calculated using PSS 78 from salinometer measurements or by conversion from older salinity values (as described by Lewis and Perkin, 1981)) together with suitable computing facilities, then the IES 80 could

TABLE 8.1(b)   *Oceanographic data, etc., and calculation of geopotential anomalies* ($\Delta\Phi$)
for station B
(using tables in *Processing Oceanographic Data*, Lafond, 1951)

Depth (m)	T°C	S	σ_t	$\Delta_{s,t}$	$\delta_{s,p}$	$\delta_{t,p}$	δ	$\bar{\delta}$	$\bar{\delta}\times\Delta p$	$\Sigma(\bar{\delta}\times\Delta p)=\Delta\Phi_B$
									Station B 41°28′, 50°09′W	Units of $10^{-8}\,\mathrm{m^3\,kg^{-1}}$ · Units of $\mathrm{m^3\,kg^{-1}\,Pa=m^2 s^{-2}}$
0	13.04	35.62	26.88	118	0	0	118			7.894
								119	0.298	
25	13.09	35.63	26.88	118	0	1	119			7.596
								119	0.298	
50	13.07	35.63	26.88	118	0	1	119			7.298
								119	0.298	
75	13.05	35.64	26.89	117	0	2	119			7.000
								120	0.300	
100	13.05	35.62	26.88	118	0	3	121			6.700
								122	0.610	
150	13.00	35.61	26.88	118	0	4	122			6.090
								122	0.610	
200	12.65	35.54	26.90	116	0	5	121			5.480
								117	1.170	
300	11.30	35.36	27.02	105	0	7	112			4.310
								98	0.980	
400	8.30	35.09	27.32	76	0	7	83			3.330
								70	1.400	
600	5.20	34.93	27.61	49	0	8	57			1.930
								52	1.030	
800	4.20	34.92	27.73	38	0	8	46			0.900
								45	0.900	
1000	4.20	34.97	27.77	34	0	10	44			0

TABLE 8.2.   *Geopotential anomalies from Tables 8.1 (a, b) and calculated mean relative
velocities between stations A and B at various depths*

Depth (m)	$\Delta\Phi_B$ ($\mathrm{m^2\,s^{-2}}$)	$\Delta\Phi_A$ ($\mathrm{m^2\,s^{-2}}$)	$(\Delta\Phi_B-\Delta\Phi_A)$ ($\mathrm{m^2\,s^{-2}}$)	Vrel ($\mathrm{m\,s^{-1}}$)	
0	7.894	6.638	1.256	0.26	Stn. A: 41°55′N, 50°09′W
25	7.596	6.273	1.323	0.27	Stn. B: 41°28′N, 50°09′W
50	7.298	5.935	1.363	0.28	Diff. = 27′, 0′
75	7.000	5.620	1.380	0.29	i.e. stations are 27 n ml apart,
100	6.700	5.315	1.385	0.29	$= 50\,\mathrm{km} = 5\times10^4\,\mathrm{m}$.
150	6.090	4.755	1.335	0.28	
200	5.480	4.300	1.180	0.24	sin 41°28′ = 0.662
300	4.310	3.470	0.840	0.17	sin 41°55′ = 0.668
400	3.330	2.820	0.510	0.11	mean sin ϕ = 0.665
600	1.930	1.780	0.150	0.03	
800	0.900	0.880	0.020	0.005	$2\Omega \sin\phi = 9.70\times10^{-5}\,\mathrm{s^{-1}}$
1000	0	0	0	0	

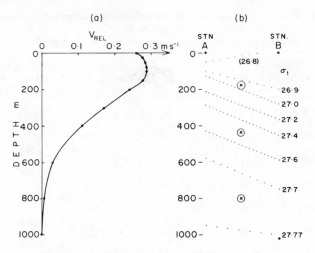

FIG. 8.6   (a) *Profile of velocity relative to that at 1000 m depth as calculated from data of Tables 8.1 and 8.2. (b) Mean slopes of isopycnal surfaces between stations A and B for the same data.*

be used for the calculations. In this case, a value for the integral of α_B (equation (8.8)) and a similar one for α_A could be calculated directly and subtracted to provide the difference inside the square brackets in equation (8.9A) (since tables for the specific anomaly terms are not yet available for the IES 80).

8.44   *An alternative derivation of the geostrophic equation*

The geostrophic equation may also be derived directly from the equations of motion (6.2) as follows. Assume

no acceleration, i.e. $$\frac{du}{dt} = \frac{dv}{dt} = \frac{dw}{dt} = 0,$$

and the vertical velocity w = small so that $2\Omega \cos \phi \, w$ may be neglected, and no other forces, i.e. $\mathbf{F} = 0$.

Justification for these assumptions for the interior of the ocean was given in Section 7.3. The z-equation becomes

$$0 = 2\Omega \cos \phi \, u - \alpha \frac{\partial p}{\partial z} - g,$$

or $$\delta p = -\rho \, \delta z (g - 2\Omega \cos \phi \, u),$$

which is the hydrostatic equation with the addition of the z-component of the Coriolis acceleration. The latter, however, is small, e.g. for $\phi = 45°$, and

$u = 2.5\,\text{m s}^{-1}$ (a high value), then the Coriolis term is about $2.6 \times 10^{-4}\,\text{m s}^{-2}$, which is negligible compared with $g = 9.8\,\text{m s}^{-2}$. Thus the hydrostatic equation (8.2) applies with all the accuracy needed even in water moving at realistic ocean speeds. This fact is fortunate as otherwise the calculation of current speeds by means of the geostrophic equation would be rendered more complicated.

Now the x- and y-equations become

$$\left.\begin{array}{l} 0 = 2\Omega\sin\phi\,v - \alpha\dfrac{\partial p}{\partial x} \\[2ex] 0 = -2\Omega\sin\phi\,u - \alpha\dfrac{\partial p}{\partial y} \end{array}\right\} \quad \text{component geostrophic equations.} \quad (8.10)$$

These say that for purely horizontal motion

$$\text{Coriolis force} = -\,\text{Pressure force.}$$

Notice that in the x- and y-equations we must not reject the Coriolis term—it is small but the only other term in each equation is the horizontal pressure term which is also small but must be balanced for steady-state flow. (One can only neglect a small term if there are other much larger terms in the equation as in the equation for the vertical component as shown above.)

Also notice that in the component equations (and see Fig. 8.7 (a, b)) the pressure gradient in the x-direction, $\partial p/\partial x$, is associated with v, while the pressure gradient in the y-direction, $\partial p/\partial y$, is associated with u. The x- and y-equations can be combined into a single one

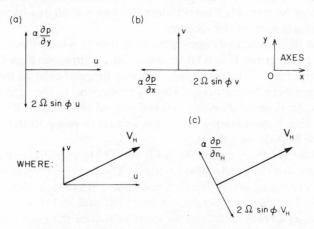

FIG. 8.7 (a, b) Directional relationships of velocity components (u, v) to pressure and Coriolis force terms (northern hemisphere), (c) directional relationship of total horizontal velocity (V_H) to pressure and Coriolis force terms (northern hemisphere).

$$2\Omega \sin \phi \, V_H = \alpha \frac{\partial p}{\partial n_H} \tag{8.11}$$

where V_H = magnitude of the vector sum of u and v
                    $= (u^2 + v^2)^{1/2}$,

and $\partial p / \partial n_H$ = the horizontal pressure term perpendicular to the
                    direction of $\mathbf{V}_H$ (see Fig. 8.7(c)).

One way to remember the relative directions of the pressure force and the velocity is to think of the sequence:

(1) the pressure gradient is initiated somehow,
(2) the fluid starts to move down the gradient,
(3) the fluid then experiences the Coriolis force to the right (in the northern hemisphere) and therefore swings to the right,
(4) the fluid eventually moves *along* the isobars, i.e. *along* the slope, not down it, with the pressure force *down* the slope balanced by the Coriolis force *up* the slope.

The equivalent situation in the atmosphere was shown in Fig. 8.2 to aid in defining the terms cyclonic and anticyclonic. It is left as a simple exercise for the reader to verify that the circulations in this figure are consistent with geostrophy.

Notice that an alternative procedure would be to start a fluid moving in some direction, Coriolis force would then make it swing to the right (in the northern hemisphere) and pile up there (slope up to the right) so developing a pressure force to the left. Therefore the geostrophic equation simply tells us that the pressure force balances the Coriolis force—it does not tell us which came first, the pressure gradient or the motion.

Equation (8.11) is actually applicable no matter in which direction we take the pressure derivatives. If n_H is taken in an arbitrary direction then V_H becomes V_1, the component of the geostrophic velocity perpendicular to the direction n_H. In the northern hemisphere, taking n_H to increase to the right the flow is away from the observer if $\partial p / \partial n_H > 0$ and toward the observer if $\partial p / \partial n_H < 0$. This is another way of stating that if the isobars slope up to the right (as in Fig. 8.4) the flow is "into the paper".

How do we get from equations (8.10) and (8.11) to (8.9A) and (8.9B), the practical forms of the geostrophic equation? The pressure derivatives in (8.10) and (8.11) are taken on surfaces of constant z which are also surfaces of constant Φ. The pressure derivatives in (8.10) and (8.11) are not directly measurable, as already noted, so we must introduce the geopotential. Now using the rule from differential calculus for implicit functions

$$\left(\frac{\partial p}{\partial x} \right)_{y,z \text{ or } \Phi \text{ constant}} = - \left(\frac{\partial \Phi}{\partial x} \right)_{y,p \text{ constant}} \bigg/ \left(\frac{\partial \Phi}{\partial p} \right)_{x,y \text{ constant}}$$

and remembering that $\partial\Phi/\partial p = -\alpha = -1/\rho$ (Section 8.3) we get $\partial p/\partial x$ $= \rho(\partial\Phi/\partial x)$ where this is the change in Φ as we go along an *isobar* in the *x*-direction. Likewise $\partial p/\partial y = \rho(\partial\Phi/\partial y)$ and $\partial p/\partial n_H = \rho(\partial\Phi/\partial n_H)$. These relations between p and Φ gradients can easily be obtained from first principles instead of the calculus rule. Suppose that one moves a small distance δn_H from the point A_1 in Fig. 8.4. Over this distance the height on the p_1 isobar will change by δz, the pressure on Φ_1 will increase by $\rho g\,\delta z$ and Φ will increase by $g\,\delta z$. Thus $\delta p = \rho\,\delta\Phi$ and dividing both sides by δn_H and taking the limit as $\delta n_H \to 0$ gives the same relation for the derivatives as does the calculus rule. Substituting the Φ term for the p terms in (8.10) and (8.11) gives an alternate form for the geostrophic equations.

Now these Φ gradients cannot be measured either, but differences from one level to another can be obtained from the density field. From equation (8.5) we have

$$\Phi_1 = \Phi_2 + \Delta\Phi_{\text{std}} + \Delta\Phi$$

and $\Delta\Phi_{\text{std}}$ is the same at every station, so its derivatives with respect to horizontal coordinates are always zero. Consider the *x*-equation at levels 1 and 2 of Fig. 8.4,

$$2\Omega\sin\phi\,v_1 = \alpha\left(\frac{\partial p}{\partial x}\right)_{\text{on }\Phi_1} = \left(\frac{\partial\Phi_1}{\partial x}\right)_{\text{on }p_1} = \frac{\partial\Phi_2}{\partial x} + \frac{\partial(\Delta\Phi)}{\partial x}, \quad \text{and}$$

$$2\Omega\sin\phi\,v_2 = \frac{\partial\Phi_2}{\partial x} \quad \text{and the difference is}$$

$$2\Omega\sin\phi\,(v_1 - v_2) = \frac{\partial(\Delta\Phi)}{\partial x}. \quad \text{Likewise}$$

$$2\Omega\sin\phi\,(u_1 - u_2) = -\frac{\partial(\Delta\Phi)}{\partial y} \quad \text{and} \quad 2\Omega\sin\phi\,(V_1 - V_2) = \frac{\partial(\Delta\Phi)}{\partial n_H} \quad (8.12)$$

where V_1, V_2 represent the horizontal velocity components perpendicular to the direction of n_H, at levels 1 and 2.

These are differential forms of the geostrophic equation written in a way which can be used with the kind of observations which we can make. Equations (8.9), the practical equations, appear to be finite difference forms of equation (8.12) but in fact they are integral forms. The average along a direction n_H from 0 to L is, by definition,

$$= \frac{1}{L}\int_0^L (\text{quantity to be averaged})\,dn_H.$$

Applying this to equation (8.12), using f for $2\Omega\sin\phi$ and an overbar to indicate an average, gives

$$\overline{f(V_1 - V_2)} = \frac{1}{L}\int_A^B \frac{\partial(\Delta\Phi)}{\partial n_H}\, dn_H = \frac{1}{L}(\Delta\Phi_B - \Delta\Phi_A). \qquad (8.13)$$

The only difference between equations (8.13) and (8.9) is that averaging is not explicitly shown in the latter and we must assume that $\overline{f(V_1 - V_2)} = \bar{f}(V_1 - V_2)$ which will be a very good approximation since over the distances used in practice f is nearly constant. In the example given in Table 8.2 with n_H in the southward direction (for which f variations with n_H are a maximum), f changed by only 1% between A and B.

8.45   The "thermal wind" equations

These are another variation of the geostrophic equations originally derived to show how temperature differences in the horizontal could lead to vertical variations in the geostrophic wind velocity, hence the term *thermal wind equations*. Consider the x-equation of (8.10) with f introduced and both sides multiplied by ρ, i.e. $\rho f v = \partial p/\partial x$. Differentiation with respect to z gives

$$\frac{\partial(\rho f v)}{\partial z} = \frac{\partial}{\partial z}\frac{\partial p}{\partial x}.$$

Changing the order of differentiation, which will be correct for a variable such as p, and using the hydrostatic equation, $\partial p/\partial z = -\rho g$, gives

$$\frac{\partial(\rho f v)}{\partial z} = \frac{\partial}{\partial x}\frac{\partial p}{\partial z} = \frac{\partial(-\rho g)}{\partial x} = -g\frac{\partial\rho}{\partial x}.$$

The same procedure can be followed for the y-equation and the thermal wind equations are

$$\frac{\partial(\rho f v)}{\partial z} = -g\frac{\partial\rho}{\partial x}, \quad \text{and} \quad \frac{\partial(\rho f V_H)}{\partial z} = -g\frac{\partial\rho}{\partial n_H}. \qquad (8.14)$$
$$\frac{\partial(\rho f u)}{\partial z} = g\frac{\partial\rho}{\partial y},$$

Again these equations show that from the density field we can only determine the vertical *variation* of velocity, i.e. the velocity shear $\partial u/\partial z$ and $\partial v/\partial z$. The horizontal density gradients are large enough to be observed. Because of depth uncertainties, the derivatives (based on finite difference approximations) will not be exactly on level surfaces but the errors are small whereas for $\partial p/\partial x$, $\partial\Phi/\partial x$, etc., they are much larger than the actual values. In practice, if one has tables for α rather than for ρ, as has been the usual case, one would use the fact that $(1/\alpha)(\partial\alpha/\partial x) = -(1/\rho)(\partial\rho/\partial x)$ and likewise for the y-derivative. In meteorology, where ρ can be expressed in terms of a virtual potential temperature, gradients of this quantity may be used in place of ρ. In the ocean, in the upper 1000 m or so, as a first approximation $\partial\sigma_t/\partial x$ and $\partial\sigma_t/\partial y$ can probably be used for $\partial\rho/\partial x$ and $\partial\rho/\partial y$, respectively. In deeper water, this is not likely to be a good approximation if temperature gradients are the dominant contribution to density gradients. (See the discussion in Chapter 5 on the use of σ_t as an approximation in the static stability equation—the terms neglected compared with $\partial\sigma_t/\partial T$ and $\partial\sigma_t/\partial S$ are the same as there, although they are now the coefficients of horizontal rather than vertical property gradients.) However, in deep water, the "thermal wind" equations are not likely to give useful results.

When we discussed the Boussinesq approximation in Section 7.41 we said that density variations could be ignored in the horizontal equations. However, they enter the thermal wind equations because the buoyancy effects do affect the pressure field, as we also noted, and these equations were derived using the hydrostatic equation in which density variations must be included. In the terms $\partial(\rho f v)/\partial z$ and $\partial(\rho f u)/\partial z$ the effect of density variation is small compared with the effect of vertical gradients of u and v and it would be consistent with the Boussinesq (and be a good) approximation to replace these terms with $\rho f(\partial v/\partial z)$ and $\rho f(\partial u/\partial z)$ respectively. Here f also comes outside the derivative because it does not depend on z.

The (northern hemisphere) rule "light water on the right" relative to the water below when looking in the flow direction comes from the thermal wind equations (8.14). Suppose ρ decreases to the east, then $\partial\rho/\partial x < 0$ and $\partial(\rho f v)/\partial z\,(=\rho f\,\partial v/\partial z$ in the Boussinesq approximation$) > 0$, that is v increases as we go upward in the water column. If ρ decreases to the south $\partial\rho/\partial y > 0$ and $\partial(\rho f u)/\partial z > 0$. The example given in Subsection 8.43 illustrates the rule very nicely. From 1000 m up to 100 m depth (Tables 8.1 (a) and (b)) $\rho_B < \rho_A$ at each depth and V_{rel} ($= u$, the eastward component) becomes more positive. Above 100 m, $\rho_B > \rho_A$ and V_{rel} decreases. Relative to the flow at 100 m, the flow above is negative (to the west) and the light water is to the right. Since the flow at 100 m is fairly large and to the east, the flow above 100 m is actually to the east because it changes only slightly relative to the flow at 100 m.

The "rule" also applies in an integral sense as already noted. That is, $V_1 > V_2$ if $\partial(\Delta\Phi)/\partial n_H > 0$ but $\partial\bar{p}/\partial n_H$ will have the opposite sign to $\partial(\Delta\Phi)/\partial n_H$ where $\bar{p}$ is the average density between pressure surfaces p_1 and p_2 or indeed between the depths of Φ_1 and Φ_2 since the difference in depth between Φ and p surfaces is 0.14 m or less. In the example of Subsection 8.43, $\Delta\Phi_B > \Delta\Phi_A$ at all depths above 1000 m, the average density is lower at B than at A and the flow at all depths is to the east relative to that at 1000 m.

8.5   Deriving absolute velocities

The geostrophic calculation gives the relative velocity component $(V_1 - V_2)$ between two depths, i.e. the velocity shear dV/dz. Therefore if we know the absolute value of either V_1 or V_2 we will know the absolute value of the other. There are several possibilities:

(1) assume that there is a *level* or *depth of no motion* (*reference level*) e.g. that $V_2 = 0$ in deep water, and then calculate V_1 for various levels above this (the classical method);

(2) when there are stations available across the full width of a strait or ocean, calculate the velocities and then apply the equation of continuity to see if the resulting flow is reasonable, i.e. complies with all facts already known about the flow and also satisfies conservation of heat and salt;

(3) use a "level of *known* motion", e.g. if surface currents are known or if the currents have been measured at some depth(s) by current meters or neutrally buoyant floats (preferably while the density measurements for the geostrophic calculations were being made). In the future, it is possible that satellite measurements of the surface slope may enable the surface currents to be calculated, at least in regions of strong currents.

Note that the above techniques are the "classical" ones for obtaining absolute velocities; a more recent technique, the "beta-spiral" approach, is described in Section 8.9.

Since surface velocities are important and can be inferred quickly from the slope of the sea surface (which is assumed to be isobaric) it is common to plot the geopotential (or "dynamic") topography of the sea surface relative to some deeper surface, if a sufficient grid of station data is available. The relative current directions will be parallel to lines of constant geopotential and relative speeds will be inversely proportional to the spacing of the lines (i.e. close spacing = steep slope = large speed, e.g. Figs. 8.8 or 11.4(b)). It is also possible to plot the geopotential topography of subsurface isobaric surfaces to deduce the motion there.

Remember that these geopotential topography plots are usually based on some assumed level of no motion, generally in the deep water, unless adequate direct current measurements are available which is rarely the case. If the average upper-layer currents are much larger than the average deep currents, which is often the case, we may get quite good values for them even if the deep currents are not exactly zero. Note, however, that although a small current which is neglected (or not known) in the deep water may not affect the calculated *velocities* in the upper layer very much, it may make a substantial contribution to the *total volume transport* when integrated over the full depth. For instance, an average current of 10 cm s^{-1} in the upper 1000 m based on zero current assumed at (and below) 1000 m would give a volume transport from surface to 4000 m of 100 m^3 s^{-1} for each metre width of the current. If the actual current from 1000 to 4000 m were 2 cm s^{-1} in the same direction as that in the upper layer, this would give rise to a 20 % error for the upper layer

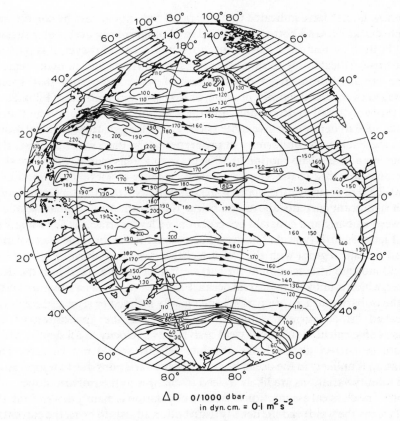

FIG. 8.8   *Annual mean dynamic topography of the Pacific Ocean sea surface relative to 1000 dbar (10 000 kPa) in dynamic centimetres (= 0.1 m² s⁻²), i.e ΔD = 0/1000 dbar); 36 356 observations. Arrows indicate flow directions deduced from topography. (Wyrtki, 1974, fig. 5)*

velocity but the total volume transport from surface to 4000 mm would be 180 m³ s⁻¹ or 80% more than that assuming zero current below 1000 m.

Notice that there is one known velocity region which cannot be used as a level of no motion—the sea bottom. The reason why this level cannot be used is that the velocity tends to zero there because of the action of friction, a force which was deliberately assumed to be negligible when deriving the geostrophic equation. Remember then that the geostrophic equation does not apply in regions where friction is important.

8.6   Relations between isobaric and level surfaces

Classically the basis for the assumption of a level of no motion was the belief that velocities are small in deep water. Observations in recent years with

Swallow floats* have indicated that this belief may not always be correct, and ripple marks on sand bottoms recorded in photographs in deep water suggest that bottom currents of 0.5 m s^{-1} or more may occur. However, it is possible that these indicate only local and/or transient currents and in many regions there are indications from distributions of water properties that speeds averaged over several months or more, or over tens or hundreds of kilometres, are probably small in deep water, and the selection of a level of no motion at about 1000 m depth may give quite good results for geostrophic calculations.

In the Pacific, the uniformity of properties in the deep water suggests that assuming a level of no motion at 1000 m or so is reasonable, with very slow motion below this. In the Atlantic, there is evidence of a level of no motion at 1000–2000 m (between the upper waters and the North Atlantic Deep Water) with significant currents above and below this depth. A selection of relations between isobaric and constant geopotential or level surfaces is shown in Fig. 8.9. Figure 8.9(a) is typical of the west Pacific and Fig. 8.9(b) of the west Atlantic (Gulf Stream region) with the characteristics described above. Figure 8.9(c) would indicate little motion at the surface but increasing speed into the deep water, which is unlikely in the real ocean. Figure 8.9(d) shows a situation where all the isobaric surfaces are parallel and equally inclined to level surfaces—the so-called "slope current" situation. In this case, the application of the geostrophic calculation would yield zero relative velocity at all depths which would be correct although the absolute velocity would not be zero. This situation is unlikely in the ocean because density variations due to temperature and salinity variations are likely to lead to changes in the isobaric slopes with depth. The classical assumption is that the circulation is mainly driven from the surface (by the wind) and the density distribution adjusts to bring the current to zero at mid-depth. However, it is possible that there may be a slope current in the deep water where T and S variations are small, plus an additional vertical variation in the upper 1000 m or so. If we add slopes like those of Fig. 8.9(d) to those of Fig. 8.9(a) we would have such a situation, and there would not be a level of no motion at any depth. Observations, though they tend to be indirect, suggest that deep-water slope currents are at least considerably smaller than near-surface currents. While the deep currents may have lower speeds they may transport large amounts of water if they extend over a large depth range (as indicated in the previous section). Surface currents based on geostrophic calculations are similar to those of Pilot Charts (obtained independently from ship navigation data), which is probably the best evidence that the average speeds of the deep waters are at least small compared with near-surface average speeds.

* These neutrally buoyant floats are sealed aluminum tubes which are ballasted to sink to and then float at a predetermined level where they then travel with the water; they have a sound source so that they can be tracked from a ship or shore station. They are named after their inventor, John Swallow (1955, see also Baker, 1981).

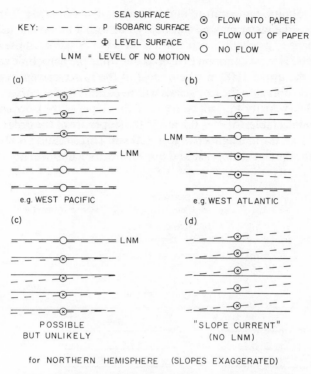

FIG. 8.9   *Relationship between isobaric and level surfaces and flow directions for northern hemisphere.*

In Fig. 8.9, (a), (b) and (c) are examples of "baroclinic" situations while (d) is a "barotropic" one. These terms will be defined in the next section.

8.7   Relations between isobaric and isopycnal surfaces and currents

An *isobaric* surface in a fluid is one on which the hydrostatic pressure is constant, while an *isopycnal* surface (sometimes called *isosteric*) is one on which the density of the fluid is constant. When the density of a fluid is a function of pressure only (i.e. $\rho = \rho\,(p)$), as in fresh water of uniform potential temperature, the isobaric and isopycnal surfaces are parallel to each other—this is called a *barotropic* field of mass. If the density is a function of other parameters as well and actually varies horizontally with them, the isobaric and isopycnal surfaces may be inclined to each other—the *baroclinic* field. This situation could occur in a freshwater lake where the density was a function of temperature as well as pressure ($\rho = \rho\,(t, p)$) or in the sea where density is a function of salinity, temperature and pressure ($\rho = \rho(s, t, p)$). With a barotropic of mass the water may be stationary but with a baroclinic field, having horizontal density gradients, such a situation is not possible. In the ocean, the barotropic case is most common in deep water while the baroclinic case is most common in the upper 1000 m where most of the faster currents occur.

In the barotropic case the isopycnals will be parallel to the isobars which are all parallel. The velocity V_{rel} will be zero and the slopes of the isopycnals will be small and undetectable; for $V = 0.1$ m s^{-1} the slopes are of the order of 10^{-6} at mid-latitudes, i.e. 0.1 m height change in 100 km. This situation is illustrated in Fig. 8.10(a) for the stationary case and Fig. 8.10(b) for a uniform flow. Note that

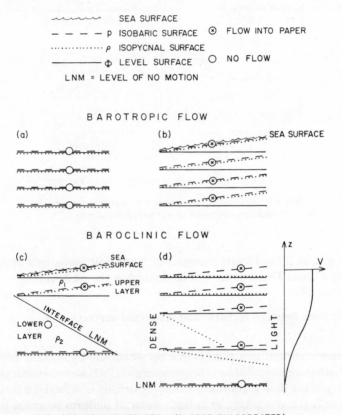

FIG. 8.10 *Schematic examples of: (a, b) barotropic and (c, d) baroclinic fields of mass and pressure with related flow directions for the northern hemisphere. Slopes are exaggerated: p and ρ by about 10⁵ in (b, c), interface by 10³ in (c), p by 10⁵ in (d), ρ by 10³ in (d). With this exaggeration, maximum speeds implied are about 0.1 m s⁻¹.*

the slopes in (b) are exaggerated! From the first thermal wind equation (8.14) and with $\partial v/\partial z = 0$ we have $fv\,(\partial \rho/\partial z) = -g\,(\partial \rho/\partial x)$; horizontal gradients of density and corresponding isopycnal slopes arise from vertical variations of density associated with compression effects. In the Boussinesq approximation we neglect density variations on the left-hand side and would get $\partial \rho/\partial x = \partial \rho/\partial y = 0$ in the barotropic case; as we shall see these gradients and corresponding isopycnal slopes are very small compared with those that occur in baroclinic cases, so the approximation is reasonable. Alternatively, we could use potential density, or σ_t when it is a suitable approximation, which would be constant in the barotropic case.

In the baroclinic case, there is no simple relation between the isobars and isopycnals. From the geostrophic equation (8.6) the slopes of isobars are proportional to the velocity. From the thermal wind equation (8.14) the horizontal density gradients and corresponding isopycnal slopes are proportional to $\partial V/\partial z$, the variation of velocity with depth change or the vertical shear. Note that if density *decreases* to the right then the isopycnals will slope *down* to the right.

Consider first the simplest idealization of baroclinic flow—a two-layer system with the upper layer moving and the lower layer stationary. The upper layer has constant potential density ρ_1 and the lower layer has constant potential density ρ_2. In Section 9.14.2 we show that the slope of the interface between the two layers (the isopycnal) $= -\rho_1/(\rho_2 - \rho_1)$ times the slope of the surface (an isobar). Thus in this simple case, the isopycnal slope is opposite in sign to the isobaric slope. The magnitude of the isopycnal slope is about 1000 times that of the isobaric slope since $(\rho_2 - \rho_1) \simeq 1\,\mathrm{kg\,m^{-3}}$ and $\rho_1 \simeq 1000\,\mathrm{kg\,m^{-3}}$ for a model of the ocean. This situation is illustrated in Fig. 8.9(c). Note again that the slopes are much exaggerated, with the isobaric slopes and isopycnal slopes in the upper layer being much more exaggerated than the interface slope. One can also consider case (c) to be a combination of cases (b) for the upper layer and (a) for the lower layer. While this model is simplistic and a bit unphysical, since both ρ and V have discontinuities at the interface, it does illustrate that in baroclinic cases isopycnal slopes will be much larger than isobaric slopes and of opposite sign, at least in part of the flow.

As a more realistic case consider the example of Section 8.43. By linear interpolation the depth of the $\sigma_t = 27.0$ surface is 130 m at station A and 280 m at station B. Thus it descends 150 m from A to B while the isobars ascend about 0.13 m from A to B. The slopes of σ_t surfaces will differ from the slopes of isopycnals (density *in situ*) by the slopes of isobars which are negligible within the observation errors, so we can use σ_t slopes to represent isopycnal slopes. The value $\sigma_t = 27.7$ is found at 570 m at station A and at 750 m at station B for a drop of about 180 m. In both cases the isopycnal slopes are opposite in sign and about 1000 times the size of the isobaric slopes. However, at 100 m the isopycnal is level within observation error and above 100 m the isopycnals slope the same way as the isobars because u decreases with height above 100 m. For example, $\sigma_t = 26.8$ is at a depth of about 50 m at station A and must reach the surface before station B, so the slope is greater than 50 m in 50 km. Overall, the isopycnal slopes are opposite in sign and much larger in size than the isobaric slopes.

We see from this example that the horizontal density gradients and corresponding isopycnal slopes are large enough to be observed whereas the pressure gradients and isobar slopes are too small to observe, except perhaps by satellite altimetry for strong currents (since the expected accuracy of ± 0.1 m is comparable to the true height variations over much of the ocean). Density differences in the ocean are small and so isopycnal slopes are magnified from isobaric slopes by the factor $\rho/\Delta\rho$. If we could measure density only to the same accuracy as pressure and depth (about 1 in 10^3) we would not be able to detect the isopycnal slopes either. Fortunately we can measure density changes, through salinity and temperature, to higher accuracy (about ± 5 in 10^6) or 0.005 in σ_t. If we could measure depth and pressure differences to 1 in 10^5 we could use the pressure field to get the relative velocity field directly. Finally, if we could establish the sea surface level to ± 1 cm in 100 km we could get the absolute velocity field to ± 1 cm s^{-1} at mid-latitudes but this ability seems unlikely in the foreseeable future.

Figure 8.10(d) shows a somewhat more realistic example similar to that of Subsection 8.43 but without the decrease of V above 100 m. Note that unlike the previous figures the exaggerations of the p and ρ slopes are different, i.e. p by $\simeq 10^5$ and ρ by 10^3, to illustrate that ρ slopes are greater although we cannot show the 1000 : 1 ratio of the ocean. In this case, ρ slopes due to pressure effects are too small to show compared with the slopes associated with the velocity shear in the vertical. In the upper part of the flow, V is large and independent of z. The pressure surfaces are parallel and slope up to the right (slope in figure $\simeq 10^{-1}$, true slope $\simeq 10^{-6}$). The ρ surfaces are level (actual slopes with this exaggeration would be 10^{-3} and true slopes 10^{-6}). When the velocity begins to decrease with depth, the isobar slopes gradually decrease to the level of no motion. The isopycnal slopes are large at first and down to the right where the shear is large; they gradually decrease to the level of no motion where $\partial V/\partial z = 0$ as well as $V = 0$, and there is no discontinuity in either.

If one has a plot of the isopycnal surfaces only (dynamic heights not yet having been computed) then one must integrate mentally starting from a level where the motion is expected to be zero or at least small. In Fig. 8.10 (d), starting at the level of no motion the isopycnal surfaces slope down to the right (light water on the right) so V_{rel} must be increasing upward until the isopycnals level out and V becomes uniform. Since V increases and is into the paper, the isobars must slope up to the right (opposite to the isopycnals in the region where they slope down).

Figure 8.11 shows a reasonably realistic and more complicated case. Again

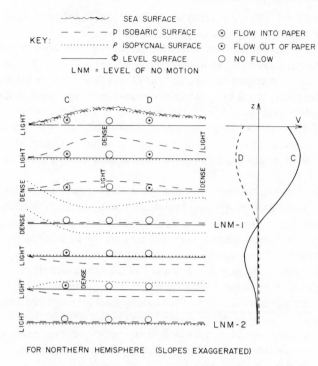

KEY:
〜〜〜〜〜 SEA SURFACE
— — — — p ISOBARIC SURFACE   ⊗ FLOW INTO PAPER
················ ρ ISOPYCNAL SURFACE   ⊙ FLOW OUT OF PAPER
————— Φ LEVEL SURFACE   ○ NO FLOW
LNM = LEVEL OF NO MOTION

FOR NORTHERN HEMISPHERE   (SLOPES EXAGGERATED)

FIG. 8.11 *A more realistic (and complicated) schematic example of relations between baroclinic fields of mass and pressure with related flow directions, and velocity profiles for stations C and D for the northern hemisphere. Slopes are exaggerated: p by about 10^5 and ρ by about 10^3.*

the isobar slope exaggeration is 10^5, the isopycnal exaggeration is 10^3. The flow on the right (region D) is similar to that of the previous figure but in the reverse direction and with a decrease in magnitude near the surface so that there the isopycnals have slopes of the same sign as the isobars. Above the upper level of no motion (*LNM* 1), region C is similar to region D with the flow reversed except that at the upper level of no motion the shear and the isopycnal slope are not zero as the flow reverses. As the flow becomes negative, the isobars slope down to the right; as the shear goes to zero the isopycnals become level. Lower down the isopycnals slope up to the right while the isobars slope down to the right; finally at the lower level of no motion (*LNM* 2) all the surfaces are level again. Overall, since the flow starts at zero and becomes larger either positive or negative upward in the end, the isobars and isopycnals tend to have slopes of opposite sign but locally they may have the same sign or one may be level while the other is sloping. Mentally integrating the isopycnal section is rather more difficult in this case. Inferring the velocity field from the density field requires considerable experience except in very simple cases.

One must also have some idea where the velocity may be zero to get started. One could imagine adding a barotropic field, such as that in Fig. 8.10 (b), so that the flow at C never reversed and the flow at D was positive at depth (although an effect this extreme is unlikely). Then one would have the correct relative field but the absolute field would be quite different. If the level of no motion is picked carefully one can hope to get a reasonable picture. Although we showed the upper level of no motion to be the same at C and D, one must remember that it might change with location and be different in the two regions.

In low- and mid-latitudes, temperature is the main factor determining density and a section of temperature (plotted directly in the field) may be used as a reasonable approximation for a density section. The rule "light water on the right (northern hemisphere) when looking in the flow direction" for the current direction relative to the water below becomes "warm water on the right" with isotherms sloping down to the right. In the example of Subsection 8.43, the water at B is warmer than that at A at each depth (except at 800 m where the difference is small). Thus the velocity increases upward and is to the east. The small velocity decrease above 100 m is not seen in the temperature field; the lower densities at A than at B above this level being caused by the lower salinities. However, the main features of the flow are shown by the temperature field.

Often in oceanography, barotropic flow is thought of as the flow due to a uniform tilt of the pressure surfaces like that of the deep water where the density essentially depends only on pressure, and the velocity is uniform with depth. Baroclinic flow is the part due to additional tilts of the pressure surfaces caused by density variations. Both are geostrophic flows in the sense that there is a balance between the pressure force and Coriolis force. For example, in Fig. 8.12 if ABC is the vertical profile of the horizontal speed, it can be regarded as made up of two parts, a barotropic part V_b and a baroclinic part V_c. In Fig. 8.12(a) the barotropic and baroclinic components are in the same direction while in (b) they are in opposite directions. The baroclinic part may be obtained from a geostrophic (velocity shear) calculation but the barotropic part will not appear in this calculation, it must be obtained by some other means. It appears that this separation is somewhat arbitrary but it is consistent with the general definition of a barotropic fluid that $\rho = \rho(p)$ which is true in deep water. However, some theoretical physical oceanographers take the surface velocity as the barotropic part and variations from this value as the baroclinic part, so it is important to make sure which system is being used in any particular discussion. Taking the barotropic motion to be that associated with the surface slope and velocity is appropriate in the context of time-dependent wave motions which will be discussed in Chapter 12.

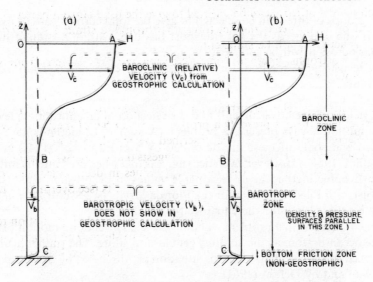

FIG. 8.12 *Horizontal speed as a combination of barotropic (V_b) and baroclinic (V_c) parts of the geostrophic velocity; V_b and V_c in (a) the same direction, (b) opposite directions. (Note that in reality V_b and V_c need not be colinear.)*

8.8 Comments on the geostrophic equation

The procedure of calculating the geostrophic currents from the oceanographic data at two stations, e.g. A and B (Fig. 8.13(a)), yields only the component V_1 of current perpendicular to the line AB. To obtain the total current it is necessary also to make calculations for a second pair of stations, e.g. BC, to get another component (V_2). These may then be added vectorially to obtain the total current ($\mathbf{V}_H$) as in Fig. 8.13(b). Generally one has a grid of stations from

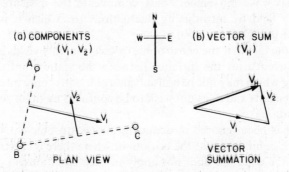

FIG. 8.13 *Stations AB yield velocity component V_1, stations BC yield component V_2; the total horizontal velocity $\mathbf{V}_H$ is the vector sum of these.*

which groups of three can be selected to give the total current pattern over an area. If one only needs the net transport through a strait, a straight line of stations across it will be sufficient.

The geostrophic method for calculating currents suffers from several disadvantages:

(1) It yields only relative currents and the selection of an appropriate level of no motion always presents a problem. (A number of methods have been used and some of these are described by Sverdrup *et al.* (1946) and by Defant (1961).) However, evidence suggests that when used intelligently the geostrophic method gives reasonable values in deep water, e.g. tests by Wüst with Pillsbury's data for the Florida Straits (see Sverdrup *et al.*) and by Knauss (1960) for the equatorial Pacific.

(2) One is faced with a problem when the selected level of no motion reaches the ocean bottom as the stations get close to shore. The (rather artificial) procedures for dealing with this situation are discussed by Sverdrup *et al.* (1946) and by Defant (1961).

(3) It only yields mean values between stations which are usually many tens of kilometres apart. It is impractical to place stations very close because:

  (a) of the limitations to the accuracy of measurement of S, T and p and hence of α (differences between station values must be significantly larger than errors of measurement at an individual station);

  (b) limited navigational accuracy means that the distance (L) between the stations may have a significant error. Of course, if the ship is equipped with accurate position-determining equipment this error will be minimized. Satellite navigation can now be used to determine position to an accuracy of the order of 100 m or better, for a stationary ship, but drift of the ship while on station may introduce significant uncertainty in the value of L to be used in the geostrophic calculation;

  (c) internal wave movements may complicate the measurement of the density field by introducing fluctuations for which it is difficult to correct (see Defant, 1950).

Actually, the fact that the geostrophic calculation only yields a mean value for the current over the distance between the stations introduces some smoothing which may not be a disadvantage if one is only interested in the bulk movements and does not wish to be confused by small-scale or short-term variations.

(4) Friction has been ignored in deducing the geostrophic equation. It may actually be significant near the bottom or where there is current shear, and therefore the equation does not apply in such situations.

(5) The equation breaks down near the equator where the Coriolis force becomes so small that the friction forces may be important. However, comparisons by Knauss (1960) with direct measurements of currents

indicate that this breakdown is only important within $\pm 0.5°$ latitude (i.e. ± 50 km) of the equator.

(6) The calculated geostrophic current will include any long-period transient currents. It is not possible to separate the transients from the "steady" ocean currents if the geostrophic current is calculated from only two stations. In principle one could repeat stations at frequent intervals and look for time-varying components of the current—it is rarely practical to do so.

Despite all these disadvantages, it must be admitted that application of the geostrophic equation has provided us with much of our present knowledge of the velocities of ocean currents. It is still the only technique for obtaining information relatively quickly from a large area. Recent instrumental developments have provided means for spot observations, e.g. Swallow floats and variants give a Lagrangian picture of deep currents directly, current meters from ships held stationary near moored buoys yield Eulerian current profiles, and strings of current meters anchored in position are now widely used to investigate the circulation in limited regions. Brief descriptions of drifting buoys, current meters and mooring techniques will be found in *Descriptive Physical Oceanography* (Pickard and Emery, 1982). All of these techniques require a large amount of effort for relatively localized returns. The data obtained from Swallow floats and current meters show that there are large transient currents with many different periods which have quite complicated variations in space so that it is difficult to get a good measure of the average flow to obtain good checks of the geostrophic equation and to establish levels of known motion for geostrophic calculations over large regions. A recently proposed method of analysis (the "beta spiral") for obtaining the depth of no motion without direct current measurements is described in the next section.

Some networks of anchored current meters, often with attached T, S, etc., recorders, have been located in areas of interest to permit fairly long-term studies of currents and to yield information on their variation with time. This approach is very expensive, and probably cannot be justified economically for ocean-sized regions; perhaps the best approach will be to use numerical modelling (as described in Chapter 11) with limited measurement programmes in the ocean for adjusting the model parameters and checking the results.

8.9 The beta-spiral

Schott and Stommel (1978) proposed a method for obtaining the absolute velocity from measurements of the density field alone. The method is applicable in regions where the geostrophic equations (8.10) $[fv = \alpha\,(\partial p/\partial x); -fu = \alpha\,(\partial p/\partial y)]$ are a good approximation. It is also assumed that the density does not change as the water flows along, i.e. $d\rho/dt = 0$; with $\partial\rho/\partial t = 0$ (the steady state),

$$u\frac{\partial \rho}{\partial x} + v\frac{\partial \rho}{\partial y} + w\frac{\partial \rho}{\partial z} = 0. \tag{8.15}$$

If we use the thermal wind equations (8.14) with the Boussinesq approximation

$$\left(\rho_0 f\frac{\partial v}{\partial z} = -g\frac{\partial \rho}{\partial x};\ \rho_0 f\frac{\partial u}{\partial z} = g\frac{\partial \rho}{\partial y} \text{ where } \rho_0 \text{ is an average density} \right)$$

to replace $\partial \rho/\partial x$ and $\partial \rho/\partial y$ in equation (8.15) we get

$$-u\frac{\partial v}{\partial z} + v\frac{\partial u}{\partial z} = -\left(\frac{g}{f\rho_0}\right)w\frac{\partial \rho}{\partial z}. \tag{8.16}$$

If we write the velocity in polar form components as $u = V\cos\theta, v = V\sin\theta$ (Appendix 1) then equation (8.16) becomes

$$\frac{\partial \theta}{\partial z} = w\frac{\partial \rho}{\partial z}\left(\frac{g}{\rho_0 f V^2}\right). \tag{8.17}$$

Now $\partial \rho/\partial z < 0$ and $f > 0$ in the northern hemisphere so, if $w > 0$ (upward), $\partial \theta/\partial z < 0$ and the current rotates to the right (θ becomes smaller) as we go upward in the water column (or to the left as we go downward). With $f < 0$ in the southern hemisphere, the rotation is to the left as we go upward for $w > 0$. Of course, if $w = 0$, there is no rotation of the flow direction as the depth changes. To see that $w \neq 0$ and rotation is likely, we derive an equation for $\partial w/\partial z$ from the geostrophic equations (8.10) and the continuity equation (4.4) ($\partial u/\partial x + \partial v/\partial y + \partial w/\partial z = 0$). In equation (8.10) we use the Boussinesq approximation, i.e. treat α as a constant. Then

$$\frac{\partial (fu)}{\partial x} = f\frac{\partial u}{\partial x} = -\alpha\frac{\partial^2 p}{\partial x\partial y};\ \frac{\partial (fv)}{\partial y} = f\frac{\partial v}{\partial y} + v\frac{\partial f}{\partial y} = \alpha\frac{\partial^2 p}{\partial y\partial x}.$$

Since the order of differentiation does not matter for p,

$$f\left(\frac{\partial u}{\partial x} + \frac{\partial v}{\partial y}\right) + v\frac{\partial f}{\partial y} = 0.$$

Using equation (4.4) and β for $\partial f/\partial y$ we get

$$\beta v = f\frac{\partial w}{\partial z}. \tag{8.18}$$

Now if $v \neq 0$, i.e. the flow is not just east–west, $\partial w/\partial z \neq 0$, i.e. w changes with z and cannot be zero everywhere. Thus the fact that f changes with latitude, so that $\beta \neq 0$, makes rotation of the geostrophic flow with depth changes likely; hence the term "β-spiral". (Note that equation (8.18) is the vorticity-conservation equation for geostrophic flow; vorticity will be discussed in Chapter 9.)

How do we use this approach to calculate absolute velocities? Let $z = -h_0 + h(x, y)$ be an isopycnal surface; $h(x, y)$ is the height above $z = -h_0$. Suppose that we go along this surface in the x-direction, then $dz = (\partial h/\partial x)dx$ and $d\rho = [(\partial \rho/\partial x)dx + (\partial \rho/\partial z)dz] = 0$. Thus $\partial h/\partial x = -[(\partial \rho/\partial x)/(\partial \rho/\partial z)]$ is the slope of the surface in the x-direction. Likewise $\partial h/\partial y = -[(\partial \rho/\partial y)/(\partial \rho/\partial z)]$. Using these values for $\partial h/\partial x$ and $\partial h/\partial y$ in equation (8.15) gives

$$u\frac{\partial h}{\partial x} + v\frac{\partial h}{\partial y} = w. \tag{8.19}$$

Taking $\partial/\partial z$ of this equation gives

$$\frac{\partial}{\partial z}\left(u\frac{\partial h}{\partial x} + v\frac{\partial h}{\partial y}\right) = \frac{\partial w}{\partial z} = \frac{\beta v}{f}$$

and rewriting the thermal wind equations (8.14) in terms of $\partial h/\partial x$, $\partial h/\partial y$ as

$$\frac{\partial v}{\partial z} = \frac{g}{\rho_0 f}\left(\frac{\partial h}{\partial x}\frac{\partial \rho}{\partial z}\right); \quad \frac{\partial u}{\partial z} = -\frac{g}{\rho_0 f}\left(\frac{\partial h}{\partial y}\frac{\partial \rho}{\partial z}\right)$$

we see that $\dfrac{\partial u}{\partial z}\dfrac{\partial h}{\partial x} + \dfrac{\partial v}{\partial z}\dfrac{\partial h}{\partial y} = 0$ and

$$u\frac{\partial^2 h}{\partial x \partial z} + v\left(\frac{\partial^2 h}{\partial y \partial z} - \frac{\beta}{f}\right) = 0. \tag{8.20}$$

Now $\partial h/\partial x$ and $\partial h/\partial y$ can be determined as functions of z from the observations of ρ (or σ_t or ρ_θ) if there are enough oceanographic stations in an area and $\partial/\partial z$ of $\partial h/\partial x$ and of $\partial h/\partial y$ can be calculated. Suppose that we have calculated u' and v', the relative geostrophic velocities based on some reference level where the velocity components are u_0 and v_0, then $u = u_0 + u'$ and $v = v_0 + v'$ gives

$$u_0\frac{\partial^2 h}{\partial x \partial z} + v_0\left(\frac{\partial^2 h}{\partial y \partial z} - \frac{\beta}{f}\right) + u'\frac{\partial^2 h}{\partial x \partial z} + v'\left(\frac{\partial^2 h}{\partial y \partial z} - \frac{\beta}{f}\right) = 0. \tag{8.21}$$

If we use N levels to determine u', v', $(\partial^2 h/\partial x \partial z)$ and $(\partial^2 h/\partial y \partial z)$ we get N equations for u_0, v_0 in the form of equation 8.21. If the observations were error free, two levels would be sufficient. However, there are likely to be errors in the observations, particularly noise from time-dependent motions, so the set of equations (8.21) will not be exactly satisfied. One uses least-squares techniques to find u_0 and v_0 such that the sum of the squares of the left-hand sides of the equations (8.21) will be a minimum.

Schott and Stommel (1978) tested the method with historical data at a number of locations. The u_0, v_0 values obtained at a particular location often depended on the range of depths used for calculating $(\partial^2 h/\partial x \partial z)$ and $(\partial^2 h/\partial y \partial z)$ in equations (8.21). Behringer (1979) reformulated the approach somewhat in terms of using the flow directions which can also be determined from the density field. He suggested least-squares minimization procedures based on rewriting the equations for u_0, v_0 in terms of the flow direction θ in two ways which differ from the Schott and Stommel (1978) procedure. These seem to give more consistent results. Also Behringer found that if he smoothed the flow directions first, to reduce the noise, all three minimization methods gave very similar results. Behringer and Stommel (1980) tested the method with data collected especially for the purpose in the mid-Atlantic. All the minimization procedures gave consistent results. The results using the same data and an inverse procedure given by Wunsch (1978) are also consistent. The "beta-spiral" approach and that of Wunsch seem to have promise but probably more detailed observations of the density field than are presently available will be needed in much of the ocean.

8.10   Justification for using the geostrophic approach to obtain the speeds of strong currents

Consider the Gulf Stream as an example. It is convenient to orient the coordinates with the stream so let us take the x-axis across the stream and the y-axis along it. (This is the one exception to our general use of a coordinate system with the x-axis positive to the east and the y-axis positive to the north.) We use scaling arguments and friction terms written in eddy viscosity form as in Chapter 7. For the width of the Stream we use $L_x = 100$ km $= 10^5$ m; for the length we use $L_y = 1000$ km $= 10^6$ m. For the along-stream, y, component of current we use $V = 1$ m s^{-1} (maximum values are up to 3 m s^{-1}); for the cross-stream, x, component we use $U = 0.1$ m s^{-1} since the stream may spread out. Then taking the depth scale $H = 10^3$ m and using continuity, the vertical velocity is of $0\,(UH/L_x = VH/L_y = 10^{-3}$ m s^{-1}). The vertical equation will reduce to the hydrostatic equation as we noted and left as an exercise in Chapter 7. Assume a steady state ($\partial u/\partial t = \partial v/\partial t = 0$) and take the maximum values for eddy viscosity $A_z = 0.1$ m^2 s^{-1} and $A_x = A_y = 10^5$ m^2 s^{-1} and examine the x-equation

$$u\frac{\partial u}{\partial x}+v\frac{\partial u}{\partial y}+w\frac{\partial u}{\partial z} = -\alpha\frac{\partial p}{\partial x}+fv-2\Omega\,\cos\,\phi\,w$$

$$+A_x\frac{\partial^2 u}{\partial x^2}+A_y\frac{\partial^2 u}{\partial y^2}+A_z\frac{\partial^2 u}{\partial z^2}.$$

Introducing our scales and taking $f = 10^{-4}$ s^{-1} (the value for $\phi = 45°$), the orders of the terms are

$$\frac{10^{-2}}{10^5}+\frac{10^{-1}}{10^6}+\frac{10^{-4}}{10^3} = ?+10^{-4}-10^{-7}+10^5\frac{10^{-1}}{10^{10}}+10^5\frac{10^{-1}}{10^{12}}+10^{-1}\frac{10^{-1}}{10^6}$$

or by dividing through by 10^{-4}, the scale for fv,

$$10^{-3} + 10^{-3} + 10^{-3} = ? + 1 - 10^{-3} + 10^{-2} + 10^{-4} + 10^{-4}.$$

The x- or cross-stream equation therefore remains geostrophic within 1%. (Remember that we have used maximum values for eddy viscosity, so 1% or so should be an upper limit for friction effects.) This equation can be used to obtain the downstream component (v) even in strong currents such as the Gulf Stream.

Now consider the y-equation

$$u\frac{\partial v}{\partial x} + v\frac{\partial v}{\partial y} + w\frac{\partial v}{\partial z} = -\alpha\frac{\partial p}{\partial y} - fu + A_x\frac{\partial^2 v}{\partial x^2} + A_y\frac{\partial^2 v}{\partial y^2} + A_z\frac{\partial^2 v}{\partial z^2},$$

i.e. $10^{-6} + 10^{-6} + 10^{-6} = -? - 10^{-5} + 10^{-5} + 10^{-7} + 10^{-7}.$

The non-linear terms are now about 10% of fu and the largest friction term is of about the same size as fu. If we use a maximum value of 3 m s^{-1} for V, the non-linear terms become of order 30% of fu and the friction terms may be up to three times fu. (Using a larger value for v in the x-equation does not change the relative importance of the terms because v also comes into the Coriolis term.) Thus the geostrophic approximation is not good for the y-equation. While we can use geostrophy to compute the downstream velocity component relative to a reference level from the density distribution, we cannot use this approximation in seeking a solution to the equations in a current as strong as the Gulf Stream. Friction, and perhaps non-linear terms, must be considered. Indeed, if friction is somewhat smaller than the maximum values (which came from estimates of friction effects in the Antarctic Circumpolar Current) non-linear terms may be comparable to or even larger than the friction terms. (In the terminology sometimes used by theoreticians, the Rossby number (Ro) and horizontal Ekman number (E_x in this case) become of order one in the y-momentum equation in a region such as the Gulf Stream.)

9

Currents with Friction;
Wind-driven Circulation

9.1 Wind-driven circulation—introduction

A notable feature of the gross surface-layer circulation of the oceans is that it is clockwise in the northern hemisphere and anticlockwise in the southern (e.g. see Fig. 8.8). This fact for the North Atlantic was known to Spanish navigators in the early 1500s and was subsequently recognized for the other oceans as navigational records accumulated. In the mid-1800s this circulation was attributed to the differential solar heating between the equatorial and polar regions but no one produced any quantitative theory of the process. About 1875 Croll became convinced that this hypothesis was incorrect and suggested that the frictional stress of the wind was the direct cause, although he did not present any theory. In 1878, Zöppritz apparently demonstrated quantitatively that the transfer of momentum and energy from wind to water was much too slow a process to account for ocean currents but his demonstration was numerically in error, although he cannot be blamed. In his calculation he used the molecular coefficient of viscosity (i.e. friction) as determined in the laboratory for laminar (smooth) flow and showed that apparently it should take months for a change of current at a depth of only a few metres to follow a change of wind at the surface. However, it was soon shown that current changes in the upper tens of metres followed wind changes in a matter of hours, not months. The reason is that in natural water bodies the flow is almost invariably turbulent and in this type of flow the turbulent or "eddy" viscosity comes into play with the effect of increasing the vertical transfer of momentum and energy to a rate of up to hundreds of thousands of times that due to molecular processes alone. This effect was not known in Zöppritz' time.

Both wind driving and the effects of density changes are important for the overall circulation but the former probably dominates in the upper 1000 m or so in most regions of the ocean. We shall discuss the wind-driven flows in this chapter and consider the differential density driving in the following chapter.

Making use of the eddy coefficient of viscosity concept, there followed a series of steps in the development of the theory of the wind-driven circulation which is now accepted, at least as a start in the right direction:

(1) about 1898 Nansen explained qualitatively why wind-driven currents flow not in the direction of the wind but at 20° to 40° to the right of it (in the northern hemisphere);

(2) in 1902 Ekman explained quantitatively for an idealized ocean how the rotation of the earth was responsible for the deflection of the current which Nansen had observed;

(3) in 1947 Sverdrup showed how the main features of the equatorial surface currents could be attributed to the wind as a driving agent;

(4) in 1948 Stommel explained the westward intensification of the wind-driven circulation;

(5) in 1950 Munk combined most of the above to obtain analytic expressions which described quantitatively the main features of the wind-driven circulation in terms of the real wind field;

(6) in recent years, numerous numerical models have been developed for the circulation of individual ocean areas and for the world ocean.

The major wind-driven circulations referred to above are relatively steady-state motions. Superimposed on these are the inertial and tidal motions which vary periodically with time. In addition, as direct observations of the ocean currents have accumulated in recent years it has become clear that there are other time-dependent motions of many space and time scales, often with amplitudes considerably larger than the mean in some regions. In this chapter we shall give the development of the theory for the steady-state or long-term mean circulation, stages (1) to (5) above, while stage (6) will be described in Chapter 11. Various aspects of the time-dependent motions will be discussed in Chapters 11, 12 and 13.

9.2 Nansen's qualitative argument

First we will present the essentials of a qualitative argument advanced by the biologist Nansen to explain why icebergs in the Arctic drifted in a direction to the right of the direction of the wind at the sea surface, not in the direction of the wind itself.

In Figure 9.1(a) is shown a perspective view of a cube of water in the surface layer, while the feathered arrow indicates the wind direction. The wind friction gives rise to a tangential force F_t on the top surface of the cube of water tending to move it in the wind direction. As soon as it starts to move, the Coriolis force F_c comes into action directed to the right. In consequence the motion will actually be in some direction between those of F_t and F_c. Also, when the surface water moves relative to that below it, there will be a retarding force of fluid friction F_b on the bottom of the cube in a direction opposite to the motion. The combination of F_t and F_c would cause the cube to accelerate but as it does so the retarding force F_b increases. Finally, a steady state is reached in which F_t, F_c

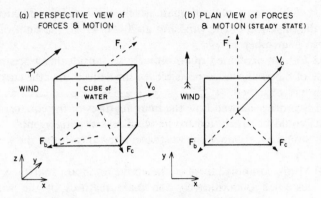

FIG. 9.1   *Forces on a cube of water in the surface layer.*

and $\mathbf{F}_b$ are in balance (Fig. 9.1(b), plan view) and the cube continues to move at a steady speed V_0 in some direction between $\mathbf{F}_t$ and $\mathbf{F}_c$, i.e. to the right of the wind direction (northern hemisphere). To determine the exact direction relative to the wind it is necessary to apply a quantitative argument from the equations of motion as Ekman did (see Section 9.4).

9.3   The equation of motion with friction included

The horizontal equations of motion become, when friction is included (and the Coriolis term involving w is omitted as noted in Section 6.34 and justified in Section 7.3),

$$\frac{du}{dt} = fv - \alpha\frac{\partial p}{\partial x} + F_x,$$

$$\frac{dv}{dt} = -fu - \alpha\frac{\partial p}{\partial y} + F_y, \tag{9.1}$$

where F_x and F_y stand for the components of friction per unit mass in the fluid.

If there are no accelerations (i.e. a steady state and zero, or at least negligible, advective accelerations), then $du/dt = dv/dt = 0$ and we are left with a balance of three forces on unit mass as

$$fv + F_x - \alpha\frac{\partial p}{\partial x} = 0,$$

$$-fu + F_y - \alpha\frac{\partial p}{\partial y} = 0, \tag{9.2}$$

i.e. Coriolis + Friction + Pressure   = 0

as shown schematically in Fig. 9.2. Remember that these are forces and must be added according to the rules for vector addition. The two equations (9.2) give the component form; we can also add them graphically as in Fig. 9.2.

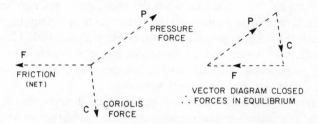

FIG. 9.2   *Three forces in equilibrium on a water parcel.*

This situation differs from the geostrophic relationship in that, with the third force (friction) acting, the pressure and the Coriolis forces are no longer directly opposed. Before we can look for solutions to these equations we must write expressions for the frictional forces F_x and F_y. Friction is essentially a force which comes into being when relative motion occurs or tends to occur between material objects. Friction between two solid bodies is well recognized; in a fluid if two parts are in relative motion friction will also occur. The two parts may be moving in opposite directions, or may be moving in the same direction with one going faster than the other (Fig. 9.3). In either case, there is said to be "velocity shear" in the fluid and the value of the friction is related to this. The amount of shear is measured, e.g. in Fig. 9.3, by: $(u_5 - u_4)/(z_5 - z_4) = \delta u/\delta z$ which tends to $\partial u/\partial z$ as δz tends to 0. Newton's Law of Friction states that in a fluid, the friction stress τ, which is the force per unit area on a plane parallel to the flow, is given by

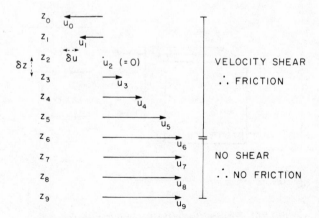

FIG. 9.3   *Illustrating velocity shear and absence of shear.*

$$\tau = \mu\frac{\partial u}{\partial z} = \rho v\frac{\partial u}{\partial z}. \tag{9.3}$$

The stress τ acts on the surface between the two layers which are moving at different speeds, tending to slow down the faster and to speed up the slower. (A fluid for which the friction law of equation (9.3) holds is called "Newtonian". Water, including sea water, and air behave in this way but molecularly more complicated substances, such as long-chain polymers, may have a more complicated behaviour and be non-Newtonian.)

The quantity μ is the coefficient of (molecular) *dynamic* viscosity, while $v = \mu/\rho$ is the coefficient of (molecular) *kinematic* viscosity. For sea water at 20°C, μ has a value of about 10^{-3} kg m^{-1} s^{-1}, so that v has a value of about 10^{-6} m^2 s^{-1}. Values vary from about 0.8 to 1.8 times these values in the ocean, with temperature variation being mainly responsible although there is a slight salinity effect. These are the molecular values and apply to water in smooth, laminar flow, as in a small diameter capillary tube, for Reynolds Numbers ($Re = UL/v$) of less than about 1000 as discussed in Section 7.13. In the ocean, where the motion is generally turbulent, the effective value of kinematic viscosity is the eddy viscosity discussed in Section 7.21 and having values of A_x, A_y of up to 10^5 m^2 s^{-1} for horizontal shear (e.g. $\partial u/\partial y$, $\partial v/\partial x$) or of A_z of up to 10^{-1} m^2 s^{-1} for vertical shear (e.g. $\partial u/\partial z$).

The eddy friction stress $\tau = \rho A_z(\partial u/\partial z)$ expresses the force of one layer of fluid on an *area* of its neighbour above or below, but for substitution in the equation of motion we need an expression for the force on a *mass* of fluid. In Fig. 9.4 a small cube of fluid is shown with shear in the z-direction and the required force would be $(\tau_2 - \tau_1)\delta s$ in the x-direction.

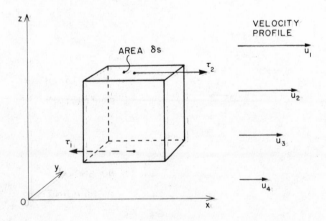

FIG. 9.4   *For derivation of the friction term in the equation of motion.*

As $\qquad \tau_2 = \tau_1 + \dfrac{\partial \tau}{\partial z}\delta z,$

$$\therefore (\tau_2 - \tau_1)\delta s = \frac{\partial \tau}{\partial z}(\delta s\,\delta z) = \frac{\partial \tau}{\partial z}(\delta V) \quad \text{where} \quad \delta V = \text{volume of cube.}$$

In the limit as $\delta s,\ \delta z \to 0$, so that $\delta V \to 0$,

the force per unit volume $= \partial \tau / \partial z$ and

$$\text{the force per unit mass} = \frac{1}{\rho}\frac{\partial \tau}{\partial z} = \alpha \frac{\partial \tau}{\partial z} = \alpha \frac{\partial}{\partial z}\left(\rho A_z \frac{\partial u}{\partial z}\right). \tag{9.4}$$

We use A_z here because we are concerned with vertical shear ($\partial u/\partial z$ or $\partial v/\partial z$). The form of equation (9.4) where A_z is inside the bracket is appropriate when the eddy-friction coefficient varies with depth z. As we have very little information on the manner in which (ρA_z) does vary with depth, we will limit ourselves to the case where (ρA_z) is assumed to be constant and we can therefore rewrite equation (9.4) as

$$\text{friction force per unit mass} = A_z \frac{\partial^2 u}{\partial z^2}. \tag{9.5}$$

In expression (9.4) the effect of ρ variations which are very small are not important compared with variations of $A_z\,(\partial u/\partial z)$ and we could have taken the ρ outside at that point, an approximation consistent with the Boussinesq approximation discussed in Section 7.41.

Then the horizontal equations of motion become

$$fv + \alpha \frac{\partial \tau_x}{\partial z} = fv + A_z \frac{\partial^2 u}{\partial z^2} = \alpha \frac{\partial p}{\partial x},$$

$$-fu + \alpha \frac{\partial \tau_y}{\partial z} = -fu + A_z \frac{\partial^2 v}{\partial z^2} = \alpha \frac{\partial p}{\partial y}. \tag{9.6}$$

The vertical equation reduces to the hydrostatic equation as justified in Section 7.3. The vertical velocity component, w, does not appear explicitly in the equations of motion in this form. It is obtained using the equation of continuity after first solving the equations of motion for u and v. In Section 7.3 we showed that the friction terms are small enough to be neglected in the interior of the ocean but we noted that they might not be negligible near the sea surface or bottom. For a term like $A_z(\partial^2 u/\partial z^2)$ to be significant in the equations of motion it must be comparable in size with the Coriolis term, i.e. $A_z(U/H^2) \simeq fU$. For instance, for $A_z = 10^{-1}\,\mathrm{m^2\,s^{-1}}$, $f = 10^{-4}\,\mathrm{s^{-1}}$, then $H^2 \simeq A_z/f \simeq 10^{-1}/10^{-4} \simeq 10^3\,\mathrm{m^2}$ or $H \simeq 30$ m. The friction term would still be about 10% of the Coriolis term at about $H \simeq 100$ m, so that we should be prepared to take friction into account within this distance from the surface or bottom. (In theoretical terminology, the vertical Ekman Number, E_z, becomes of order one near the top and bottom of the ocean.)

9.4   Ekman's solution to the equation of motion with friction present

A difficulty with equations (9.6) is that there are two causative forces for motion, the distribution of mass (i.e. density) which gives rise to the pressure terms, and the wind friction term. Note that we can think of the velocity as having two parts, one associated with the horizontal pressure gradient and one with vertical friction. Each part can be solved for separately and the two added together, i.e.

$$fv = f(v_g + v_E) = \alpha \frac{\partial p}{\partial x} - A_z \frac{\partial^2}{\partial z^2}(u_g + u_E) \qquad (9.7)$$

where

$$fv_g = \alpha \frac{\partial p}{\partial x}, \qquad u_g, v_g \quad \text{being the geostrophic velocity components,}$$

and

$$fv_E = -A_z \frac{\partial^2 u_E}{\partial z^2}, u_E, v_E \quad \begin{array}{l} \text{being the Ekman velocity components associated} \\ \text{with vertical shear friction (non-geostrophic).} \end{array}$$

The term $-A_z(\partial^2 u_g/\partial z^2)$ is neglected because it is $\lesssim 10^{-3}\alpha(\partial p/\partial x)$ as shown in Section 7.3.

This separation is possible because the equations are linear. It provides an example of the principle of superposition, i.e. for a linear system the sum of two solutions is also a solution. If non-linear effects become important this separation scheme does not work.

To simplify the problem, Ekman (1905) assumed the water to be homogeneous and that there was no slope at the surface, so that the pressure terms would be zero and v_g therefore also zero, i.e. he solved for v_E only. He also assumed an infinite ocean to avoid the complications associated with the lateral friction at the boundaries and the diversion of the flow there.

At Nansen's suggestion, Ekman first studied the effect of the frictional stress at the sea surface due to the wind blowing over it. Altogether he assumed:

(1) no boundaries;
(2) infinitely deep water (to avoid the bottom friction term);
(3) A_z constant;
(4) a steady wind blowing for a long time;
(5) homogeneous water and the sea surface level so that $\partial p/\partial x = \partial p/\partial y = 0$ since density depends only on pressure, i.e. a barotropic condition and therefore there is no geostrophic flow;
(6) f to be constant, i.e. the f-plane approximation.

The reason for assumption (2) was because there was reason to believe that the wind-driven current would decrease as depth increases and therefore in very deep water the speed would become negligible. Hence the shear would also be negligible and so the fluid friction would vanish and there would be only the friction near the surface to take into account. The reason for assumption (3) was partly to simplify the problem and partly because so little was known about the variation of A_z with z.

The equations then became

$$\left. \begin{aligned} fv_E + A_z \frac{\partial^2 u_E}{\partial z^2} &= 0, \\[2mm] -fu_E + A_z \frac{\partial^2 v_E}{\partial z^2} &= 0, \end{aligned} \right\} \quad \text{the Ekman equations,} \qquad (9.8)$$

i.e. Coriolis + Friction = 0, as in Fig. 9.5(a).

If, for simplicity, we assume the wind to be blowing in the y-direction (Fig. 9.5(b)), the solutions to Ekman's equations are

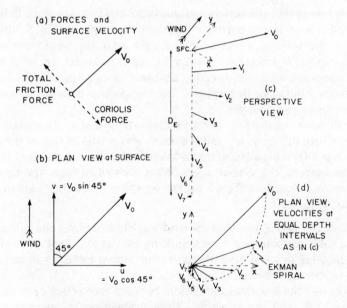

FIG. 9.5   *Wind-driven currents from Ekman analysis: (a) net frictional stress balances Coriolis force with surface current* $\mathbf{V}_0$ *perpendicular to both; (b) wind in y-direction, surface velocity* $\mathbf{V}_0$ *and components; (c) perspective view showing velocity decreasing and rotating clockwise with increase in depth; (d) plan view of velocities at equal depth intervals, and the "Ekman spiral" (all for northern hemisphere).*

$$u_E = \pm V_0 \cos\left(\frac{\pi}{4} + \frac{\pi}{D_E}z\right)\exp\left(\frac{\pi}{D_E}z\right), \quad \begin{array}{l}(+\text{for northern hemisphere})\\(-\text{for southern hemisphere})\end{array}$$

$$v_E = V_0 \sin\left(\frac{\pi}{4} + \frac{\pi}{D_E}z\right)\exp\left(\frac{\pi}{D_E}z\right), \quad (9.9)$$

where $V_0 = (\sqrt{2}\pi\tau_{yn})/(D_E\rho|f|)$ is the total Ekman surface current,   (9.10)

τ_{yn} = magnitude of the wind stress on the sea surface (approximately proportional to the wind speed squared and acting in the direction of the wind),

$|f|$ = the magnitude of f,

$D_E = \pi(2A_z/|f|)^{1/2}$ is the *Ekman depth* or *depth of frictional influence* (discussed below).

We will interpret these solutions:

(1) At the sea surface where $z = 0$, the solutions become

$$u = \pm V_0 \cos 45°, \quad v = V_0 \sin 45°,$$

which means that the surface current flows at 45° to the right (left) of the wind direction in the northern (southern) hemisphere (Fig. 9.5(b)).

(2) Below the surface, where z is no longer zero, the total current speed $= V_0 \exp(\pi z/D_E)$ becomes smaller as depth increases, i.e. as z becomes more negative, while the direction changes clockwise (anticlockwise) in the northern (southern) hemisphere. The perspective drawing of Fig. 9.5(c) shows these two changes.

(3) The direction of the flow becomes opposite to that at the surface at $z = -D_E$ where the speed has fallen to $\exp(-\pi) = 0.04$ of that at the surface. The depth D_E is usually arbitrarily taken as the effective depth of the wind-driven current, the *Ekman layer*. When viewed in plan, the tips of the current vector arrows form a decreasing spiral called the "Ekman current spiral" (Fig. 9.5(d)).

We made the assumption that the wind was blowing along the y-direction to keep the solutions relatively simple in form for our first look at them. If the wind is blowing in some other direction the current pattern will be the same relative to the wind direction.

In order to obtain numerical relations between the surface current, V_0, the wind speed, W, and the depth D_E, Ekman made use of two experimental observations:

Obs. 1:   The wind stress magnitude $\tau_n = \rho_a C_D W^2$ where ρ_a = the density of air, the drag coefficient $C_D \simeq 1.4 \times 10^{-3}$ (non-dimensional), and W is the wind speed in m s^{-1}. Then $\tau_n = 1.3\,\text{kg m}^{-3} \times 1.4 \times 10^{-3}$

$\times W^2 = 1.8 \times 10^{-3} W^2$ Pa. If we substitute this expression in equation (9.10) we obtain

$$V_0 = \frac{\sqrt{2} \times \pi \times 1.8 \times 10^{-3} \times W^2}{D_E \times 1025 \text{ kg m}^{-3} \times |f|} = 0.79 \times 10^{-5} \frac{W^2}{D_E |f|} \text{ m s}^{-1}. \quad (9.11)$$

Obs. 2:   Field observations analysed by Ekman indicate that the surface current and the wind speed are related as

$$\frac{V_0}{W} = \frac{0.0127}{(\sin |\phi|)^{1/2}} \text{ outside } \pm 10° \text{ latitude from the equator.} \quad (9.12)$$

Substituting this expression in equation (9.11) (and remembering that $f = 2\Omega \sin \phi$) we get

$$D_E = \frac{4.3 \, W}{(\sin |\phi|)^{1/2}} \text{ metres (with } W \text{ in m s}^{-1}). \quad (9.13)$$

Therefore, if we know W at latitude ϕ we can calculate V_0 and D_E, and the velocity at any depth below the surface. The fact that D_E depends on W suggests that the eddy viscosity A_z increases as W increases; if we know D_E we can estimate a value for A_z. Some numerical values from the above relations are given in Table 9.1.

TABLE 9.1.   *Some values for D_E and estimated A_z*

Latitude:	$\phi = 10°$	45°	80°	
V_0/W	= 0.030	0.015	0.013	A_z
Wind speed				
(W)				(estimated)
10 m s^{-1}	$D_E = 100$	50	45 m	0.014 m^2 s^{-1}
20	= 200	100	90	0.055

9.41   *Comments on the experimental observations used by Ekman*

It should be noted that while Obs. 1 and 2 are believed to be reasonable, they are not exact. For instance, there is variability in the value of C_D, the present estimates suggesting values from 1.3 to 1.5 × 10^{-3} ± 20%, for wind speeds up to about 15 m s^{-1}. (A discussion of the recent estimates of C_D will be given in Section 9.9.)

The three figure accuracy implied by the constant in Ekman's expression (9.12) for V_0/W probably overstates its accuracy because more extensive data yield values ranging from 2 to 5% in mid-latitudes. In addition, time-dependent effects and the mixed-layer depth are probably important.

Very often, values for D_E have been estimated from or compared with the depth of the upper mixed layer, although the assumption that this "mixed-layer depth" is identical with the Ekman depth is not correct very often. The mixed-layer depth depends on the past history of the wind in the locality rather more than on the wind speed at the time of observation. It also depends on the stability of the underlying water and on the heat balance through the surface (which determines convective effects). The formation of the mixed layer is a complicated time-dependent process which is still not fully understood—it is an active area of research in physical oceanography at the present time. (A brief discussion will be given in Section 10.5.) One would expect the Ekman depth to be rather less than the mixed-layer depth in most cases, because the latter may be much influenced by even short periods of strong winds. It follows that values for A_z which are calculated from the apparent D_E are, in general, probably too large.

Finally, all these detailed results depend on the assumptions that A_z is constant with depth and that the wind is constant, neither of which is likely. Thus, although the main features of the current turning to the right and decreasing with depth are probably correct, the details are not to be taken too seriously. You will notice that we say "probably correct" because there are very few measured current profiles which are adequate to test the theory. It is difficult to make accurate current measurements in the open, deep ocean, the only region where the Ekman theory applies, and difficult to get sufficiently steady wind conditions. At the same time, it must be recognized that a version of the Ekman theory also applies to the velocity structure in the atmosphere above the earth's surface and there are some observations to show that the theory applies fairly well in this case. Time-dependent effects remain a problem because the wind does not often blow with a steady speed and direction for a long enough period (a few pendulum days) to produce the steady-state situation. Stability may also be important.

As we mentioned in Section 7.21, in the atmosphere near the ground A_z increases linearly with height at first. Higher up it is probably constant for some distance and then decreases to zero as the shear and frictional effects diminish. This layer is spoken of as the *planetary boundary* or *Ekman layer*, or sometimes as the *region of frictional influence*. Assumption of a similar variation for A_z in the ocean might lead to more realistic results although the presence of surface waves may make the region near the surface behave differently in the ocean than in the atmosphere. The surface waves also make measurement of currents near the surface extremely difficult and one would have to separate the Ekman flow from the geostrophic and time-varying flows (e.g. tidal, inertial), so verifying theoretical results in detail would be a non-trivial problem to say the least.

Actually, provided that the flow is reasonably uniform in the horizontal, the time-dependent problems can be solved. If u_E and v_E do not vary with horizontal position x and y, then by the continuity equation $w = 0$ and all the non-linear terms for the mean flow are zero. While the real flows in the ocean and atmosphere will not be exactly horizontally uniform they should be sufficiently so that the non-linear terms are negligible in most cases. Thus the equations including time variations remain linear and can be solved (by numerical methods if analytic solutions cannot be found). The problem remains of choosing a suitable behaviour for A_z. Solutions for the atmosphere and also for the ocean have been obtained using various dependencies of A_z on z but such details are beyond the scope of this book.

9.42   *Transport and upwelling—the effect of boundaries*

The wind-driven Ekman current has its maximum speed at the surface and the speed decreases with depth increase. Because the strongest currents are to the right (or left) of the wind direction, it is easy to appreciate that the net transport will be to the right (or left) of the wind direction, in fact it will be shown to be at *right angles* to the wind direction.

The basic form of the equations for horizontal motion (equations (9.6)) in the absence of any pressure gradient is

$$\rho f v_E + \frac{\partial \tau_x}{\partial z} = 0 \qquad -\rho f u_E + \frac{\partial \tau_y}{\partial z} = 0$$

which we can write as

$$\rho f v_E dz = -d\tau_x \qquad -\rho f u_E dz = -d\tau_y. \tag{9.14}$$

Now $\rho v_E dz$ is the mass flowing per second in the y-direction through a vertical area of depth dz and width one metre in the x-direction, and $\int_z^o \rho v_E dz$ will be the total mass flowing in the y-direction from the level z to the surface for this strip 1 m wide, while $\int_z^o \rho u_E dz$ will be the total mass transport per unit width in the x-direction. If we choose the lower level deep enough, then the integrals will include the whole wind-driven current. We choose a value $z = -2D_E$ where the speed will be $\exp(-2\pi)$ $= 0.002$ of that at the surface, i.e. substantially zero. If we use the symbols M_{xE} and M_{yE} to represent the Ekman (i.e. wind-driven) mass transports in the x- and y-directions respectively, then

$$f M_{yE} = f \int_{-2D_E}^{o} \rho v_E dz = -\int_{-2D_E}^{o} d\tau_x = -(\tau_x)_{sfc} + (\tau_y)_{-2D_E},$$

$$f M_{xE} = f \int_{-2D_E}^{o} \rho u_E dz = \int_{-2D_E}^{o} d\tau_y = (\tau_y)_{sfc} - (\tau_y)_{-2D_E} \tag{9.15}$$

Now $(\tau_x)_{-2D_E}$ and $(\tau_y)_{-2D_E}$ will be essentially zero because the velocity below the wind-driven layer is substantially zero and therefore there can be no shear and therefore no friction. So we have

$$fM_{xE} = \tau_{y\eta} \quad \text{and} \quad fM_{yE} = -\tau_{x\eta} \tag{9.16A}$$

where we use the subscript η to indicate surface values. (We use η rather than 0 (zero) because in this and following discussions the sea surface may be above or below (by an amount η) the level of the origin of the coordinate system which is taken to be at *mean* sea level.)

The variations of ρ are small and so the ρ's may be taken outside the integrals in equation (9.15) with negligible error; the value then used for ρ would be a typical one, e.g. a vertical average over $2D_E$ in the region being considered.

The volume transport (per unit width) $Q_y = \int_z^0 v_E dz$ is often used as an alternative to the mass transport. Then $M_{yE} = \rho Q_{yE}$ and $M_{xE} = \rho Q_{xE}$ and alternative forms of equations (9.16) are

$$fQ_{xE} = \alpha\tau_{y\eta} \quad \text{and} \quad fQ_{yE} = -\alpha\tau_{x\eta}. \tag{9.16B}$$

These results are correct even if the details of the Ekman spiral are not.

In our example, where the wind is entirely in the y-direction, $\tau_{x\eta} = 0$ and therefore $M_{yE} = 0$, but $M_{xE} > 0$ because $\tau_{y\eta} > 0$, showing that the net transport is to the right of and at right angles to the wind direction (in the northern hemisphere and vice versa in the southern hemisphere). This result remains correct for wind in any direction.

The equation of continuity then requires that there must be inflow from the left of the wind direction to replace the flow away to the right. For Ekman's infinite ocean there is no trouble in supplying this inflow in the surface layer. However, if the wind is blowing parallel to a coastline which is on the left of the wind (in the northern hemisphere) a difficulty arises. The wind causes the surface or Ekman layer to move to the right, i.e. away from the coast, but because of the coast there is no supply of *surface* water on the left of the wind for replacement. What happens in nature is that, as the Ekman layer is skimmed away from the coast, water from *below* the surface comes up to replace it. This behaviour is called *upwelling* and the region near the coast is one of divergence. This phenomenon occurs at times along many regions of the eastern sides of the oceans. In the northern hemisphere, the wind must blow along the coast in a southerly direction, which usually happens during the summer. In the southern hemisphere, the transport is to the left of the wind and so it must blow in a northerly direction for upwelling to occur. In general we can say that upwelling will occur when the wind blows equatorward along an eastern boundary of an ocean in either hemisphere or poleward along a western boundary, although this latter situation is less common.

Examples of regions which have been studied where upwelling occurs due to equatorward winds are off the west coasts of North and South America in the Pacific Ocean (British Columbia to California and off Peru) and off West and South Africa in the Atlantic. One example of coastal upwelling due to poleward winds is off the Somali Republic and Arabia in the Indian Ocean during the southwest monsoon.

The upwelled water does not come from great depths. Studies of the properties of the upwelled water indicate that it comes from depths not greater than 200–300 m. When the upwelled water has high nutrient content, plankton production may be promoted and the process is therefore important biologically. Some 90 % of the world's fisheries are in 2–3 % of the ocean's area, mostly in coastal upwelling regions. However, not all subsurface waters are high in nutrient content and so upwelling does not invariably promote biological production.

If the wind blows away from the equator along the eastern boundary of an ocean, then water will be forced toward the coast and the level will rise. This process may then give rise to a surface slope and a consequent geostrophic current. In upwelling regions also, a surface slope is usually caused, in this case down toward the coast. The induced geostrophic currents along the coast generally have considerably higher speeds than the wind-induced (Ekman) onshore–offshore currents, making the latter difficult to measure. On the eastern side of an ocean, the downward slope toward the coast requires an equatorward flow at the surface if the pressure gradient is to be balanced, at least in the main, by the Coriolis force. We say "in the main" because near shore and/or in shallow water friction is likely to become important and the current may not be purely geostrophic. As the density of the water near the coast is higher than that offshore at the same level, baroclinic compensation will occur, i.e. the longshore flow will decrease with depth. Sometimes "overcompensation" may occur and the offshore pressure gradient changes sign at depth, requiring a poleward undercurrent to provide a balancing (or partially balancing) Coriolis force. (A number of papers on upwelling observations and theory are presented in *Coastal Upwelling*, F. A. Richards (Ed.), 1981.)

9.43   *Upwelling and downwelling away from boundaries*

Over the real ocean, the wind is not uniform as assumed by Ekman but varies with position. For example, if the wind remains constant in direction but varies in speed across the direction of the wind, then the Ekman transport perpendicular to the wind will vary and the upper-layer waters will be forced toward or away from each other, i.e. convergences or divergences will develop. Continuity then requires that a convergence be accompanied by downward motion (downwelling) while a divergence will be accompanied by upward motion (upwelling).

For instance, in the North Atlantic the general direction of the wind is to the east at higher latitudes and to the west at lower latitudes. (The former are called "westerlies" because they come *from* the west to the observer while the latter are called "easterlies".) Figure 9.6(a) shows the main winds in simplified form, the

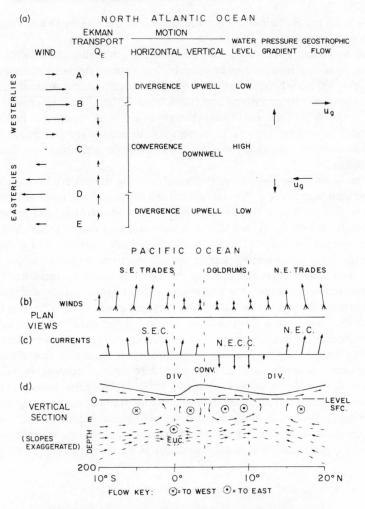

FIG. 9.6   (a) Convergences and divergences related to wind shear and associated geostrophic flow, North Atlantic Ocean; schematic representations in plan view for eastern equatorial Pacific Ocean of (b) wind system and (c) currents; (d) vertical section for same region showing circulation, convergences and divergences.

lengths of the wind arrows indicating the wind speed. The Ekman transport due to the wind will be in the southward direction from the westerlies and in the northward direction from the easterlies, and the transport will be greater for greater wind speeds. The result is that the Ekman transport to the south will increase from A to B. To supply the increase, water must upwell from below the Ekman layer and there will be a zone of divergence. From B to C the southward

Ekman flow will decrease to zero and from C to D it will be in a northward direction, increasing as one goes toward D. In consequence, the region around C will be one of convergence and water must descend below the surface. Between D and E there will be a region of divergence and upwelling. By setting up such pressure gradients the wind causes flows much deeper than those it drives directly in the Ekman layer. Thus the wind drives flows in the Ekman layer directly to 100 to 200 m depth, and geostrophic flows associated with convergence and divergence of the Ekman flows, and the pressure gradients which they cause, to depths of 1000 to 2000 m.

In a region of convergence the surface level will tend to be high while in a divergence region it will tend to be low, and there will be consequent pressure gradients and geostrophic flows (u_g) set up as shown on the right of Fig. 9.6(a).

A wind blowing to the west along the equatorial zone (Fig. 9.6(b)), even without horizontal shear, will cause divergence and upwelling at the equator (Fig. 9.6(d)) because the Ekman layer transport will be to the right north of the equator and to the left south of the equator, i.e. away from the equator in both cases. (Other aspects of Fig. 9.6(b,c,d) will be discussed in Sections 9.13.1 and 9.13.2.)

9.44   *Bottom friction and shallow water effects*

If a current is flowing over the sea bottom, friction there will generate an Ekman spiral current pattern above the sea bottom but with the direction of rotation of the spiral reversed relative to the wind-driven near-surface Ekman layer. The current pattern is shown in Fig. 9.7 for friction acting at the sea bottom, in perspective and plan views.

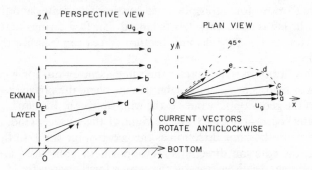

FIG. 9.7   *Frictional effects on a geostrophic current near the bottom of the ocean (northern hemisphere).*

Assuming that A_z is constant, Ekman's equations (9.8) still apply but the boundary conditions are different. The tangential velocity must vanish at the bottom (i.e. $u = v = 0$) and must go to a constant value above the region of friction effects (the Ekman layer), assuming that the geostrophic flow above this layer is independent of z. If, as a specific example, we take $u = u_g$, $v = 0$, in the geostrophic region (although again the general results of rotation relative to the direction of the geostrophic current do not depend on its direction), the solution (for the northern hemisphere) is

$$u_E = u_g [1 - \exp(-\pi z/D_E) \cos(\pi z/D_E)],$$
$$v_E = u_g \exp(-\pi z/D_E) \sin(\pi z/D_E), \qquad (9.17)$$

where $z = 0$ is taken at the bottom (assumed level) in this case to make the formulae simpler, and $D_E = \pi \sqrt{(2A_z/|f|)}$ as before. Equations (9.17) satisfy (9.8) (as the reader may verify by substitution). At $z = 0$, $u_E = v_E = 0$ as required. As z becomes large compared with D_E/π, $\exp(-\pi z/D_E)$ goes to zero and $u_E = u_g$, $v_E = 0$ as required. Near, but not right at, the surface $(\pi z/D_E) \ll 1$ and expansion of exponential, sine and cosine terms, keeping terms proportional to $(\pi z/D_E)$ but neglecting higher powers, gives $u_E = \pi z u_g/D_E = v_E$. Thus near the bottom u and v vary linearly with z and the near-bottom current direction is $45°$ to the left of the geostrophic current (in the northern hemisphere and vice versa in the southern hemisphere). The current rotates from the geostrophic direction to $45°$ to the left of it and the speed goes to zero at the bottom.

Using a qualitative argument similar to Nansen's for the swing of the surface current to the right, it is easy to see why the current near the bottom swings to the left. Before friction begins to act we have a geostrophic current with the Coriolis force acting to the right and the pressure gradient force to the left. With a barotropic case (reasonable near the ocean bottom) the pressure gradient is independent of depth. As the bottom is approached, friction slows the flow; the Coriolis force (proportional to speed) decreases and the pressure gradient to the left is not completely balanced. The flow swings to the left until the vector sum of the Coriolis and friction forces can balance the pressure gradient force.

The same solution is valid (under the same assumption) for wind blowing over the sea or land. Since in the northern hemisphere the surface wind is at $45°$ to the left of the geostrophic wind, i.e. the wind above the Ekman layer, and the surface (water) current is at $45°$ to the right of the surface wind, the surface current will be in the same direction as the geostrophic wind. In the southern hemisphere, the rotation directions are opposite in both cases and the final result is the same. Again because of the simple form chosen for A_z the details should not be taken too seriously. The direction of rotation to the left in the atmosphere is usually less than $45°$, $10-20°$ is more commonly observed over the ocean. This discrepancy may be due to neglect of time-dependent and stability effects as well as to the simple form for A_z. Likewise, the wind-driven surface (water) current is likely to be to the right of the wind direction but not by exactly $45°$

It is worth noting that the near-surface wind speed is still an appreciable fraction of the geostrophic speed. At 10 m height it is 60–70% of the geostrophic speed; most of the reduction to zero occurs very close to the surface. The Ekman layer thickness in the atmosphere is typically 10 times that in the ocean. Thus the kinematic eddy viscosity based on this depth will be about 100 times the eddy viscosity for the ocean surface Ekman layer. This difference is a consequence of the greater speeds of flow in the atmosphere leading to greater shears and stronger turbulent friction effects, at least as evidenced by the value of A_z.

A more complicated situation might consist of a combinaton of a geostrophic current with a wind-driven Ekman spiral superimposed at the surface (and with the Ekman bottom layer if the water were shallow and the geostrophic current extended near to the bottom). Now imagine a tidal current superimposed, the direction of which might also be rotating, and it will be appreciated that things can get quite complicated in the real ocean. It may be very dificult to analyse into its components a current system consisting of all three, geostrophic, wind-driven and tidal, particularly if they are all changing with time.

If you visualize the water becoming shallow and the depth decreasing to the order of D_E or less, you can see that the surface Ekman layer and the bottom Ekman layer will close up and even overlap. In shallow water the two spirals tend to cancel each other so that the total transport is more in the direction of the surface wind rather than at right angles to it. When the water depth decreases to about $D_E/10$ then the transport is essentially *in* the wind direction, the effect of Coriolis force being swamped by the friction.

9.45   *Limitations of the Ekman theory*

The above theory is quite elegant in its way but in fact it is doubtful if anyone has actually observed a well-developed Ekman spiral current distribution in the sea as even Ekman admitted in one of his last papers. However, this is not to say that the theory is incorrect—the Ekman spiral is well known and clearly observed in the laboratory where the viscosity is molecular and constant, and there is evidence for such behaviour in the atmosphere as already discussed. Furthermore, some of the integrated effects, such as the upwelling consequence, are well known and common phenomena which support the Ekman theory on broad grounds. Then why is the Ekman spiral so elusive in the sea?

The first reason is that the problem in the form solved by Ekman is very much idealized. Commenting on his assumptions:

(1) No boundaries—not realistic, but probably not too bad an assumption away from the coast, and the consequences near the coast do support the solution obtained.
(2) Infinitely deep water—again not exactly true but presents only a small

source of error in the open ocean, i.e. D_E values of the order of 100–200 m compared with the average ocean depth of 4000 m.

(3) A_z constant—probably not true but at present we do not really know enough about it to say whether or not this assumption leads to much error. It probably does not, because Rossby and Montgomery (1935) have solved the equations with $A_z = f(z)$ in likely ways and found only detailed differences from Ekman's solution, e.g. the angle between the wind direction and the current at the surface became slightly smaller and a function of latitude and wind speed.

(4) Steady-state solution and steady wind—probably a real source of difficulty, since neither wind nor sea is really steady (except approximately in the trade wind zones).

(5) Furthermore, there are other sources of motion in the sea (thermohaline, tidal, surface and internal waves) and a current meter placed in the sea cannot distinguish one from another. It tries to record the sum and the oceanographer has to try to sort them out. (The surface wave velocities and their effects are usually neither measured nor averaged out fully, which may cause large errors.) To sort out the individual velocities it is necessary to have long series of measurements (say hourly, or even more frequently, perhaps for months); these we lack in sufficient detail to test Ekman's theory adequately. Added to this deficiency are the practical difficulties of measuring currents in deep water, the only region where it is reasonable to seek the Ekman spiral.

(6) Homogeneous water—distinctly unreal and one assumption that should be criticized although as noted the wind fricton part of the flow can be calculated separately. Sverdrup was probably the first to try to do something to correct this fault, as will be described in the next section.

(7) The f-plane assumption—a minor source of error for zonal bodies of water and for areas of limited latitudinal extent.

Despite its idealized nature this theory of wind-driven currents, stimulated by Nansen and worked out by Ekman, opened the way to the understanding of the mechanism giving rise to the upper-layer currents. The key to Ekman's success here was the use of the large eddy coefficient of viscosity rather than the much smaller molecular one which rendered Zöppritz' earlier attempt sterile.

Finally we should mention that there are extensions of the simple Ekman theory to water of finite depth (as described qualitatively in Section 9.44) and to the transient state, i.e. the development of the Ekman current system when a wind suddenly starts to blow. Summaries of these may be found in *Principles of Physical Oceanography*, Neumann and Pierson, 1966.

9.5   Sverdrup's solution for the wind-driven circulation

The equations of motion assuming negligible accelerations and friction from horizontal gradients of velocity are

$$\alpha \frac{\partial p}{\partial x} = fv + \alpha \frac{\partial \tau_x}{\partial z},$$

$$\alpha \frac{\partial p}{\partial y} = -fu + \alpha \frac{\partial \tau_y}{\partial z}, \tag{9.6'}$$

i.e. Pressure = Coriolis + Friction (forces).

Ekman simply ignored the pressure gradient terms on the left side, assuming an unrealistic homogeneous ocean with level isobars, i.e. non-geostrophic flow. In this section we ignore horizontal friction terms which would be important in currents such as the Gulf Stream, so the solutions will not be valid there. However, we have added wind driving and can examine its possible effects away from coastal boundaries.

Essentially what Sverdrup did was to retain the pressure terms but abandon any attempt to determine the details of the velocities u and v as functions of z. He was satisfied to determine the total transport in the x and y directions in the whole layer affected by the wind (i.e. M_x and M_y when expressed as mass transport). He integrated the equations from the surface (taken as $z = 0$ because η will be small compared with the depth of frictional influence) to $z = -h$ (assumed to be above the ocean bottom) where the wind-driven motion has become zero. Such motion would include not only the Ekman but any geostrophic flows caused by divergence of the Ekman flow, so $h \gg D_E$. In the first stage of integration the equations take the form

$$\int_{-h}^{0} \frac{\partial p}{\partial x} dz = \int_{-h}^{0} \rho f v \, dx + \tau_{x\eta} = f M_y + \tau_{x\eta},$$

$$\int_{-h}^{0} \frac{\partial p}{\partial y} dz = - \int_{-h}^{0} \rho f u \, dz + \tau_{y\eta} = -f M_x + \tau_{y\eta}. \tag{9.18}$$

Here, $\tau_{x\eta}$ and $\tau_{y\eta}$ represent the wind-friction stress at the sea surface, all that remains of the friction terms in the previous pair of equations (9.6'). The reason is that when the derivative of a continuous function is integrated between limits, the values *at* the two limits determine the value of the integral. In this case the value of the friction stress in the water is equal to the wind stress at the surface $z = 0$ (the τ values), and is zero at $z = -h$ because it was selected to be where the motion had become zero, and with no motion in a fluid there is no friction. To simplify the notation we shall omit the η subscript in the rest of this section and use τ_x and τ_y for the surface stress components.

We shall carry out a more general derivation which includes Sverdrup's simpler one as a special case presently, so we shall use some results without proof here. If we differentiate the first of equations (9.18) with respect to y and the second with respect to x, then the differentiation of the pressure terms can be taken inside the integrals since the limits are constants. (The surface is not exactly level but this variation does not matter, as we will show later, and we ignore it.) The two pressure terms are then the same except for the order of differentiation which can be interchanged for a variable such as p, so the pressure terms are the same. Then subtracting the two equations, noting that the term $M_x(\partial f/\partial x)$ which appears is zero (because the Coriolis term, f, does not change in the x-direction (east–west) and that $(\partial M_y/\partial y)+(\partial M_x/\partial x)=0$ by continuity, the resulting equation is

$$M_y\frac{\partial f}{\partial y}=\frac{\partial \tau_y}{\partial x}-\frac{\partial \tau_x}{\partial y} \qquad (9.19)$$

and, together with the equation of continuity for mass transport

$$\frac{\partial M_y}{\partial y}+\frac{\partial M_x}{\partial x}=0, \qquad (9.20)$$

these form a pair of equations describing the mass transports M_x and M_y.

Sverdrup's procedure assumes either that there is zero velocity in the deep water or that the bottom is level and that the friction there is small compared with that at the surface. If there is a barotropic current and a variable bottom depth where the water is moving there will be additional terms which we shall discuss later. The interesting feature of equation (9.19) is that it is not the components themselves, τ_x and τ_y, of the wind surface stress τ_η which appear but their horizontal *gradients* $\partial \tau_x/\partial y$ and $\partial \tau_y/\partial x$. In equation (9.19) the combination $(\partial \tau_y/\partial x-\partial \tau_x/\partial y)$ is the vertical component $(\text{curl}_z\tau_\eta)$ of the curl of the wind stress $(\nabla\times\tau_\eta)$, the only component which is non-zero for a horizontal wind. The symbol β is often used for $\partial f/\partial y$ and with these changes of notation, equation (9.19) becomes

$$\beta M_y=\text{curl}_z\tau_\eta \qquad (9.21)$$

which is called the *Sverdrup equation*. (Note that as $\beta\,(=\partial f/\partial y)$ represents the variation of the Coriolis parameter f with *latitude* it is essential that the y-axis be north–south when using this β-plane notation.)

At some places $\text{curl}_z\tau_\eta$ will vanish (equal zero) and there will be no north–south transport (although there may be flows which when added up will cancel). Lines along which $\text{curl}_z\tau_\eta=0$ provide natural boundaries which separate the circulation into "gyres".

The quantities M_x and M_y are the total mass transports in the wind-influenced layer, defined as $M_x=\int_{-h}^{0}\rho u\,dz$ and $M_y=\int_{-h}^{0}\rho v\,dz$. We write

$$M_x=M_{xE}+M_{xg}$$

where the first term on the right is the Ekman wind-driven transport (non-geostrophic) while the second is the geostrophic transport (due to the distribution of mass), and similarly for M_{yE} and M_{yg} (just as we separated v into v_E and v_g in equation (9.7)).

Then equations (9.18) become

$$fM_{yE} = f \int_{-h}^{0} \rho v_E dz = -\tau_x,$$

$$fM_{yg} = f \int_{-h}^{0} \rho v_g dz = \int_{-h}^{0} \frac{\partial p}{\partial x} dz, \qquad (9.22)$$

and similarly for M_x.

9.51   Orders of magnitude of the terms

It is useful to look at the magnitudes of some of the terms. We use a position in the North Atlantic at about $35°N$ with a wind of $7-8$ m s^{-1} (about 15 knots) from the west. Then $\tau_x \simeq 10^{-1}$ N m^{-2} (or Pa) and $\tau_y = 0$,

$$\text{curl}_z \tau_\eta \simeq -\frac{\partial \tau_x}{\partial y} \simeq -\frac{10^{-1} \, \text{N m}^{-2}}{1000 \, \text{km}} \simeq -10^{-7} \, \text{N m}^{-3}$$

$$f \simeq 10^{-4} \, \text{s}^{-1}, \quad \beta \simeq 2 \times 10^{-11} \, \text{m}^{-1} \text{s}^{-1}.$$

From these values we get, using equation (9.16),

$$M_{yE} = -\frac{\tau_x}{f} = -10^{-3} \, \text{kg m}^{-1} \text{s}^{-1}$$

(where the $-$ sign indicates flow to the south) and, using equation (9.21),

$$M_y = M_{yE} + M_{yg} = \frac{\text{curl}_z \tau_\eta}{\beta} = -\frac{10^{-7}}{2 \times 10^{-11}} = -5 \times 10^3 \, \text{kg m}^{-1} \text{s}^{-1}.$$

We see that $M_{yg} = -4 \times 10^3$ kg m^{-1} s^{-1} which is caused by the north–south variation of the wind and consequent convergence of M_{yE} and is considerably larger than M_{yE} as is often the case.

The value for M_y above is for only a 1-m-wide strip, so that for the width of an ocean of 5000 km $= 5 \times 10^6$ m, the southward flow would be 25×10^9 kg s$^{-1} \simeq 25 \times 10^6$ m^3 s^{-1} in volume $= 25$ sverdrups. (Here, as is common practice in oceanography, we use 1000 kg m^{-3} for ρ when converting from mass to volume transport since the error, $\sim 3\%$ relative to using the actual density, is negligible compared with the uncertainty in the estimates of transport.)

9.52   Application of the Sverdrup equations

Sverdrup applied these equations to the trade-wind zones in lower latitudes where τ_y and $\partial \tau_y / \partial x$ can be assumed to be negligible (i.e. much smaller than the terms retained), and τ_x variations with x are averaged out. Substituting

$f = 2\Omega \sin \phi$ and noting that $y = R\,\mathrm{d}\phi$ where $R =$ radius of the earth, then $\beta = \mathrm{d}f/\mathrm{d}y = \mathrm{d}(2\Omega \sin \phi)/R\,\mathrm{d}\phi = 2\Omega \cos \phi/R$ and equation (9.19) gives M_y while using (9.20) gives

$$\frac{\partial M_x}{\partial x} = -\frac{\partial M_y}{\partial y} = \frac{1}{2\Omega \cos \phi}\left(R\frac{\partial^2 \bar{\tau}_x}{\partial y^2} + \frac{\partial \bar{\tau}_x}{\partial y}\tan \phi \right)$$

which can be integrated from $x = 0$, the coast, where $M_x = 0$. Finally we have

$$M_x = \frac{x}{2\Omega \cos \phi}\left(\frac{\partial \bar{\tau}_x}{\partial y}\tan \phi + \frac{\partial^2 \bar{\tau}_x}{\partial y^2}R \right),$$

$$M_y = -\frac{R}{2\Omega \cos \phi}\frac{\partial \bar{\tau}_x}{\partial y}. \tag{9.23}$$

Here, x is the distance from a north–south coastline at the east side of the ocean westward to a point P in the ocean as in Fig. 9.8, so that the numerical value entered in the expression will be negative. The bars over the stress terms indicate that mean values are taken over the distance x and the values for M_x and M_y are for the point P.

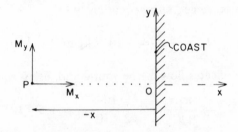

FIG. 9.8 For derivation of the Sverdrup transport on the east side of an ocean.

Comparing the Ekman and the Sverdrup solutions, Sverdrup lost the details of the current velocities with depth but gained the possibility of having a coastal boundary at one side of the ocean, a step toward a more realistic situation than Ekman's horizontally infinite, i.e. boundaryless, ocean. Sverdrup's solution also is no longer bound by the homogeneous ocean assumption and the solutions therefore have this additional feature of the real oceans.

Referring to the expression for M_x in equation (9.23) above, it turns out in practice that in the trade wind and equatorial zones the important term of the two on the right is $\partial^2 \bar{\tau}_x/\partial y^2$. Figure 9.9 shows for the eastern Pacific the character of the mean x-component of the wind as the full line while the

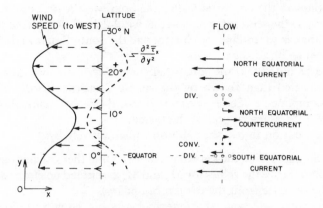

FIG. 9.9 *Smoothed representation of the east–west com-*
ponents of wind speed and stress terms, and related currents
at low latitudes, eastern Pacific Ocean. Regions of diver-
gence (DIV) and convergence (CONV) indicated.

corresponding character of $\partial^2 \bar{\tau}_x / \partial y^2$ is shown by the dashed line. By "character" we mean schematically—the actual wind variation with latitude is not as regular as shown.

It will be seen that:

(a) north of about 15°N and south of about 2°N, the value of $\partial^2 \bar{\tau}_x / \partial y^2$ is + and x is $-$, $\therefore$ M_x is $-$, i.e. the flow is to the WEST (North Equatorial Current and South Equatorial Current), and

(b) between 15°N and 2°N (the doldrums), the value of $\partial^2 \bar{\tau}_x / \partial y^2$ is $-$ and x is $-$, $\therefore$ M_x is +, i.e. the flow is to the EAST (North Equatorial Countercurrent).

The figure shows qualitatively how Sverdrup's solution explains the existence of an equatorial current system consisting of two westward flowing currents (N.E.C. and S.E.C.) with an eastward flowing current (N.E.C.C.) between them. It will be noted that this system is not symmetrical about the equator but is displaced to the north of it, because the trade-wind system is displaced this way in the Pacific. The current distribution is similar in the Atlantic. However, in the Indian Ocean the wind pattern changes seasonally and the current pattern changes with it. During the Northeast Monsoon (November to March), the wind pattern is similar to those in the Pacific and Atlantic but is situated somewhat further south; the westward North Equatorial Current straddles the equator and the eastward Equatorial Countercurrent and the westward South Equatorial Current are both south of the equator. During the Southwest Monsoon (May to September), the Southeast Trade Winds continue south of the equator but north of the equator the winds blow to the northeast (called the Southwest Monsoon). As a result there is only a two-current system—the wind-driven Southwest Monsoon Current to the east straddling the equator and the South Equatorial Current to the west well south of the equator.

It should be noted that the equatorial current system is now known to be more complicated than shown by the simple Sverdrup theory, with more east–west components. (See *Descriptive Physical Oceanography*, Pickard and Emery, 1982, for a description of the system.)

Sverdrup went on to test the solutions quantitatively by:

(a) calculating τ_x from the known mean winds and then calculating the curl, etc., and thence the values for M_x and M_y at selected positions defined by x (the distance from the eastern boundary),

(b) determining M_{xg} and M_{yg} independently by the geostrophic method from oceanographic field data and adding the Ekman transport to get the total transport, and

(c) comparing these two independent calculations.

The result of this calculation as revised by Reid (1948) is shown in Fig. 9.10 where the values for M_x are shown on the left and for M_y on the right. Because M_y is smaller, the relative errors in obtaining it from the density field are greater which may be at least part of the reason for the poorer agreement in the two methods for M_y than for M_x. Note that $\partial^2 \bar{\tau}_x / \partial y^2$, which is proportional to M_x,

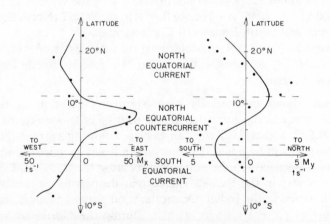

FIG. 9.10 *Mass transport components in the eastern Pacific calculated from the mean wind stress (lines) compared to those from geostrophic calculations from oceanographic data (dots). M_x and M_y in tonnes per second through a vertical area 1 m wide and 1000 m deep (approximately equivalent to 0.1 Sv per degree of latitude). (From Reid, 1948.)*

has a more complicated structure when the actual winds are used than that in the schematic picture shown in Fig. 9.9.

Note that $M_x \simeq 10M_y$ which is fairly typical, particularly for equatorial regions. The reason lies in the difference between the east–west and the north–south length scales of the gyre systems. The east–west scale (L_x) is determined by continental barriers, the north–south (L_y) by the lines of $\text{curl}_z \tau_\eta$ = 0. Typically there is approximately a 10 to 1 ratio of lengths L_x/L_y. Because by continuity

$$\frac{\partial M_x}{\partial x} + \frac{\partial M_y}{\partial y} = 0, \text{ then } \frac{M_x}{M_y} \simeq \frac{L_x}{L_y} \simeq \frac{10}{1}.$$

Looking at it another way, water which goes north or south in a gyre must then go east or west to close the gyre. Thus the total transport north–south equals that east–west, i.e. $M_y L_x = M_x L_y$. This gives the same result using the idea of continuity of volume in an integral rather than a differential sense.

Figure 9.11 shows an analytic solution by R. O. Reid (1948) calculated from a simplified form for the wind stress but for the real coastline in the eastern equatorial Pacific. Presentation of flow patterns in this way is discussed in Appendix 1. The stream function ψ will be defined presently. For the moment, take it to be a horizontally integrated mass transport. The water flows in the

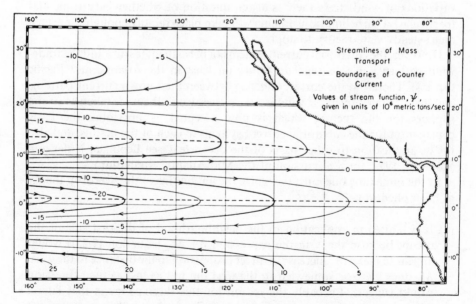

FIG. 9.11   *Streamlines of mass transport in the eastern Pacific from the mean wind stress.*
(From Reid, 1948.)

direction indicated by the arrows on the lines; between any two lines the total transport is 5×10^6 tonnes per second (or about 5 Sv).

The calculation (a) above in Sverdrup's test was from the climatological mean wind for October/November (the cruise data period) while the calculation (b) was for specific cruise data. It should be pointed out that Leetmaa, McCreary and Moore (1981) have expressed reservations about the success of the Sverdrup test for the following reasons. In recent years, studies of the equatorial region wind stress have shown that it has considerable variation in space and time. In the eastern Pacific, the magnitude of the zonal wind stress varies by a factor of about 5 from a maximum in the northern summer to a minimum in winter with shorter period (smaller) variations superimposed. The question then is whether Sverdrup's assumption of a steady state is applicable. Leetmaa *et al.* (1981) repeated the Sverdrup calculation using more recent wind-stress values for the calculation of transport, and oceanographic data from the EASTROPAC expedition for geostrophic calculations for a number of meridional sections from 2° to 12°N during a period of about a year in 1967/68. The transports calculated from the wind stress and the geostrophic ones show some similarity to the single Sverdrup section calculation but also show seasonal variations in the strengths of the three equatorial currents, with the North Equatorial Countercurrent being near the equator in the northern winter but moving northward in the summer. For calculations relative to a depth of no motion of only 500–1000 m and with the observed temporal variations of wind stress there is also a question of whether barotropic and (particularly) baroclinic adjustments will take place rapidly enough for even a quasi-steady-state theory to apply.

However, the equatorial current system is a permanent feature and the trade-wind pattern is also a regular feature (at least in the Atlantic and Pacific Oceans). Therefore the correspondence between the known current patterns and that obtained from the wind stress via Sverdrup's theory is taken as strong support for the theory, which is now accepted as providing the basic explanation for the equatorial current systems in each of the oceans, and also for its application to other parts of the ocean (e.g. see Leetmaa, Niiler and Stommel, 1977, for a discussion of the Mid-Atlantic circulation).

At the same time one must recognize the limitations of the Sverdrup theory as he applied it, i.e.:

(1) It is limited in application to the neighbourhood of the east side of the ocean, because the x in the expression for M_x (equation (9.23)) would appear to make M_x increase in direct proportion to the distance to the west. M_x does increase somewhat to the west but not as fast as the expression would suggest. Probably the reason for this discrepancy is that lateral friction (between the currents) has been ignored. It will increase as the currents increase and therefore in the real ocean M_x does not go on

increasing to the west as rapidly as the Sverdrup relation suggests. The stress terms τ_x and $\partial^2 \tau_x / \partial y^2$ no doubt also have some variation with x which was not included.

(2) The differential equations allow only one boundary condition to be satisfied; in the solution given it is that there shall be no flow through the coast. To be able to apply more boundary conditions (e.g. no slip at the eastern boundary and perhaps conditions at a western boundary) it is necessary to go to more complicated equations, as will be described later.

(3) The solutions give the depth integrated mass transport but no details of the velocity distribution with depth.

9.6   The general form of the Sverdrup equation—convergence and divergence

As we have already seen when using mass transports, we need an equation of continuity for them. Consider a column of fluid with sides δx and δy extending from the sea surface to $z = -h$. To be completely general we denote the value of z at the surface by η because we know that the surface may have small variations from the level surface $z = 0$ which we may take to be the average level of the sea surface (mean sea level) over the region being examined. Mass transports in the x and y directions respectively are

$$M_x = \int_{-h}^{\eta} \rho u \, dz \quad \text{and} \quad M_y = \int_{-h}^{\eta} \rho v \, dz.$$

The mass flow into the column in the x-direction is $M_x \delta y$ and the flow out is $[M_x + (\partial M_x / \partial x) \delta x] \delta y$. (This calculation is just like that used in the derivation of the equation of continuity in Chapter 4, except that now we have a column of height $(h + \eta)$ instead of height δz.) The net flow out in the x-direction is $(\partial M_x / \partial x) \delta x \delta y$. Similarly in the y-direction it is $(\partial M_y / \partial y) \delta x \delta y$. At the bottom of the column ($z = -h$) there may be flow out if the vertical velocity is not zero; it will be $-\rho w_{-h} \delta x \delta y$ (where the minus sign is required because we want flow *out* while w positive is *up*). At the top there may be an effective outflow velocity w_n if there is a net difference between evaporation and precipitation. We shall neglect this effect but it could be included as a driving term. If the value of η is changing with time then the equivalent mass flow out is $\rho(\partial \eta / \partial t) \delta x \delta y$. As we are considering steady-state cases we shall take it to be zero. (It would be important in time-dependent calculations for such phenomena as tides and storm surges.) Because mass must be conserved the net flow of mass out must be zero, i.e.

$$\left(\frac{\partial M_x}{\partial x} + \frac{\partial M_y}{\partial y} - \rho w_{-h} \right) \delta x \delta y = 0$$

and dividing through by $(\delta x \delta y)$ gives

$$\frac{\partial M_x}{\partial x} + \frac{\partial M_y}{\partial y} - \rho w_{-h} = 0$$

or

$$\nabla_H \cdot \mathbf{M} - \rho w_{-h} = 0 \tag{9.24}$$

where $\nabla_H \cdot = [\mathbf{i}(\partial/\partial x) + \mathbf{j}(\partial/\partial y)]$ is the horizontal divergence operator and $\mathbf{M}$ is the vector mass transport.

If the velocity is zero at $z = -h$ and deeper, then the last term of equation (9.24) is zero and we get the form used by Sverdrup. If $h = -z_B$ is the total depth (which in general will be variable) then the last term also vanishes because there can be no flow through the bottom.

Sometimes it is convenient to take the point of view that the vertical velocity associated with the divergence of one type of transport provides a driving force for the convergence of another type of transport. For example, consider the Ekman flow, then $(\partial M_{xE}/\partial x) + (\partial M_{yE}/\partial y) - \rho w_E = 0$. Here, w_E is the vertical velocity at the bottom of the Ekman layer associated with convergence or divergence of the Ekman transport. If $w_E \neq 0$, it requires a corresponding divergence or convergence of the flow below. Denoting this flow by $\mathbf{M}_g$ (g for geostrophic) and assuming for the present that the flow vanishes below $z = -h$, so that $w_{-h} = 0$, then $\nabla_H \cdot \mathbf{M} = \nabla_H \cdot (\mathbf{M}_E + \mathbf{M}_g) = 0$. Thus $\nabla_H \cdot \mathbf{M}_g = -\nabla_H \cdot \mathbf{M}_E = -\rho w_E$. The process is sometimes spoken of as *Ekman pumping*. If the flow does not go to zero before approaching the bottom, there may be a bottom Ekman layer (denoted by the subscript B). Then $\nabla_H \cdot \mathbf{M}_{EB} + \rho w_{EB} = 0$ where ρw_{EB} is added now since it is flow up through the top and $\nabla_H \cdot \mathbf{M}_g = -\rho(w_E - w_{EB})$. We expect w_{EB} to be rather small compared with w_E as a general rule. Note that w_E is a good approximation to the total vertical velocity at the base of the layer. There may be divergence of the part of $\mathbf{M}_g$ occurring in the relatively thin Ekman layer. The total divergence $\nabla_H \cdot \mathbf{M}$ is only $-\rho w_E$ and, since the Ekman layer is only a small fraction of the region of geostrophic flow, the divergence of $\mathbf{M}_g$ in the Ekman layer will be a small part of the total and may be neglected.

If we use volume transports, $Q_x = \int_{-h}^{\eta} u\,dz$ and $Q_y = \int_{-h}^{\eta} v\,dz$, and assume incompressibility, the net volume flow out must vanish. By the same sort of derivation and neglecting the possible flows through the surface (e.g. precipitation or evaporation) we have

$$\frac{\partial Q_x}{\partial x} + \frac{\partial Q_y}{\partial y} - w_{-h} = 0. \tag{9.25}$$

Let us now integrate the equations of motion (9.6) vertically from the bottom $z = z_B$ to the surface $z = \eta$. In the equations (9.6) accelerations and friction from velocity variations in the horizontal have been assumed small; this should be a good approximation except in strong currents (usually near the western boundary) such as the Gulf Stream. We obtain

$$\int_{z_B}^{\eta} \frac{\partial p}{\partial x}\, dz = fM_y + \tau_{x\eta} - \tau_{xB},$$

$$\int_{z_B}^{\eta} \frac{\partial p}{\partial y}\, dz = -fM_x + \tau_{y\eta} - \tau_{yB}, \qquad (9.26)$$

where we have retained the possibility of friction occurring at the bottom although we do not expect it to be important. The symbol η in the subscript for a stress component indicates a surface value while B denotes a value at the bottom.

We cannot directly follow Sverdrup's procedure of cross-differentiating the pressure terms because the limits of integration are not independent of x and y now. To get around this problem we must write the pressure terms in a somewhat different form. Here we follow Fofonoff's article in *The Sea*, Vol. 1 (M. N. Hill, Ed., 1962) but do not derive all the results in detail; the interested reader should consult the article which also contains other valuable information. We define a new function $E_p = \int_{z_B}^{\eta} p\, dz$. It can be shown (by integrating by parts taking $p = 0$ at the surface and using the hydrostatic equation) that

$$E_p = \int_{z_B}^{\eta} \rho g\,(z - z_B)\mathrm{d}z = \text{work done to pile up the water above the bottom,}$$

i.e. it is a measure of the potential energy. Now

$$E_p = \frac{1}{g} \int_{z_B}^{\eta} (p\alpha)\rho g\,\mathrm{d}z \text{ is the original definition rewritten.}$$

Using the hydrostatic equation $\mathrm{d}p = -\rho g\,\mathrm{d}z$ and $p = 0$ at $z = \eta$, $p = p_B$ at $z = z_B$, we have

$$E_p = \frac{1}{g} \int_0^{p_B} p\alpha\,\mathrm{d}p = \frac{1}{g} \int_0^{p_B} p\alpha_0\,\mathrm{d}p + \frac{1}{g} \int_0^{p_B} p\delta\,\mathrm{d}p = E_p^0 + \chi. \qquad (9.27)$$

We have removed the minus sign using $\int_{p_B}^{0} = -\int_0^{p_B}$. E_p^0 is a function of the bottom pressure only; it is equal to the potential energy of sea water of $S_0 = 35$, $T = 0°\mathrm{C}$. χ is the potential energy anomaly, i.e. the difference from E_p^0 associated with the difference between the reference specific volume and the actual specific volume; it is sometimes a useful function to consider in synoptic physical oceanography. This separation is similar to the one made when discussing the geopotential in the previous chapter and χ can be calculated from oceanographic observations in a similar way to that used to obtain $\Delta\Phi$. Now consider $\partial E_p / \partial x$:

$$\frac{\partial}{\partial x} \int_{z_B}^{\eta} p\,dz = \int_{z_B}^{\eta} \frac{\partial p}{\partial x}\,dz + p_\eta \frac{\partial \eta}{\partial x} - p_B \frac{\partial z_B}{\partial x}.$$

The terms $p_\eta(\partial \eta/\partial x)$ and $p_B(\partial z_B/\partial x)$ occur because η and z_B vary with x. Since we take $p_\eta = 0$, the first of these vanishes. (It can be retained if important, e.g. in storm surge calculations, but it is not normally important for the large-scale, steady-state circulation.) Rearranging

$$\int_{z_B}^{\eta} \frac{\partial p}{\partial x}\,dz = \frac{\partial E_p}{\partial x} + p_B \frac{\partial z_B}{\partial x} = \frac{\partial \chi}{\partial x} + \frac{\partial E_p^0}{\partial x} + p_B \frac{\partial z_B}{\partial x}. \tag{9.28}$$

The last two terms may be combined (see Fofonoff, 1962) to give

$$\int_{z_B}^{\eta} \frac{\partial p}{\partial x}\,dz = \frac{\partial \chi}{\partial x} + \frac{p_B \alpha_B}{g}\left(\frac{\partial p_B}{\partial x}\right)_z \tag{9.29}$$

where $(\partial p_B/\partial x)_z$ means the change of pressure on a level surface at the bottom. The same equation will hold with y replacing x, so we can substitute these equations in (9.26) to give

$$-fM_y = -\frac{\partial \chi}{\partial x} - \frac{p_B \alpha_B}{g}\left(\frac{\partial p_B}{\partial x}\right)_z + \tau_{x\eta} - \tau_{xB},$$

$$fM_x = -\frac{\partial \chi}{\partial y} - \frac{p_B \alpha_B}{g}\left(\frac{\partial p_B}{\partial y}\right)_z + \tau_{y\eta} - \tau_{yB}. \tag{9.30}$$

The first terms are associated with variations in the density distribution; they arise from the baroclinic part of the geostrophic velocity, i.e.

$$fM_{yc} = \frac{\partial \chi}{\partial x}, \quad fM_{xc} = -\frac{\partial \chi}{\partial y}.$$

The stress terms may be associated with Ekman transports as we did earlier

$$fM_{yE} = -\tau_{x\eta}; \quad fM_{yEB} = \tau_{xB}; \quad fM_{xE} = \tau_{y\eta}; \quad fM_{xEB} = -\tau_{yB}.$$

The terms involving bottom pressure are the barotropic transports times f.

$$M_{yb} = \int_{z_B}^{\eta} \rho v_b\,dz = v_b \int_{z_B}^{\eta} \rho\,dz \text{ because } v_b \text{ does not vary with } z.$$

Now $\rho\,dz = -\dfrac{dp}{g}$, so $\displaystyle\int_{z_B}^{\eta} \rho\,dz = -\int_{p_B}^{\eta} \frac{dp}{g} = \frac{p_B}{g}$, and $M_{yb} = \dfrac{v_b p_B}{g}$.

The barotropic flow is the geostrophic flow in the deep water, so

$$fv_b = \alpha_B\left(\frac{\partial p_B}{\partial x}\right)_z \quad \text{and} \quad \frac{p_B \alpha_B}{g}\left(\frac{\partial p_B}{\partial x}\right)_z = \frac{fv_b p_B}{g} = fM_{yb}.$$

Likewise, the p_B term in the second equation of (9.30) is $-fM_{xb}$. Thus equation (9.30) may be written as

$$M_y = M_{yc} + M_{yb} + M_{yE} + M_{yEB}.$$
$$M_x = M_{xc} + M_{xb} + M_{xE} + M_{xEB}. \tag{9.30'}$$

These equations simply divide the total transport into components which are related respectively to the density field, bottom pressure gradient and stress terms of equation (9.30).

Now, following Sverdrup's approach we take

$$\frac{\partial}{\partial x}(fM_x) + \frac{\partial}{\partial y}(fM_y) = f\left(\frac{\partial M_x}{\partial x} + \frac{\partial M_y}{\partial y}\right) + \beta M_y = \beta M_y \tag{9.31}$$

because by continuity $(\partial M_x/\partial x + \partial M_y/\partial y) = 0$ for the total transport from surface to bottom. Following the same procedure for the right-hand sides of equations (9.30) gives

$$\beta M_y = \mathrm{curl}_z \tau_\eta - \mathrm{curl}_z \tau_B - \rho'f\left(u_b \frac{\partial z_B}{\partial x} + v_b \frac{\partial z_B}{\partial y}\right) \tag{9.32}$$

or in vector form

$$\beta M_y = (\nabla \times \tau_\eta)_k - (\nabla \times \tau_B)_k - \rho' f(\mathbf{V}_b \cdot \nabla_H z_B)$$

where $\nabla_H = \mathbf{i}\,(\partial/\partial x) + \mathbf{j}\,(\partial/\partial y)$ and $\rho' = \rho_B\left[1 + \dfrac{p_B}{\alpha_B}\left(\dfrac{\partial \alpha}{\partial p}\right)_B\right]$

(see Fofonoff (1962) for a partial derivation of the last term). The χ terms cancel out just as did the pressure gradient terms in Sverdrup's derivation because he assumed the deep flow (barotropic part) to be zero. With no flow near the bottom there is no stress there either, and equation (9.32) reduces to the simpler form of (9.21) derived by Sverdrup. If the bottom is level and the bottom stress negligible, one also gets the simpler form (9.21). Finally, if the flow is along the bottom contours, i.e. is entirely horizontal so that $\mathbf{V}_b$ is perpendicular to $\nabla_H z_B$, then the terms involving z_B also vanish. It has been quite common to neglect the bottom stress; this approximation is probably good in most cases although, as we shall see in Chapter 11, some numerical model results suggest that bottom friction is not completely negligible.

It has also been quite common to neglect the $(\mathbf{V}_b \cdot \nabla_H z_B)$ term based on the idea that the velocities vanish at great depths since the driving for the flow is from the surface. It is only possible to include this effect in analytic treatments (with suitably idealized bottom topography) if the baroclinicity is also excluded, i.e. the geostrophic part of the flow is depth independent, another reason for leaving it out. Even if we take the density field as given from observations, the difficulty remains that we cannot obtain adequate information about $\mathbf{V}_b$ because we cannot obtain it from geostrophic calculations, and deep-current meter observations are too few and of too short duration to help much. As we shall see in Chapter 11 on numerical models, the $(\mathbf{V}_b \cdot \nabla_H z_B)$ term may be important in the real ocean.

9.7   The mass transport stream function

By integrating along the vertical we have derived the mass transport per unit width of current (the density times the vertically averaged velocity times the depth) which depends on x and y but not on z. When we have such a flow, which depends on only two space variables (and is either incompressible or steady state), it is possible to use a scalar function called a *stream function* from which the velocity may be derived. This approach can provide a useful simplification because it may be easier to find a single scalar function than the two components of a vector. We put

$$M_x = \frac{\partial \psi}{\partial y} \quad M_y = -\frac{\partial \psi}{\partial x} \qquad (9.33)$$

where ψ is the stream function. The flow is parallel to lines on which ψ is a constant, which means that plots showing such lines are convenient for display purposes as mentioned earlier in connection with Fig. 9.11 and described somewhat more fully in Appendix 1.

The reader should note that some writers introduce the minus sign in the M_x equation of (9.33)—we have chosen to follow what we believe is the more common practice in fluid mechanics although not, unfortunately, in physical oceanography which is rather divided on this matter. If the opposite sign convention to ours is used and a $\psi = 0$ streamline is used, as is common, the signs of the ψ values in our convention must be multiplied by -1 to convert. Normally there is no problem in interpreting the plots because arrows are usually placed on the lines to show the direction of flow.

$$\text{Consider} \quad \frac{\partial M_x}{\partial x} + \frac{\partial M_y}{\partial y} = \frac{\partial^2 \psi}{\partial x \partial y} - \frac{\partial^2 \psi}{\partial x \partial y}.$$

Now by continuity, if we take the total transport this quantity should vanish and the order of differentiation of ψ can be interchanged (which is a mathematical condition for ψ to be well behaved, as the interested reader may see by consulting a text on fluid mechanics, e.g. Batchelor, 1967). In practice, if we can find an equation for ψ from the equations of motion, usually in the form of the Sverdrup equation (9.21), or the more general form of equation (9.32), or extensions discussed presently, the solution will automatically be such that continuity is satisfied. To get an equation for ψ in the case where the Sverdrup equation holds we simply replace the (βM_y) term by $(-\beta (\partial \psi / \partial x))$.

While the total transport is non-divergent, i.e. continuity takes the form

$$\frac{\partial M_x}{\partial x} + \frac{\partial M_y}{\partial y} = \nabla_H \cdot M = 0$$

the individual transports, Ekman, baroclinic, barotropic (if any) and bottom Ekman (if any) may not be so; it may not be possible to represent them by stream functions. The negative of the potential energy anomaly $(-\chi)$ is almost a stream function for the baroclinic transport. Actually, contours of $(-\chi)$ are "streamlines" for fM_c but since f varies slowly the baroclinic flow will be almost along the contours. The relation between the contour spacing and the strength of the transport will vary with f and hence with latitude.

Similarly, a contour plot of $\Delta\Phi$ (or ΔD in the mixed units system) shows the pattern of the horizontal geostrophic flow (relative to that at the reference level). It is not a true stream function unless $(\partial u/\partial x) + (\partial v/\partial y) = 0$ is also true, but it provides the same useful display features as a stream function.

9.8   Westward intensification—Stommel's contribution

A feature of the ocean circulation seen as a whole is the so-called *westward intensification*, for example as shown in Figs. 8.8 (Pacific) and 11.4(b) (North Atlantic) where the flow lines are close together in the west or northwest off Japan and off the United States respectively, whereas they are more widely spaced over most of the rest of the ocean. Where the flow lines are close, the flow must be swift, and vice versa. Similar flow patterns are evident in the South Atlantic and probably in the Indian Ocean, but are less evident in the South Pacific where the flow is complicated by the islands in the west. Stommel (1948) was the first to present an adequate explanation for this feature.

His demonstration was done with a simplified theoretical model of an ocean and wind pattern. He took a rectangular ocean of constant depth and all on one side of the equator and, for convenience, assumed the earth to be flat, i.e. he used the tangent plane approximation mentioned in Section 6.4. He also assumed a wind stress which varied with latitude as shown in Fig. 9.12, so that it was to the west at the south and to the east at the north. This is a reasonable approximation to the real wind stress which will be discussed shortly. He also included a simple friction term to prevent acceleration so that he could investigate the steady state. Basically he used Sverdrup's equation (9.21) with friction added. He calculated the flow patterns in this ocean for three conditions:

(1) non-rotating ocean (i.e. non-rotating earth);
(2) rotating ocean but Coriolis parameter f constant (the f-plane approximation);
(3) rotating earth with Coriolis parameter varying with latitude ϕ in a simple

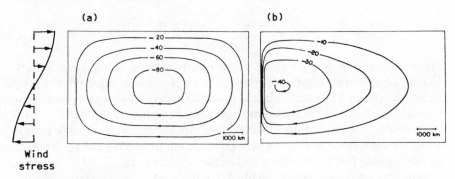

FIG. 9.12   *Flow patterns (streamlines) for simplified wind-driven circulation with:* (a) *Coriolis force zero or constant,* (b) *Coriolis force increasing linearly with latitude. (From Stommel, 1948.)*

but realistic fashion, i.e. linearly with latitude ϕ from $10°$ to $50°$ latitude (the β-plane approximation).

The flow patterns which he obtained then appeared as in Fig. 9.12, the second, (b), being most like the flow pattern in the real oceans (e.g. as in Fig. 8.8). In case (1) the surface remained nearly level. In case (2), to balance the Coriolis effect a higher water level was found at the centre to provide the necessary pressure gradient (Fig. 9.12(a)). A similar high level was found in case (3) but it was not symmetric in the east–west direction as in case (2) but displaced to the west (Fig. 9.12(b)).

It was therefore clear that the variation with latitude, $\partial f/\partial \phi$, of the Coriolis parameter was responsible for the westward intensification. Nowadays, this result is discussed in terms of vorticity as will be described shortly. (Basically, wind stress puts vorticity, i.e. spin, into the ocean and friction is required to take it out. When f varies with latitude, strong friction in the west is needed to take out the vorticity and for this strong friction to occur strong currents with strong shear are needed.)

This result of Stommel's was obtained for a very simplified version of the real ocean, but it is clear that the variation of the Coriolis parameter is a fundamental feature of the dynamics which must be taken into account in any study of large ocean circulations. In addition, Stommel's addition of a friction term permitted a solution with a closed circulation, which Sverdrup's assumptions did not. Stommel's model was not intended to represent a real ocean but to illustrate a principle, as is often the case with models which are much idealized from the real world. The mathematical details are not important, provided that one believes the result (as we do in this case), and we have not given them here. For a fuller exposition, the interested reader may consult Stommel's book *The Gulf Stream* (1965) which also contains much other interesting information.

9.9   The planetary wind field and the drag coefficient C_D

Having indicated the development of the wind stress explanation for the upper-layer currents we should look for a few moments at the general character of the winds on a global scale. The main features are shown in Fig. 9.13. On the left are shown the main wind systems which are distributed in a zonal fashion. The graph of the east–west component of the wind on the right shows that Stommel's form for the wind stress as a function of latitude is a reasonable approximation to the real stress between about 10° and 50°N.

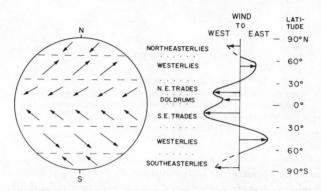

FIG. 9.13   *General pattern of global winds and mean east–west wind components.*

This graph represents an average across the whole width of the ocean, i.e. east–west variations are ignored. In more recent treatments, particularly the numerical ones (Chapter 11), values for the wind stress for a grid of points over the ocean are used so that a better approximation to the real wind stress both north–south and east–west is applied. Of course, it is still a time-averaged distribution.

The presently available wind stress results are probably not very representative in detail. The procedure is to use the relation that wind stress $\tau = \rho C_D W^2$ in the direction of the wind W. The problem is what value to use for the drag coefficient C_D. As far as we know, most stress calculations for oceanographic purposes have used a step or smoothed step function for C_D as in Fig. 9.14(a). More recent measurements (Smith, 1980; Large and Pond, 1981) which are also shown in Fig. 9.14 (curves b and c) do show C_D increasing with wind speed but not so abruptly. The Large and Pond results are based on an extensive set of open sea observations and are in good agreement with other open sea measurements, in particular those of Smith (1980) which are the only other set containing much data for wind speeds above $15 \, \mathrm{m \, s^{-1}}$. Note that C_D is a function of height since the wind speed varies with height. The values shown are for a measurement height of 10 m above sea level; for other heights a

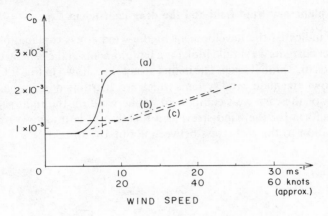

FIG. 9.14   *Form of drag coefficient* (C_D) *for wind over water as a function of wind speed:* (a) *as used for many calculations of wind-driven circulation; measured values for* 10 m *height for neutral stability by* (b) *Smith* (1980), (c) *Large and Pond* (1981).

correction is easily applied (see Large and Pond, 1981). C_D is also a function of the static stability of the air; the values shown are for neutral stability. For mid-latitudes, stability effects are usually small on average but for the tropics and subtropics they need to be included. An approximate correction can be made using the air–sea temperature difference (Large and Pond, 1981). Note also that C_D is not a constant at a given wind speed but has considerable variability which is not surprising since the momentum transfer process is turbulent. The standard deviation is about 20 % of the mean, so for any single observation there is about 68 % probability that the true stress is within 20 % of the calculated value. As one averages observations together the statistical fluctuations should average out so that a mean stress based on a few days or more should be correct to about 10 % which is the overall uncertainty provided that the wind observations are made accurately. For winds below 10–11 m s^{-1}, C_D seems to be approximately constant on average. In fact, there is a range of somewhat lower values when the previous winds have been weak and somewhat higher values when the previous winds have been considerably stronger or in a much different direction. The constant value in this range should represent average conditions in the open ocean. Since winds above 20 m s^{-1} or so are usually too infrequent to make much contribution to the average stress, the Large and Pond C_D formulation should allow stresses *averaged over a few days or more* to be calculated to about $\pm 10\%$ provided that good wind data are available. Note that the Large and Pond formulation is *not* applicable during transient conditions (rapid increase of speed or large direction shifts) and will probably underestimate the stress under such conditions. For example, in storm-surge calculations at about $15 \, \text{m s}^{-1}$, C_D ~ 0.0025 is required to obtain water-level changes comparable to observed values (Donelan, 1982). Under such circumstances one must take the development of the wave field into account using a scheme such as that proposed by Donelan (1982).

The indirect measurements at high wind speeds on which the step or smoothed step function is based are now considered to be unreliable. They were based on "set-up" of the water level, i.e. rise above predicted tidal level. Unfortunately the problem is more complicated than a simple balance between the pressure gradient associated with the set-up and the wind stress at the surface. Time variations, the effect of bottom topography and non-linear effects in shallow water on the waves associated with the strong winds where the set-up measurements were made are all important factors. The C_D values from surface tilts are larger than the Smith and the Large and Pond values, particularly in the $15\text{–}20 \, \text{m s}^{-1}$ wind-speed range. They may be reasonably representative of the conditions under which they were collected but not "typical" of open sea conditions. The step function for C_D is based on the apparent large values of C_D from set-up measurements and the lower values at light wind speeds from more direct measurements obtained initially. These low values may be too low because it is the difference in air–water velocity that is important. Under light winds, a relatively thin layer of water may be moving relatively quickly due to the wind and the value of W used to calculate C_D may be too large. Also these early values were probably obtained under conditions which tended to produce smaller C_D values, i.e. steady conditions with no previous strong winds.

The jump of C_D as shown in Fig. 9.14 was rationalized on the basis that wave breaking or "white-capping" becomes *obvious* at winds of about $7.5 \, \text{m s}^{-1}$ (15 knots) and the water surface becomes "rougher" leading to higher values for C_D. The use of the step function persisted long after the time when it was known not to be realistic. This provides an example of the unfortunate use, by people not fully conversant with the development of a particular field, of preliminary results as though they were settled matters.

While the new estimates of C_D allow the average stress to be calculated they do not extend to high enough speeds to allow the dramatic effects of very strong winds (e.g. hurricanes) to be calculated with certainty. Direct measurements of the stress to estimate C_D for winds in the $30\text{–}50 \, \text{m s}^{-1}$ range will be very difficult to obtain both because of the experimental difficulties and because such winds are infrequent. There are indirect estimates based on hurricane momentum budgets and on extrapolations of laboratory wind flume experiments which are reported by Garratt (1977). These two data sets agree reasonably well with each other and with both Smith and Large and Pond in the region of overlap ($15\text{–}25 \, \text{m s}^{-1}$) and with extrapolations of the same authors' results to $50 \, \text{m s}^{-1}$. Until more direct measurements become available the present C_D formulation should give reasonable estimates of the stress under extreme conditions.

Another difficulty is that the wind speed appears in the stress expression as a power of W. If we have frequent measurements of **W** (e.g. daily or more frequently) we can calculate the stress for each measurement and add these *stress* values vectorially to obtain the total effect. Unfortunately, in some early calculations mean values ($\overline{W}$) for the wind speed over periods of a month or more were used. If the values of **W** change much during the averaging period, the square of this mean value (i.e. $(\overline{W})^2$) is smaller than the mean of the squared values (i.e. $(\overline{W^2})$), so that the use of long-term mean values tends to underestimate the effect of the wind stress. Averaged winds should not be used for stress calculations unless one is prepared to work out corrections, appropriate to the region of interest, for the stress reduction which will occur.

Climatological wind data are often summarized in the form of wind roses which show the percentage of the time for which the wind blows in particular directions (usually eight) in each of a number of speed ranges, as well as the overall frequency distribution of speeds regardless of direction. Using wind roses one can calculate the contribution to the stress from each direction in each speed range and add them vectorially to obtain the climatological average *stress*. Some tests, using detailed wind data to calculate the stress both from the detailed data and from the same data put into wind-rose form, show that both methods give the same results within a few percent. Thus calculations of the climatological average stress over the ocean should be possible with reasonable accuracy. Unfortunately, it appears that the spatial resolution of the climatological data (5° in latitude at best) is insufficient to calculate the curl of the wind stress. A test by Saunders (1976) for the North Atlantic in the region of maximum wind curl (20–30°N, 60–70°W) shows that with 5° resolution in latitude the value of the curl is only about two-thirds of that using 1° resolution. However, calculations using the old step function C_D overestimate the stress which tends to compensate for the lack of spatial resolution. In the region tested by Saunders, the two effects approximately balance. To get good stress and curl estimates it will be necessary to use the accumulated body of ship reports and redo the calculations using the latest C_D estimates and high resolution. Of course, not all areas of the ocean can be covered because not all regions are well travelled. Another possibility which may give better coverage when it is better verified is to use satellite observations of sea surface roughness which should be reasonably well related on average to the stress and the wind speed.

9.10 Munk's solution

Munk combined the basic features contributed by Ekman, Sverdrup and Stommel to provide the first comprehensive solution for the wind-driven circulation, using the real wind field, albeit with less detailed wind roses than are now available and using the step function C_D. He used two friction terms:

(1) vertical, associated with vertical shear to convey momentum from the wind stress applied at the surface into the Ekman layer,
(2) lateral, associated with horizontal shear so that the ocean would remain in a steady state of circulation.

He finished up with a fourth-order differential equation describing the circulation as

$$A\nabla^4\psi - \left(\beta\frac{\partial\psi}{\partial x}\right) - \text{curl}_z\tau_\eta = 0 \qquad (9.34)$$

where A = the eddy viscosity coefficient for lateral friction for mass transport,

∇^4 = the two-dimensional biharmonic operator = $\dfrac{\partial^4}{\partial x^4} + 2\dfrac{\partial^4}{\partial x^2\partial y^2} + \dfrac{\partial^4}{\partial y^4}$,

ψ = the mass transport stream function which describes the stream lines (really trajectories in the steady state) of flow around the ocean.

The equation must be solved for ψ and then

$$M_x = \frac{\partial\psi}{\partial y}, \qquad M_y = -\frac{\partial\psi}{\partial x}.$$

In words, equation (9.34) can be written

Vorticity from lateral stress − Planetary vorticity − Wind stress curl = 0.

(Vorticity will be discussed shortly.)

The three terms in the equation are not equally important all over the ocean. In the west, where the currents are strong, the first and second are the important ones while in the remainder of the ocean the second and third are important. The lateral stress is determined by the lateral shear in the currents and is large in the west because the currents and shear are large there. Elsewhere the currents are so much less that the terms arising from the shear can never be large as we showed in Section 7.3.

Munk obtained an analytic solution to the differential equation (9.34), i.e. obtained an expression for ψ in terms of the dimensions of the ocean, of β and of τ_η. For the latter he used values for the east–west wind stress component (averaged across the ocean) only, ignoring the north–south component. For the solution, ψ is best shown in the form of the flow pattern as in Fig. 9.15(a). This is the solution for a rectangular ocean and is somewhat stylized. Later, Munk and Carrier solved for a triangular ocean which more nearly represents the real shape of the North Pacific as shown in Fig. 9.15(b).

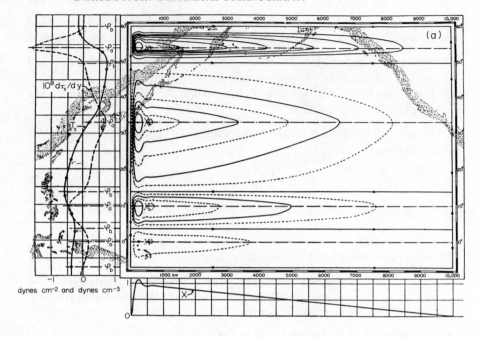

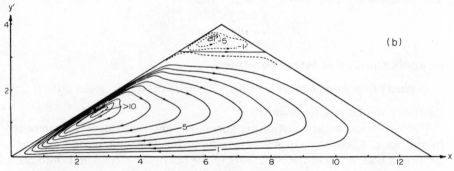

FIG. 9.15   (a) On left—mean annual wind stress over the Pacific, τ_x (full line) and its curl, $10^8\,\partial\tau_x/\partial y$ (dashed line), on right—computed mass transport streamlines (ψ) for a rectangular ocean (from Munk, 1950); (b) computed transport streamlines for a triangular ocean from $15°$ to $60°N$ (from Munk and Carrier, 1950). (1 dyne cm^{-2} ≡ 0.1 Pa; 1 dyne cm^{-3} ≡ 10 Pa m^{-1}.)

Note that the second and third terms are just the Sverdrup equation (9.21) with $-\partial\psi/\partial x$ substituted for M_y. The additional terms in the generalized Sverdrup equation (9.32) do not appear because Munk assumed that the currents went to zero above the bottom or that the bottom was level and the bottom stress was negligible because the currents ought to be very small in the deep water.

Munk assumed that in the west the friction terms associated with horizontal shears in the currents would become important but that non-linear terms would remain small. In Section 8.10 we showed that this assumption may be reasonable if the horizontal eddy viscosity is sufficiently large.

To see how Munk's equation (9.34) is obtained we write down the vertically integrated equations (9.30) but add the lateral friction terms from equation (7.6). For simplicity we omit the bottom pressure and stress terms, as under Munk's assumptions they are taken to be zero, to get

$$-fM_y = -\frac{\partial \chi}{\partial x} + \tau_{x\eta} + \int_{z_B}^{\eta} \rho A_x \frac{\partial^2 u}{\partial x^2} dz + \int_{z_B}^{\eta} \rho A_y \frac{\partial^2 u}{\partial y^2} dz,$$

$$fM_x = -\frac{\partial \chi}{\partial y} + \tau_{y\eta} + \int_{z_B}^{\eta} \rho A_x \frac{\partial^2 v}{\partial x^2} dz + \int_{z_B}^{\eta} \rho A_y \frac{\partial^2 v}{\partial y^2} dz.$$

Now assume that $A_x = A_y = A_H$ and that

$$\int_{z_B}^{\eta} \rho A_H \frac{\partial^2 u}{\partial x^2} dz \simeq A \frac{\partial^2 M_x}{\partial x^2}$$

and likewise for the other terms. Now if $z_B = $ constant as Munk assumed and A_H does not vary with z, the result is nearly exact because the η variations are too small to matter and density variations can also be ignored in this case (by the Boussinesq approximation discussed in Section 7.41). Recall that we have already assumed that A_x and A_y variations with x and y may be ignored. If some of the assumptions are not exactly correct then we argue by analogy that friction for transports may be represented by $A(\partial^2 M_x/\partial x^2)$, etc., with $A \simeq A_H$. Recall also, however, that the use of the eddy viscosity, particularly a constant value, is a crude way to represent the effects of turbulence. Any results which depend strongly on this assumption must be viewed with suspicion until verified by observations.

Using the eddy viscosity representation gives friction terms $A\nabla_H^2 M_x$ in the first equation and $A\nabla_H^2 M_y$ in the second (where $\nabla_H^2 = \partial^2/\partial x^2 + \partial^2/\partial y^2$). Now we take $(\partial(fM_x)/\partial x + \partial(fM_y)/\partial y)$ as we did earlier. Except for the friction term we already know the result $(\beta M_y = \text{curl}_z \tau_\eta)$ so we just work out its form, assuming that A is constant, as

$$A\frac{\partial}{\partial x}(\nabla_H^2 M_y) - A\frac{\partial}{\partial y}(\nabla_H^2 M_x).$$

Using $M_x = \partial \psi/\partial y$, $M_y = -\partial \psi/\partial x$ and writing out in full we have

$$-A\left[\frac{\partial}{\partial x}\left(\frac{\partial^2}{\partial x^2} + \frac{\partial^2}{\partial y^2}\right)\frac{\partial \psi}{\partial x} + \frac{\partial}{\partial y}\left(\frac{\partial^2}{\partial x^2} + \frac{\partial^2}{\partial y^2}\right)\frac{\partial \psi}{\partial y}\right]$$

$$= -A\left[\frac{\partial^4 \psi}{\partial x^4} + 2\frac{\partial^4 \psi}{\partial x^2 \partial y^2} + \frac{\partial^4 \psi}{\partial y^4}\right] = -A\nabla^4 \psi$$

and we have finally

$$\beta M_y = -\beta \frac{\partial \psi}{\partial x} = \text{curl}_z \tau_\eta - A \nabla^4 \psi$$

which is Munk's equation.

Note that this is a fourth-order equation (i.e. it contains fourth partial derivatives) and therefore its solution can satisfy four boundary conditions—no flow through and no-slip along both east and west boundaries. The vanishing of $\text{curl}_z \tau_\eta$ at certain latitudes breaks the flow into gyres as in Fig. 9.15. In Stommel's model, because of the simpler form which he assumed for the friction term, his equation was only of second order and he could not satisfy the no-slip condition. Also because of the higher order, Munk's solution allows for the countercurrent (a fairly strong southward flow observed to the east of the Kuroshio and the Gulf Stream). In Fig. 9.15, the western boundary current is where the streamlines are close together indicating strong currents. The countercurrent is indicated in the largest gyre by the swing to the south of the streamlines as they enter and leave this region. Stommel's model (Fig. 9.12(b)) does show the western intensification but not the countercurrent. Munk's solution is more realistic in these regards.

9.10.1   *Comments on Munk's solution*

These solutions show a series of "gyres" which include the equatorial current system and the westward intensification. Quantitatively, for the larger currents such as the Gulf Stream and the Kuroshio, Munk's calculated values for the transports were only about one-half of the then accepted values from geostrophic calculations. Subsequent studies of the geostrophic transports in the North Atlantic (e.g. Leetmaa, Niiler and Stommel, 1977) have suggested that this apparent discrepancy arose because the "accepted" values were for the strong northeastward-flowing part of the Gulf Stream system and that large return flows (to the southwest) in the western boundary region just to the east of the Gulf Stream had been ignored. When these return flows are allowed for there is reasonably good agreement between the circulation derived from the curl of the wind stress and those from geostrophic calculations and from direct measurements, at least for the North Atlantic, and it appears that the Sverdrup relation is a good approximation in the interior of the ocean.

At the same time, it is worth pointing out several detailed aspects of the wind-stress calculation. One point is that the wind-stress curl may be underestimated when it is obtained using finite differences. The stress used is usually calculated at 5° intervals of latitude and longitude and the curl may be underestimated using such large separations, particularly where it has maxima or minima. The wind-stress estimates are based on wind data over many years, the curl being estimated from these stresses. In addition to the possible error because the finite differences used are based on large separations, the locations of the regions of maxima and minima of wind-stress curl will vary seasonally and also from year to year for a given season. Thus the maxima and minima of north–south transport integrated across the ocean as calculated from climatological data will be of smaller magnitude than the values which one would obtain by averaging the maxima or minima of the integrated transport (regardless of latitude) for a particular gyre based on daily values of the stress. How important this effect might be is unknown because, as far as we know, such calculations have not been made. The agreement noted above between geostrophic calculations and transports calculated from the Sverdrup equation suggests that this effect is unlikely to cause a large increase in the mean southward transport. Daily stress values could only be estimated by extrapolating the geostrophic wind from surface pressure maps to a surface wind. To do such a calculation for a few years would be a large task. Also there is uncertainty both in the maps and in the extrapolation. However, such a calculation over the range of latitudes in which the largest southward integrated transport in the Gulf Stream gyre is likely to occur could prove to be quite interesting.

In addition, when Munk made his calculations there was doubt about the value to use for C_D. He noted that using $C_D = 2.6 \times 10^{-3}$ everywhere would give better results. This value seems high, particularly in view of more recent estimates. Also, the observations are from ship's reports and, as merchant and passenger ships try to avoid strong winds, this leads to some bias toward low wind-speeds in the data and consequent underestimate of the stress. The use of the step C_D function (Fig. 9.14) has already put in some compensation for the poor resolution and possible bias in ship reports. Using a more realistic C_D (about 20 % higher than the Large and Pond formula) and $2° \times 5°$ resolution in latitude and longitude, Leetmaa and Bunker (1978), using the curl of the wind stress and the Sverdrup relation, calculated a southward transport of 32 Sv in the Atlantic gyre which includes the Gulf Stream. Munk's value was 36 Sv which is remarkably close. The Leetmaa and Bunker value is probably realistic since the slightly high C_D and the somewhat limited resolution will tend to compensate each other. Their total southward transport also is comparable to the directly observed northward flow in the Straits of Florida (37 Sv) and comparison of their transports with geostrophic calculations at several latitudes shows reasonable agreement (Leetmaa et al., 1977; Stommel et al., 1978).

More details of the theories of the wind-driven circulation may be found in *Wind-driven Ocean Circulation* (A. R. Robinson, Ed., 1963) and in *Ocean Circulation Physics* (M. E. Stern, 1975).

Munk neglected the thermohaline circulation entirely but did not think that this was a significant source of error; Stommel has suggested that the thermohaline circulation may provide a significant contribution to the total flow. Some features of this aspect of the circulation will be discussed in Chapter 10.

The neglect of the non-linear terms in the acceleration, such as $[u(\partial u/\partial x) + v(\partial u/\partial y) + w(\partial u/\partial z)]$, may not be justifiable in the western boundary current region. Including these terms makes the equations non-linear but attempts have been made by Morgan and by Charney to develop analytic theories in which the inertial terms are included. No fully satisfactory theory has yet developed but it appears that the inertial or non-linear terms must be taken into account in some circumstances and may then be just as important in the west as the lateral friction term. More recent attempts at calculating the ocean circulation have been made with numerical models in which all the effects, including the inertial ones, can be included. Friction is still approximated with constant eddy viscosity in most cases although other formulations are possible (e.g. larger values where velocity gradients are larger which, from what we know of turbulent motion, seems more realistic). These more "sophisticated" eddy viscosities have been used more in models of the atmosphere which are somewhat more advanced than models of the ocean. Numerical models have shown that inertial effects may increase the northeastward boundary current transport above that of the interior Sverdrup transport (i.e. that calculated using the Sverdrup equation (9.21)). The effect of bottom topography in a realistic baroclinic ocean according to some model results may also be responsible for the enhanced boundary current flow (in the sense of being larger than predicted by the simple wind-stress theory just described). Observations of salinity and temperature in the deep water are not accurate enough to show clearly whether or not this mechanism is really important. There are other possible enhancement mechanisms also. Features of numerical models will be discussed in Chapter 11.

9.11   Vorticity

9.11.1   *Relative vorticity—ζ*

Expressed simply, *vorticity* is a characteristic of the kinematics of fluid flow which expresses the tendency for portions of the fluid to rotate. It is directly associated with the quantity called "velocity shear". To illustrate this relationship, Fig. 9.16(a) shows on the left a plan view of some fluid which is flowing to the right (east) with velocity $u(y)$, the variation with y being such that from top to bottom of the figure, (A) the velocity first increases, then (B) is constant for a space and then (C) decreases. A small object floating in the water in zone A would tend to rotate in an anticlockwise direction as it drifted to the right as shown successively for times $t = t_1, t_2, t_3$. An object in zone C would tend to rotate clockwise, and one in the centre zone B would not tend to rotate either way. The rotation of the fluid, in this case measured by $\partial u/\partial y$, is called the

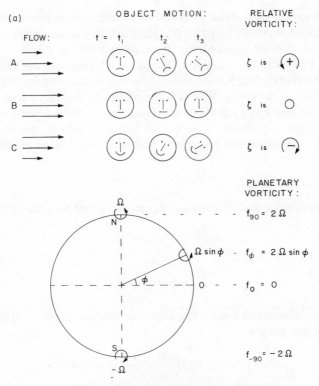

FIG. 9.16 *For discussion of vorticity: (a) relation between relative vorticity (ζ) and velocity shear, (b) planetary vorticity (f) at various latitudes on a rotating earth.*

vorticity. When it is measured relative to the earth it is called *relative vorticity* (ζ, zeta). When it is measured relative to axes fixed in space it is called "absolute vorticity" (discussed later).

The sign convention for direction is that the vorticity is positive when it is anticlockwise (zone A) as viewed from above (the same direction as the rotation of the earth as viewed from above the North Pole) and negative when clockwise (zone C).

In the general case, the relative vorticity in the horizontal plane (the vertical component) is $\zeta = \mathrm{curl}_z \mathbf{V} = (\partial v/\partial x - \partial u/\partial y)$.

9.11.2 *Planetary vorticity—f*

For a rotating solid object, the vorticity = 2 × angular velocity. By virtue of the rotation of the earth in space, at latitude ϕ a portion of its surface has angular velocity $\Omega \sin \phi$ about a vertical axis and therefore has vorticity $2\Omega \sin \phi$. This is called *planetary vorticity*. It is the quantity f appearing in some of the Coriolis terms and we will continue to use the symbol $f = 2\Omega \sin \phi$. A body of water which is stationary relative to the earth will then automatically possess planetary vorticity f; Fig. 9.16(b) shows how this quantity varies with position on the surface of the earth. Note that the planetary vorticity varies only with latitude ϕ and therefore a parcel of water at latitude ϕ can have only $f = 2\Omega \sin \phi$, no other value is possible. Also notice that f is zero at the equator, increasing to $+2\Omega$ at the north pole and decreasing to -2Ω at the South Pole.

9.11.3   *Absolute vorticity*—$(\zeta + f)$

The equations for the horizontal components of motion without friction are

$$\frac{du}{dt} - fv = -\alpha \frac{\partial p}{\partial x},$$

$$\frac{dv}{dt} + fu = -\alpha \frac{\partial p}{\partial y}. \tag{9.35}$$

If we cross-differentiate these equations and subtract them, to eliminate the pressure terms, we get

$$\frac{d}{dt}(\zeta + f) = -(\zeta + f)\left(\frac{\partial u}{\partial x} + \frac{\partial v}{\partial y}\right) = -(\zeta + f)\nabla \cdot \mathbf{V}_H \tag{9.36}$$

where $\mathbf{V}_H$ stands for the horizontal velocity and $\nabla \cdot \mathbf{V}_H$ will be recognized as a measure of the tendency for horizontal flow to diverge (if $\nabla \cdot \mathbf{V}_H$ is $+$) or to converge (if $\nabla \cdot \mathbf{V}_H$ is $-$). The quantity $(\zeta + f)$, the sum of the relative and planetary vorticities, is called the *absolute vorticity*. Equation (9.36) expresses the Principle of Conservation of Absolute Vorticity for flows on the earth when frictional effects are neglected. Here we have neglected derivatives of α with respect to x and y as usual. Consistent with this approximation, terms involving vertical shear are neglected. (In the ocean, for vorticity one may treat the flow as barotropic, taking u, v and ζ as independent of z to a good approximation. Note also that $df/dt = v(\partial f/\partial y) = \beta v$ since f is independent of x, z and t.)

In a divergence, where $\nabla \cdot \mathbf{V}_H$ is positive, the magnitude of the absolute vorticity decreases with time, whereas in a convergence where $\nabla \cdot \mathbf{V}_H$ is negative, the absolute vorticity magnitude increases with time. We consider the magnitude because $(\zeta + f)$ may be positive or negative. As f is usually much larger than ζ, positive values for $(\zeta + f)$ will usually be found in the northern hemisphere and negative values in the southern. With $\nabla \cdot \mathbf{V}_H > 0$, if $(\zeta + f) > 0$ then $d(\zeta + f)/dt < 0$, so $(\zeta + f)$ decreases with time, but if $(\zeta + f) < 0$, $d(\zeta + f)/dt > 0$, i.e. $(\zeta + f)$ becomes more positive with time but as $(\zeta + f) < 0$ to start with, its magnitude decreases.

To get a physical picture of this process, imagine a body of water in the form of a vertical cylinder of small height which is initially stationary relative to the earth so that it has planetary vorticity f only, as in Fig. 9.17(a). If the fluid now starts to flow inward (converges) toward the axis of the cylinder it must also elongate because volume is conserved; because $\nabla \cdot \mathbf{V}_H$ is negative, the absolute vorticity must increase. In this case the water will acquire some relative vorticity ζ, so that its absolute vorticity increases from f to $(f + \zeta)$ as the cylinder shrinks and elongates. In Fig. 9.17(b) is shown the opposite situation in which a tall, narrow cylinder expands (diverges) to form a low, wide one. If its initial vorticity was simply f its final vorticity will decrease to $(f - |\zeta|)$. In both cases we have assumed that the cylinder of water remains at the same position on the earth's surface so that the planetary vorticity f does not change. Note also that the curved arrows showing the motion of the water as it converges or diverges are drawn for the northern hemisphere case.

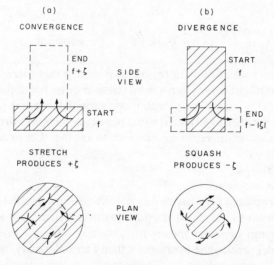

FIG. 9.17 *Change of absolute vorticity associated with (a) convergence, (b) divergence (both for northern hemisphere).*

If one is more used to thinking in terms of conservation of angular momentum, then in the first case the moment of inertia is decreased; as the angular momentum (moment of inertia × angular velocity) is not changing, there being no applied torques in this case, the angular velocity must increase. Similarly, in the second case the moment of inertia increases and the angular velocity must decrease.

9.11.4   *Potential vorticity—$(\zeta + f)/D$*

Let us consider a layer of thickness D in the sea whose density is uniform so that the horizontal velocity components are independent of depth. This thickness D is not specifically identified with the Ekman depth D_E. The depth D would more likely be the whole layer from the surface to the permanent thermocline or from the thermocline to the bottom. This is an idealization of the real situation in the ocean to one of two homogeneous layers. Clearly this "two-layer" ocean is an approximation but the general features of the results will be correct and so it is often a useful one to make. Then the equation of continuity of volume for the layer is

$$\frac{1}{D}\frac{dD}{dt} + \left(\frac{\partial u}{\partial x} + \frac{\partial v}{\partial y}\right) = 0. \tag{9.37}$$

If we combine this equation with (9.36), eliminating the horizontal divergence term, we get

$$\frac{d}{dt}\left(\frac{\zeta + f}{D}\right) = 0, \text{ i.e. } \left(\frac{\zeta + f}{D}\right) = \text{constant} \tag{9.38}$$

for the motion of a water body in the ocean provided that there is no input of vorticity (such as might come from a wind stress or other frictional effects). The quantity $(\zeta + f)/D$ is called the *potential vorticity* of the water.

This relation permits us to make some predictions about vorticity changes when a parcel of water moves from one place to another. Let us consider some possibilities:

(1) if D remains constant
  (a) then if a column of water moves zonally (along a parallel of latitude so that ϕ remains constant), then f remains constant and so must ζ;
  (b) if a column of water moves meridionally (along a line of constant longitude) toward the north pole, then f automatically increases and ζ must decrease to keep $(\zeta + f)$ constant, i.e. the water acquires more negative (clockwise) rotation relative to the earth;
  (c) conversely, if a column of water moves toward the south pole then it acquires more positive (anticlockwise) rotation;
(2) if D increases, then $(\zeta + f)$ must increase if it is positive initially,
  (a) so if the water moves zonally, then f remains constant so ζ must increase, i.e. the water acquires more positive (anticlockwise) rotation;
  (b) if the water moves meridionally toward the north pole, then f automatically increases, and it is not immediately obvious what ζ will do;
  (c) if the water moves meridionally toward the south pole, f decreases and ζ must increase, i.e. the column acquires more positive (anticlockwise) rotation;

(3) if D decreases, then $(\zeta + f)$ must decrease if it is initially positive. The reader is left to work out what will happen to ζ for zonal and for meridional flow in this case and also for D changes when $(\zeta + f)$ is initially negative.

In the interior of the ocean, for large-scale processes, ζ is negligible compared with f. In consequence, conservation of potential vorticity becomes f/D = constant. For instance, if a water column stretches (D increases) then f must increase in magnitude; the water must move toward the nearest pole, north or south (because f is a function of latitude only), and vice versa. One way in which a water column can stretch is for it to pass over a trough in the sea bottom, or it can contract by passing over a ridge. The condition f/D = constant permits us to predict which way a current will swing on passing over bottom irregularities—equatorward over ridges and poleward over troughs in both hemispheres. The deflection of the flow required to keep f/D constant is sometimes called *topographic steering*.

Note finally that several of the equations derived earlier in the chapter are vertically integrated forms of the vorticity equation (9.36). The Sverdrup equation in both simple and general form is a vertically integrated form and includes wind friction. Stommel's equation (not given explicitly) and Munk's equation are of the same form and include lateral friction as well as wind friction, e.g. in equation (9.32) with v taken as independent of z (a good approximation as noted), $\beta M_y = \rho \beta v D = \rho D (\mathrm{d}f/\mathrm{d}t); D = \eta - z_B$ and variations in D will be dominated by z_B variations, so that the final term of equation (9.32) is $\rho' f(\mathrm{d}D/\mathrm{d}t)$. Because non-linear terms were neglected, ζ terms do not appear in equation (9.32); in equations (9.36) and (9.38) they arise from the non-linear terms of equation (9.35). The curl terms of equation (9.32) do not appear in equations (9.36) and (9.38) because we left out friction.

9.12 Westward intensification of ocean currents explained using conservation of vorticity

Here we must include changes in vorticity due to frictional effects such as the wind which is important in driving the upper layer. As we have a steady state the total vorticity over the whole gyre must be constant and as each particle makes a circuit around the gyre it must arrive back at its starting point with no net change in vorticity. Changes in D do not affect these steady-state requirements so, for simplicity, take D = constant; then $\mathrm{d}(\zeta + f)/\mathrm{d}t$ = the sum of the frictional effects. Consider a northern hemisphere ocean with winds to the west in the south and to the east in the north, causing the upper-layer circulation to be clockwise (i.e. the angular rotation is negative). Then on the west side of the ocean the flow will be to the north (Fig. 9.18(a)) and the relative vorticity will become more negative (i.e. $-\zeta_p$) due to the northward movement which causes f to increase, and also more negative $(-\zeta_\tau)$ due to the wind stress which

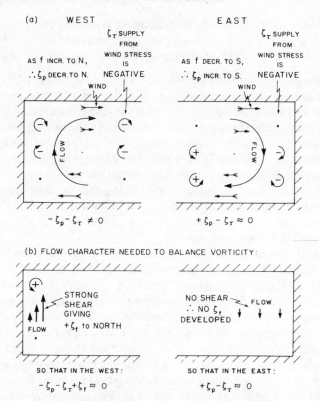

FIG. 9.18 *Use of conservation of vorticity to explain westward intensification of ocean currents: (a) relative vorticity supplied by wind stress, (b) relative vorticity supplied by frictional stress at boundary.*

provides clockwise, i.e. negative vorticity. There is, therefore, a net decrease of relative vorticity ($-\zeta_p - \zeta_\tau < 0$) on the west side. On the east side (Fig. 9.18(b)) the flow is to the south and so the relative vorticity increases ($+\zeta_p$) due to the decrease of f and decreases ($-\zeta_\tau$) due to the wind stress. Now Sverdrup demonstrated that on the east side of the ocean, $\beta M_y = \text{curl}_z \tau_\eta$ (equation (9.21)), i.e. the increase and decrease terms for relative vorticity, will be about equal in the east and there will be no change during the southward flow, i.e. $+\zeta_p - \zeta_\tau \simeq 0$ in the east.

Therefore, for the complete circulation it is necessary to supply vorticity to make up for the decrease in the west, in order to keep the total vorticity constant. One way to do so is through lateral friction on the west side only, so that the decrease of planetary vorticity and decrease of vorticity due to the wind stress are made up by increase of vorticity from lateral shear in the water currents, i.e. $-\zeta_p - \zeta_\tau + \zeta_f \simeq 0$. For this balance we need a velocity structure such as in Fig. 9.18(b) where there are strong currents with much shear in the west, but slower currents with little shear in the east, i.e. westward intensification of the ocean currents. Strong currents in the middle or east of the ocean do not introduce vorticity of the correct sign to provide a balance. (Strong currents in the west and friction acting at the bottom would also work but this possibility seems less likely than lateral friction.)

Note that there are two essential features; f must be allowed to vary with latitude (as Stommel showed) and ζ must be small compared with f. If the winds were very much stronger so that ζ became large compared with f, then f could be ignored and we would return to a symmetrical circulation with wind-vorticity input balanced by some other frictional effect (lateral or bottom) throughout the whole ocean. However, in the earth's oceans $\zeta \ll f$ for the large-scale flow in the interior and so westward intensification occurs.

Where the wind stress causes an anticlockwise circulation (the largest ones being in the southern hemisphere), similar arguments again lead to westward intensification.

9.13   The equatorial current system

9.13.1   *The surface equatorial currents*

As has been stated earlier, the most conspicuous feature of the equatorial current system is its zonal character with westward-following North and South Equatorial Currents and, between them, the eastward North Equatorial Countercurrent, all of which are evident at the surface (Fig. 9.6(c)). The earliest explanation offered for the existence of the eastward Countercurrent, which flows opposite to the prevailing westward winds, was that the wind stress piled up water at the west and that the Countercurrent was a pressure-gradient driven flow in the zone of the "doldrums", the weaker (though still westward) winds between the stronger Northeast and Southeast Trade Winds (Fig. 9.6(b)). However, later studies showed this explanation to be inadequate. Subsequently, the curl of the wind-stress theory of Sverdrup, which has been discussed in some detail, has been accepted as the proper explanation, i.e. it is the meridional *change* of wind stress between the stronger trade winds and the weaker doldrums which is responsible rather than the smaller wind stress itself in the doldrums.

The plan view and vertical section of Fig. 9.6(c, d) show the zonal flows in relation to the wind stress Fig. 9.6(b). The related meridional slopes of the sea surface (much exaggerated in the figure) and the convergences and divergences associated are also shown in Fig. 9.6(d).

In addition to these three major currents, evident at the surface, there are other surface currents, e.g. a weaker South Equatorial Countercurrent between 5° and 10°S and possibly other minor components which are not yet well described. However, there is a fourth major component of the equatorial system–the subsurface Equatorial Undercurrent.

9.13.2 *The Equatorial Undercurrent*

Beneath the surface and embedded in the westward-flowing Pacific South Equatorial Current there exists a most remarkable current—the Equatorial Undercurrent. This current flows eastward, centred on the equator. Maximum speeds are 1.5 m s^{-1} or more at a depth rising from 200 m in the west to 50 m or less in the east; the annual average transport has recently been estimated at about 40 Sv with maximum transports up to 70 Sv. The current is like a thin ribbon, about 0.2 km thick and 300 km wide; it can be followed for some 14 000 km across most of the Pacific! Its existence has been recognized for only 25 years, although once it was known to exist, evidence for it was found in earlier data.

A similar current exists in the Atlantic, and was actually first recognized in 1886 by Buchanan but his accounts were ignored. There is also evidence for an equatorial undercurrent in the Indian Ocean during the northeast monsoon period (November–March). More complete descriptions of these currents may be found in Philander (1973) and in *Descriptive Physical Oceanography* (Pickard and Emery, 1982).

Before the observational discovery of these currents there were no theoretical predictions to suggest their existence. How do they arise? Consider first what is happening right at the equator where the Coriolis effect is zero and where there is a well-defined upper, mixed (constant density) layer above the thermocline. The wind is blowing with a westward component and the surface current is to the west. Because there are land barriers at the western side the water tends to pile up there producing a surface slope up toward the west and therefore a pressure gradient force develops toward the east. The surface layer will tend to be thicker in the west with the thermocline (and pycnocline) sloping opposite to the surface slope, i.e. upward to the east. At some considerable depth the opposite slope of the isopycnals relative to the surface can lead to baroclinic compensation and a reduction of the east–west pressure gradient to zero. At the surface the westward component of the wind stress can balance the eastward pressure gradient. As one goes below the surface, the stress caused by the surface wind decreases and the eastward pressure gradient can no longer be balanced by the downward momentum transfer and produces the eastward-flowing undercurrent at the bottom of the mixed layer. Because of the strong eastward current there will be frictional forces (and perhaps inertial effects) which will balance the pressure gradient and allow a steady-state flow to occur.

As one goes away from the equator, Coriolis effects become important. In the surface layer with a westward-directed wind, the Ekman transport will be away from the equator on both north and south sides so there will be a surface divergence (Fig. 9.6(d)) with consequent upwelling and biological production. Below the surface layer there must be convergence (flow toward the equator) to balance the surface divergence. North and south of about 1.5° latitude the Coriolis force associated with this equatorward flow can balance the eastward pressure gradient. It is only very near to the equator that the Coriolis effect is so weak that the Undercurrent is possible.

Note that although the Coriolis force vanishes at the equator its variation with latitude is a maximum there and tends to stabilize the current. (The Coriolis factor $f = 2\Omega \sin \phi$ which tends to zero as $\phi \to 0°$ but its rate of change with latitude is $\partial(2\Omega \sin \phi)/\partial\phi = 2\Omega \cos \phi$ which tends to its maximum value of 2Ω as $\phi \to 0°$.) If the eastward current wanders southward the Coriolis force will be to the left pushing it back to the equator; if it goes northward the Coriolis force is to the right, again bringing it back toward the equator. In fact, the variation of the Coriolis force with latitude will tend to stabilize eastward-flowing currents anywhere in the ocean and to accentuate meandering of westward-flowing currents with the strongest effects at the equator where the variation is a maximum.

It is only in the long term that the Undercurrent is "tied" to the equator. Evidence for north–south oscillations in equatorial currents has been found. They are probably associated with the variation of the Coriolis effect with latitude and may be thought of as Rossby waves superimposed on the average current. These waves, which are possible because of the Coriolis effect variation, will be discussed in Chapter 12.

A weakness of the constant density upper-layer models is that they predict an undercurrent at the bottom of the mixed layer whereas observations indicate that the strongest current is actually below this in the thermocline/pycnocline region below the mixed layer (Fig. 9.6(d)). More elaborate theories, which developed from studies of temporal variations in the equatorial region, e.g. Rossby waves, and incorporate density stratification rather than a simple two-layer model, are able to describe the existence of the undercurrent *in* the thermocline, but these are beyond the scope of this book. However, it must be admitted that no fully satisfactory theory of the Equatorial Undercurrent is yet available.

Note that the discussion above is no more than an outline of some theories of the undercurrent. However, it does suggest how such a current can occur and what physical effects must be included in a complete theory which should be able to predict, for example, the detailed velocity distribution in the current. Both horizontal and vertical friction terms would need to be included in the equations. Because of the strength of the current and its small vertical and lateral (north–south) scales, non-linear terms might well be important too. The reader wanting more details could consult Philander (1973) and Leetmaa *et al.* (1981).

9.14 The boundary-layer approach

When we examined the sizes of the terms in the equations of motion in Chapter 7 we found that many terms were negligible. The equation for the vertical component became the hydrostatic equation; the local time derivatives were shown to be small for motions with periods greater than a few days. It appeared that friction from turbulence and non-linear terms might be important in some regions. If we divide through by the magnitude of the Coriolis term we find that the non-linear terms and friction terms have small coefficients—the Rossby number ($Ro = U/f_0 L$) and Ekman numbers $[E_H = A_H/(f_0 L^2); E_z = A_z/(f_0 H^2)]$ respectively, which were defined in Section 7.3.

Whenever higher-order derivatives in the equations exist but have small coefficients we expect *boundary layers* to occur. These are relatively thin regions near the boundary where some of the terms with higher-order derivatives will become important. Simple equations will hold in the interior of the fluid; in the oceanographic case they are the geostrophic equations. (The same approach is used in other fluid mechanics problems where rotation effects do not occur but we shall only look at the oceanographic case.) However, solutions to the geostrophic equations do not usually satisfy all the boundary conditions. Therefore near the boundaries some higher-order term or terms must become large because the length scale normal to the boundary becomes small. Eventually some extra term or terms become large enough to allow the boundary conditions to be satisfied.

To use the boundary-layer approach we re-scale the equations, changing the appropriate length scale to one relevant to the boundary-layer thickness. The advantage of this approach is that only the higher-order terms which matter will become important. Unimportant terms remain small and the "boundary-layer" equations will be simpler than the full equations and hopefully easier to solve. We expect the boundary-layer effects to disappear a few boundary-layer scale lengths from the boundary. We make the boundary-layer solution satisfy the boundary conditions and then approach the interior solution as the boundary-layer coordinate becomes very large. Because the boundary layer is very thin relative to the whole ocean and the interior solution is slowly varying, we can match to the interior solution effectively right at the boundary. Once one becomes convinced that this "connecting" of boundary layer and interior regions is possible one can concentrate on examining a range of boundary-layer solutions without worrying about the matching. This approach has been a very useful tool in analytical studies and has helped to improve our understanding of what dynamical effects may be important even when the solutions are too idealized to be applied in detail to the real ocean. The boundary-layer approach may also show how the thickness of the boundary layer depends on the parameters of the system. Here we shall not give an exhaustive treatment of this approach but illustrate the method with a few fairly simple examples.

The Ekman solution for the wind-driven currents given in Section 9.4 is a boundary-layer solution. The geostrophic equations describe the part of the motion associated with the horizontal pressure gradients. However, these solutions cannot satisfy the surface boundary condition (that the shear stress at the surface must be continuous) when there is an applied wind stress. Near the surface the vertical length scale for the vertical friction term becomes small and these vertical friction terms become large enough to balance the Coriolis term associated with the directly wind-driven flow. With the stress in the y-direction, as in the solution presented, the boundary condition at the surface is

stress in the water at the surface in the y-direction
= wind stress at the surface in the y-direction,

i.e. $\rho A_z (\partial v / \partial z)_{z=\eta} = \tau_{y\eta}$.

The reader may verify that the solution given satisfies this equation and that $(\partial u / \partial z)_{z=\eta} = 0$ since there is no stress in the x-direction. (In the more general case, $\rho A_z (\partial u / \partial z)_{z=\eta} = x$-component of the wind stress must hold also.) Indeed the speed of the surface flow V_0 in the solution is determined by this boundary condition.

The solution has the proper boundary-layer character. In addition to satisfying the surface boundary condition, the speed of flow in the solution decays rapidly with depth and is essentially zero at the bottom of a very thin layer relative to the total ocean depth.

The matching with the interior is very straightforward in this case. If the Ekman flow is non-divergent then the Ekman and geostrophic flows are completely independent and may be simply added together as noted earlier. If the Ekman flow is convergent or divergent then the vertical velocity at the bottom of the Ekman layer provides one of the boundary conditions which the geostrophic flow must satisfy as described in the section on the generalized Sverdrup equation earlier in this chapter. Because the Ekman layer is so thin one may, with negligible error, take this boundary condition on the geostrophic flow to be at $z = \eta$ rather than at the "bottom" of the Ekman layer.

9.14.1   *The use of the boundary-layer approach to obtain a solution to Munk's equation*

Munk's equation (9.34) looks quite formidable. Although he did solve the complete equation for the rectangular ocean model, it is much easier to obtain an approximate, but sufficiently accurate, solution using the boundary-layer approach. For the solution with the triangular ocean shape of Munk and Carrier the boundary-layer approach had to be used because a solution to the full equation could not be found. Here we shall obtain the solution for the rectangular basin case. As noted earlier, in the interior $A\nabla^4\psi$ can be neglected but in the side boundary layers $\partial/\partial x$ terms (derivatives with respect to the coordinate normal to the boundary) become important so that boundary conditions of no flow through and no slip at the side boundaries can be satisfied.

Now instead of just looking at the sizes of terms we shall put Munk's equation (9.34) in non-dimensional form. This is a very useful procedure, often done in theoretical studies (both in geophysical fluid dynamics, i.e. ocean-ography and meteorology, and in other branches of fluid mechanics). The terms involving non-dimensional variables are then of order 1 and their non-dimensional coefficients, based on the scales of the system being studied, determine the importance of various terms. In the non-dimensional equations of motion the Rossby and Ekman numbers are important as mentioned earlier. Here we are going to non-dimensionalize Munk's form of the vertically integrated vorticity equation. We consider finding a solution applicable to a region such as the Gulf Stream gyre system. Now in the interior away from the western boundary (under Munk's assumptions given earlier) the simple Sverdrup balance (equation (9.21)) holds. Integrating this equation with respect to x gives

$$\psi_i = -\frac{1}{\beta}\int \text{curl}_z \tau_\eta \, dx + C \tag{9.39}$$

where the subscript i indicates the stream function in the interior, C is a constant of integration and $\text{curl}_z \tau_\eta$, as before, stands for $(\partial \tau_{y\eta}/\partial x - \partial \tau_{x\eta}/\partial y)$. Now the size of ψ from equations (9.39) or (9.21), or more precisely the change of ψ from the outer edge of the western boundary layer to the eastern side of the ocean, will be of the order of τ_0/β, where τ_0 is a typical wind-stress magnitude (or the total change in $\tau_{x\eta}$ from north to south in the gyre) and β, taken to be constant in this β-plane model, is the value of df/dy at the centre of the gyre. We note that the change in the value of ψ across the western boundary must also be of the same size because the boundary current is returning the interior flow and the change in ψ in each region gives a measure of the total transport. We put

$$\psi = (\tau_0/\beta)\psi' \quad \text{and} \quad \tau_\eta = \tau_0 \tau'_\eta \tag{9.40}$$

where the primes indicate non-dimensional variables with a range of order 1. The magnitude of $\text{curl}_z \tau_\eta$ will be τ_0/L where L is the distance over which τ_η changes by τ_0, in this case the north–south dimension of the gyre. For x and y we put

$$x = \xi W \quad \text{and} \quad y = \gamma L \tag{9.41}$$

where ξ and γ are non-dimensional variables. In the interior, W is O(L). (For simplicity we shall assume a square basin so x and y both go from 0 to L but the east–west width could be L times a constant of order 1 without changing the results.) In the western boundary, $W \ll L$ because the boundary current is long and narrow. As x goes to a small fraction of L, ξ will become very large if W is taken to be the width of the western boundary region. If we substitute equations (9.40) and (9.41) into (9.34) we get

$$\frac{A\tau_0}{\beta}\left[\frac{1}{W^4}\frac{\partial^4\psi'}{\partial\xi^4} + \frac{2}{W^2 L^2}\frac{\partial^4\psi'}{\partial\xi^2\,\partial\gamma^2} + \frac{1}{L^4}\frac{\partial^4\psi'}{\partial\gamma^4}\right] - \frac{\tau_0}{W}\frac{\partial\psi'}{\partial\xi}$$

$$= \tau_0\left[\frac{1}{W}\frac{\partial\tau'_{y\eta}}{\partial\xi} - \frac{1}{L}\frac{\partial\tau'_{x\eta}}{\partial\gamma}\right].$$

Multiplying by W/τ_0 and some rearranging gives

$$\frac{A}{\beta W^3}\left[\frac{\partial^4\psi'}{\partial\xi^4} + 2\left(\frac{W}{L}\right)^2\frac{\partial^4\psi'}{\partial\xi^2\,\partial\gamma^2} + \left(\frac{W}{L}\right)^4\frac{\partial^4\psi'}{\partial\gamma^4}\right] - \frac{\partial\psi'}{\partial\xi}$$

$$= \frac{W}{L}\left[\frac{L}{W}\frac{\partial\tau'_{y\eta}}{\partial\xi} - \frac{\partial\tau'_{x\eta}}{\partial\gamma}\right]. \qquad (9.42)$$

Note that $(L/W)(\partial\tau'_{y\eta}/\partial\xi)$ remains of $O(1)$ both in the interior, where $W \simeq L$, and in the boundary layer because the magnitude of $\mathrm{curl}_z\tau_\eta$ is the same in both regions. In the western boundary region where $W \ll L$ and ξ changes rapidly, $\partial\tau'_{y\eta}/\partial\xi$ becomes very small, of $O(W/L)$. Recall that $\partial\tau_{x\eta}/\partial y$ is the dominant term in the wind-stress curl everywhere, so the term coming from $\partial\tau_{y\eta}/\partial x$ cannot become important no matter how we non-dimensionalize the equation.

Following Munk we take $A = 5 \times 10^3\ \mathrm{m}^2\,\mathrm{s}^{-1}$ and $\beta = 1.9 \times 10^{-11}\ \mathrm{m}^{-1}\,\mathrm{s}^{-1}$. For the interior take $W = 5 \times 10^6\ \mathrm{m} = 5000$ km. Then $A/(\beta W^3) = 2 \times 10^{-6}$, so the friction terms are negligible and the simple Sverdrup balance holds. To make the friction terms of $O(1)$ to balance $\partial\psi'/\partial\xi$ we must take

$$W = (A/\beta)^{1/3} = (5 \times 10^3/1.9 \times 10^{-11})^{1/3} \simeq 6 \times 10^4\ \mathrm{m} = 60\ \mathrm{km}.$$

This is an example of how the scaling can be used to find out how the "width" of the western boundary current must depend on the other parameters in the system. Now with $W = 60$ km, the friction term is $O(1)$ and so is $\partial\psi'/\partial\xi$. However, the wind-stress term is now $O(W/L)$ or $O(0.01)$ and may be neglected to a good approximation. This result should not be too surprising. The wind distribution is quite uniform and symmetric; thus the local wind-driving in the western boundary region will be similar to that in the interior. With the much higher velocities and transport/unit width in the western boundary current the relatively small local wind-induced transports can be ignored. Furthermore, not all of the higher-order terms are equally important. With $W/L \simeq 0.01$, only the first term is $O(1)$, the next largest one being $O(10^{-4})$ and therefore negligible. Thus in the western boundary we have, to a good approximation (about 1 % which is rather good by geophysical fluid dynamics standards),

$$\frac{\partial^4 \psi'}{\partial \xi^4} - \frac{\partial \psi'}{\partial \xi} = 0. \tag{9.43}$$

The use of the boundary layer approach has simplified equation (9.34) to equation (9.43) which shows the strength of the method.

Equation (9.43) has a quite simple solution (a sum of exponential functions which is usually the first type of solution tried for a linear differential equation with constant coefficients because it often works) as

$$\psi' = C_0 + \sum_{n=1}^{n=3} A_n \exp(a_n \xi) \tag{9.44}$$

where the a_n's are the roots of $a_n^3 = 1$ which are $a_1 = 1$, $a_2 = -1/2 + j\sqrt{3}/2$, and $a_3 = -1/2 - j\sqrt{3}/2$ where $j = \sqrt{-1}$. Now $A_1 = 0$ must be chosen because exp ξ is divergent (becomes very large) as ξ becomes large (mathematically as $\xi \to \infty$). The solution with $a_1 = 1$ can be used on the eastern boundary to satisfy the no-slip condition as we shall show presently. Using $\psi = 0$ (no flow through the boundary) and $\partial \psi / \partial x = 0$ ($M_y = 0$ or no slip or flow along the boundary) at $x = 0$ determines A_2 and A_3 and we have

$$\psi' = C_0 \left[1 - \exp(-\xi/2)\left(\cos(\sqrt{3}\xi/2) + \frac{\sin(\sqrt{3}\xi/2)}{\sqrt{3}} \right) \right] \tag{9.45}$$

$$= C_0 T(\xi).$$

$T(\xi)$ goes to 1 as ξ becomes very large (as we go from the boundary layer to the interior).

Thus $\psi' \to C_0$ as we move to the interior and, in dimensional units, C_0 must be the value of the interior transport stream-function at the edge of the boundary layer. Because $W/L \ll 1$, we can, to a good approximation, use ψ_i at $x = 0$. Thus $\psi = \psi_i(x, y) T(x/W)$ and from before

$$\psi_i = -\frac{1}{\beta} \int_0^x \text{curl}_z \tau_\eta \, dx + C.$$

Using $\psi_i = 0$ at $x = L$, the eastern boundary, gives

$$C = \frac{1}{\beta} \int_0^L \text{curl}_z \tau_\eta dx \qquad \text{and finally}$$

$$\psi = \left(\frac{1}{\beta} \int_x^L \text{curl}_z \tau_\eta dx \right) T(x/W) \tag{9.46}$$

where $T(x/W)$ is given by the expression multiplying C_0 in equation (9.45) with ξ replaced by x/W. To complete the solution fully, we add to the right-hand side of equation (9.46) the quantity (W'/β) $(\mathrm{curl}_z \tau_\eta)_{x=L}[\exp\{(x-L)/W'\}$ $-T(x/W)]$ which is negligible except near the eastern boundary and which makes $\partial\psi/\partial x$ vanish at $x = L$. W' is an eastern boundary width and $W' \ll L$. Since A in the eastern boundary will probably be smaller than in the west, friction probably being weaker here, W' (proportional to $A^{1/3}$) is likely to be smaller than W of the western boundary.

Now from the solution, the actual width of the western boundary current is three to four times $(A/\beta)^{1/3}$ or, for the values chosen by Munk, about 200 km. This width is probably reasonable for the climatological average Gulf Stream or Kuroshio. However, at any one time these streams are only 50–60 km wide. Using the Munk theory with $A = (\beta W^3)$ suggests for the short-term average stream that A is of order 10^2 m^2 s^{-1}. Now the inertial or non-linear terms are of order $10^3/A = 10$ times the turbulent friction terms. Thus for the short-term average western boundary current, inertial or non-linear effects are probably not negligible.

9.14.2   *A simple inertial theory by Stommel; the Rossby radius of deformation*

This is an idealized model used to see if a predominantly inertially controlled Gulf Stream is a reasonable approximation. Stommel assumes a two-layer system. The upper layer has density ρ_1 and is moving; the lower layer has density ρ_2 and is at rest. The thickness of the upper layer, D, is 0 at the coast ($x = 0$) and increases to D_0 at the outer edge of the western boundary layer. The x-axis is taken across the stream and the y-axis along it.

Before proceeding with Stommel's model we need to develop expressions for the pressure gradients in terms of the layer thickness. Consider the pressure in the lower layer

$$p = -\int_\eta^z \rho g\,dz = -\int_\eta^d \rho_1 g\,dz - \int_d^z \rho_2 g\,dz$$

where η is the surface elevation from the rest state with $z = 0$ at the surface and d is the level of the interface between the layers measured from the $z = 0$ reference. Now ρ_1 and ρ_2 are constants and may be taken outside the integrals giving

$$p = \rho_1 g(\eta - d) + \rho_2 g(d - z) = \rho_1 g\eta + (\rho_2 - \rho_1)gd - \rho_2 gz.$$

Then $\dfrac{\partial p}{\partial x} = \rho_1 g\dfrac{\partial\eta}{\partial x} + (\rho_2 - \rho_1)g\dfrac{\partial d}{\partial x}$ in the lower layer.

But in this layer there is no flow (by assumption) and therefore the horizontal pressure force must be zero, i.e.

$$\frac{\partial d}{\partial x} = -\frac{\rho_1}{\rho_2 - \rho_1}\frac{\partial \eta}{\partial x}.$$

If we wish to use the total upper layer thickness, taken to be positive, $D = \eta - d$ and

$$\frac{\partial D}{\partial x} = \frac{\partial \eta}{\partial x} - \frac{\partial d}{\partial x} = \left(1 + \frac{\rho_1}{\rho_2 - \rho_1}\right)\frac{\partial \eta}{\partial x} = \left(\frac{\rho_2}{\rho_2 - \rho_1}\right)\frac{\partial \eta}{\partial x}.$$

Now $(\rho_2 - \rho_1) \ll \rho_1$, e.g. in the Gulf Stream $(\rho_2 - \rho_1) \simeq 2 \times 10^{-3}\rho_1$. Thus the slope of the interface is much larger than the slope of the surface and is of the opposite sign. Here we have derived the result for an idealized two-layer system. In the more general case of continuous density variation, the results would be similar. If the horizontal pressure gradients go to zero at depth, the isopycnal slopes will be mainly opposite to the surface slope and much larger.

In the upper layer above $z = d$

$$p = \rho_1 g(\eta - z) \quad \text{and} \quad \frac{\partial p}{\partial x} = \rho_1 g\frac{\partial \eta}{\partial x}.$$

Also as we showed in Section 8.10, the x-momentum equation remains geostrophic to a good approximation as

$$-fv = -\frac{1}{\rho_1}\frac{\partial p}{\partial x} = -\frac{\rho_2 - \rho_1}{\rho_2}g\frac{\partial D}{\partial x} = -g'\frac{\partial D}{\partial x} \qquad (9.47)$$

where $g' = g(\rho_2 - \rho_1)/\rho_2$ is termed the "reduced gravity". The total transport of the stream is

$$Q = \int_0^W Q_y \, dx$$

where W is the value of x at the seaward edge of the stream. As the upper layer is homogeneous, v is independent of depth within the upper layer and zero below it, so $Q_y = vD$. Substituting this expression and using equation (9.47) gives

$$Q = \int_0^W \frac{g'}{f}D\frac{\partial D}{\partial x} \, dx = \int_0^W \frac{g'}{f}\frac{1}{2}\frac{\partial D^2}{\partial x} \, dx = \frac{g'}{f}\frac{D_0^2}{2}$$

since $D = 0$ at $x = 0$ and $D = D_0$ at $x = W$.

Now following Stommel we assume that the potential vorticity is essentially constant. This assumption should be valid if friction effects are small enough to be neglected. The important inertial terms are retained. Observations in the Gulf Stream do show that the potential vorticity remains nearly constant. For the relative vorticity $\partial u/\partial y$ is negligible compared with $\partial v/\partial x$ both because $v \gg u$ and because the Stream is long and narrow. Thus constant potential vorticity (equation (9.38)) reduces to $(f + \partial v/\partial x)/D = \text{constant} = f/D_0$ since at the "edge" of the Stream $\partial v/\partial x$ will be small. Taking $\partial/\partial x$ of equation 9.47* and substituting for $\partial v/\partial x$ in the potential vorticity equation gives

$$\left(f + \frac{g'}{f} \frac{\partial^2 D}{\partial x^2}\right) \bigg/ D = f/D_0. \text{ Rearrangement gives } \frac{\partial^2 D}{\partial x^2} = \frac{(D - D_0)}{\lambda^2}$$

where $\lambda = \sqrt{(g' D_0)}/f$ is called the *Rossby radius of deformation* and gives a length scale based on the parameters of the system. (Since it depends on density differences, λ is completely termed an *internal* radius of deformation (λ_i); the *external* radius of deformation (λ_e) is defined in Section 12.10.3 and compared with the internal radius.) The solution is

$$D = D_0[1 - \exp(-x/\lambda)], \quad v = \sqrt{g' D_0} \exp(-x/\lambda).$$

With $D_0 = 800$ m, $f = 10^{-4}$ s^{-1}, and $(\rho_2 - \rho 1)/\rho_2 \simeq 2 \times 10^{-3}$, Q $\simeq 63$ Sv and the maximum value of $v = 4$ m s^{-1}. Also $\lambda = 40$ km which gives a length scale for the "width" of the Stream.

This simple inertial boundary layer model gives a transport similar to that from observations for the northeastward flow of the Gulf Stream and, in the outer part of the Stream, the velocity calculated from the model solution fits the velocity calculated using the geostrophic equation and observations of temperature and salinity reasonably well. Near the inshore edge, the observed velocity decreases whereas the model velocity continues to increase.

While the model is too simple to represent the Gulf Stream in detail it does indicate that inertial effects need to be included, particularly on the outer side of the Stream south of where the transport is a maximum. On the inshore edge, friction probably becomes important and beyond the latitude of maximum transport the stream shows much more meandering so a more complicated model is needed. As mentioned earlier, Morgan and Charney developed more complete inertial models but these do not work north of the latitude of maximum transport either.

A frictionless, purely inertial, model is not likely to be completely satisfactory. As originally noted by R. W. Stewart, the fact that the no-slip boundary condition is not satisfied allows the Stream to transport relative vorticity. Thus in the inertial models a considerable amount of relative vorticity is moved to the northwest corner of the gyre, probably making the dynamics used incorrect there and perhaps throughout much of the gyre. Analytical models with both inertial and frictional effects seem very difficult to deal with so, as noted before, recent attempts at complete ocean models have been done with numerical techniques which are discussed in Chapter 11.

The Rossby radius, λ, is named after C.-G. Rossby who first introduced it in his wake stream theory of the Gulf Stream, an attempt to explain the

---

* If the Stream does not flow to the north so that x is not east–west, a term involving the variation of f with latitude would occur but it is negligible.

countercurrents both inshore and offshore of the Gulf Stream. Stommel's model indicates that it is an important scale in the inertially dominated part of the main stream itself. It also appears to be the relevant size scale for "meso-scale" eddies, the transient motions with quite large velocities which will be described briefly in Chapter 11.

10

Thermohaline Effects

10.1 The deep circulation

This is much less well known and less well described dynamically than the upper-layer circulation. Because of the difficulty of making direct measurements of velocity in the deep water, even today, most of the information on the deep circulation has been deduced from observations of distributions of water properties. Such information as is available on deep currents suggests that the property distributions are less variable than the velocity fields; however, they essentially provide only a long-term mean picture of the circulation (which may not be very suitable for studying the dynamics). The property distributions suggest that the main sources of deep water are in the North and South Atlantic. For the North Atlantic, it has been accepted since the late 1950s that the major source of deep water is overflow from the Norwegian Sea over the sills between Greenland and Scotland. There is some evidence of sinking due to winter cooling in the Labrador Sea south of Greenland but it is probably very localized in both space and time. In the South Atlantic, the main source is probably the Weddell Sea where sinking results from density increase due to freezing out of ice. Other probable sources around Antarctica are the Ross Sea in the South Pacific and off the coast around 50°E and 140°E. The processes in both north and south are thermohaline and, of course, seasonal. In addition, there is evidence that formation of deep water is intermittent even during the cooling season.

In addition to the deep water formation due to winter cooling, there are also contributions of thermohaline origin at mid-depth from the Mediterranean to the Atlantic and from the Red Sea and Persian Gulf to the Indian Ocean. These are waters rendered dense by evaporation at the surface which then sink and flow out of the seas into the neighbouring oceans.

Stommel has put forward ideas for a model of the circulation of the deep waters. He brings in another feature of the ocean structure—that the depth of the thermocline at any locality remains substantially constant. Because in low latitudes there is a net annual inflow of heat through the surface into the water, the upper warm layer, and its boundary the thermocline, should deepen with time. As this deepening does not happen, some mechanism must be opposing the tendency, and Stommel suggests that this mechanism is slow upward flow of cool deep water. Continuity requires that the sinking water in the North and South Atlantic must be balanced by rising, and Stommel suggests that while the sinking is very localized, the rising is spread over most of the low and middle latitude areas of the oceans. His model of the character of the deep circulation is shown in Fig. 10.1.

The sinking regions (S_1, S_2) are shown feeding relatively intense western boundary currents (required by conservation of vorticity in a situation in which the relative vorticity, ζ, is known to be small in the interior). Outward from these flow gentler geostrophic currents into the bodies of the oceans to supply the slow upward flow to maintain the thermocline depth constant. In the interior, upward motion causes D (ref. Section 9.11.4) to increase; water moves poleward and the magnitude of f increases; ζ stays small. To get back south or north as necessary with ζ small requires input of vorticity of the appropriate sign. This input may be achieved with a strong flow and shear on the west again, just as for the upper layer circulation as discussed in Section 9.12. Figure 10.2 shows that the strong flow and shear must be on the west rather than on the east when the return boundary flow is to the south, since southward flow requires input of negative vorticity to keep ζ small.

In discussing the western boundary surface currents in the last chapter we showed that a northward return flow also had to be on the west. Observations reported in 1977 by Warren indicate an additional feature of the deep flow in the Indian Ocean. Earlier observations had shown the northward flow

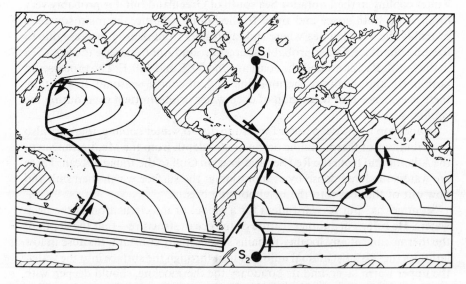

FIG. 10.1   *Model for deep ocean circulation. (After Stommel, 1958)*

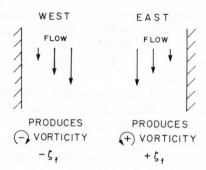

FIG. 10.2 *Relative vorticity supplied by velocity shear at west and east sides of an ocean.*

expected from Stommel's model along the west side of the ocean off Madagascar while the new observations showed a northward flow along the Ninetyeast Ridge which rises to about 4000 m depth along that meridian which is to the east of the centre of the Indian Ocean. The inference is that a western boundary-type flow can be associated with the mid-ocean ridges, if high enough above the bottom, as well as with the western boundary itself.

An aspect of this model relates to the different strengths of the surface layer western boundary currents. Where the deep currents are opposite in direction to the surface flows, the latter may be expected to be stronger to maintain continuity. The sinking water must be replaced by water which has come up through the thermocline and returns in the upper layer. Conservation of vorticity again requires western boundary currents in the return flow. Above the thermocline the upward flow causes D to decrease. To keep ζ small the flow is equatorward, i.e. opposite to the deep flow and likewise to conserve vorticity the western boundary flows in the upper layer will be opposite to those in the lower layer. The strong Gulf Stream to the northwest in the upper layer is consistent with the strong southwest flow in the deep water from S_1 (Fig. 10.1), while the less strong Kuroshio is associated with the weaker deep flow in the Pacific, although the thermohaline flow does still enhance the Kuroshio in the surface region compared with the purely wind-driven values. (Note that there is no large source of deep water in the Pacific. There is now believed to be some outflow from the Ross Sea in the Antarctic but the volume appears to be much smaller than that from the Weddell Sea.) Again, the relatively weak southward Brazil Current in the upper layer in the South Atlantic is consistent with the southward deep flow below it. Stommel suggests that the flow in the deep current under the Gulf Stream is about 30 Sv. The equal surface return flow would then almost double the Gulf Stream transport associated with wind driving. Even the addition of such a thermohaline flow to the wind-driven flow does not make the transport large enough to match values from fairly recent direct measurements of the poleward flow in the Gulf Stream itself although, as noted in Section 9.10.1, if all the counter flows in the western boundary are subtracted the agreement is reasonable considering the uncertainties in the experimental results.

Stommel makes it clear that he does not regard the above as a *theory* but as a *model* for quantitative study which might form the basis for a theory. A limited number of deep current measurements with Swallow floats have in some cases supported the model and in some cases opposed it. However, the number of measurements yet available is small and must be much increased before deciding whether it would be profitable to develop a more detailed theory from the model or to develop a new model and theory.

To produce a theory with thermohaline effects included may require taking T and S (and hence ρ) not as given by observations but as unknowns which must be solved for, in the problem, along with the velocity. If T and S are to be unknowns then we require equations for them.

10.2   Equations for salt and temperature (heat) conservation

The differential equations for salinity and temperature (representing heat) are:

$$\frac{dS}{dt} = \kappa_S \nabla^2 S \tag{10.1}$$

$$\frac{dT}{dt} = \kappa_T \nabla^2 T + Q_T \tag{10.2}$$

where κ_S and κ_T are molecular kinematic *diffusivities* for salt and for temperature (or heat) respectively, and $\nabla^2 = \partial^2/\partial x^2 + \partial^2/\partial y^2 + \partial^2/\partial z^2$. The diffusivities have the same units as kinematic viscosity ($m^2\ s^{-1}$). κ_T is about $\nu/10$ and κ_S is about $\nu/1000$. [These differences arise from the molecular nature of a liquid (water in this case). It is harder for the molecules to exchange kinetic energy (which determines the temperature) than momentum (for frictional stress). It is even harder to move a different type of molecule (e.g. salt) through the fairly closely packed water molecules. For gases at standard temperature and pressure, which are loosely packed, the kinematic viscosity and the diffusivities are approximately equal.]

It has been assumed that κ_S and κ_T vary with position slowly enough that such variations can be ignored, i.e. terms such as $\partial[\kappa_S(\partial S/\partial x)]/\partial x$ which arise in deriving these equations are approximated by $\kappa_S(\partial^2 S/\partial x^2)$. Q_T represents a source function. For example, solar radiation is absorbed over a significant depth range and causes heating and Q_T represents such effects. A similar function for S is not required because processes affecting salinity occur only at boundaries, e.g. river inputs, effects of freezing or the difference between evaporation and precipitation. Such effects would be included as boundary conditions, i.e. for the river one would specify the incoming velocity, salinity (if any) and temperature of the river water, for the other two processes one would specify the equivalent salt flux at the boundary as mass of salt per unit area per unit time. Q_T and boundary conditions will be taken to be given. (In practice, detailed specification might be difficult. If we were considering vertical average between a number of levels, as in a finite difference numerical model, and the solar radiation were all absorbed in the top layer, then the solar radiation could be included as part of the net heat flux at the surface.)

Note that on the left-hand side we have the total derivative. These equations apply to an individual, small (mathematically infinitesimal) fluid element. Indeed they may be easily derived by considering the net flux of the quantity of interest in all directions for a small element and equating it to the total rate of change of the quantity inside the element. For example, the flux of salt in the x-direction is $-\kappa_S (\partial S/\partial x)$ and the net flux in (as $\delta x \rightarrow 0$) is $\partial (\kappa_S \partial S/\partial x)/\partial x$ and similarly for the other directions. These equations are similar to the Navier–Stokes equations for momentum but are rather simpler in that many of the types of terms occurring in the latter do not appear (e.g. terms like those involving pressure, Coriolis force and gravity).

Note also that a fluid element can only change its salinity by molecular processes. Except where Q_T is important, the same is true for temperature (and thus heat). Molecular viscous effects are usually important for fluid elements too, although pressure, Coriolis and gravitational forces are also acting. When we deal with the averaged equations, as we are forced to do because we cannot solve the equations for the instantaneous quantities, such details get obscured. Nevertheless, it remains true that molecular effects are always important and often dominant for fluid elements. When these elements are in turbulent flow, the turbulence, in "stirring" the fluid, makes the instantaneous property gradients very large and greatly increases the rate of change of properties of the elements compared with the rate in a non-turbulent fluid with a comparable mean gradient. In the averaged equations the mixing is described by the Reynolds stresses for momentum (and Reynolds fluxes for salt and temperature as we shall show presently). To get actual solutions these Reynolds stresses and fluxes are related to the mean gradients by introducing eddy viscosity (or diffusivity for salt and temperature) as a necessary approximation for rather detailed processes which we cannot deal with completely.

Equations like (10.1) or (10.2) also apply to other scalar properties. For example, dissolved oxygen concentration would be represented by an equation of the form of (10.2) with oxygen concentration replacing T. Q_{ox} could be either a source or sink depending on what biological processes are occurring.

When we regarded S and T as given, we had four equations in four unknowns (Section 6.2). We have now added equations for S and T. With S and T

unknown, ρ (or α) must also be regarded as unknown but it can be obtained from S, T and p through the equation of state. (The most recent form, the International Equation of State 1980, is given in Appendix 3 as a polynomial in S, T and p.) Thus we now have seven equations in seven unknowns. With appropriate boundary conditions (which may include initial conditions if necessary) it is possible in principle to obtain solutions. In practice, the non-linear terms in the Navier–Stokes equations present difficulties as we have already seen. The addition of three more equations and unknowns is not likely to help and indeed things get worse as we shall see.

10.3   Equations for the average salinity and temperature

Equations (10.1) and (10.2) apply to the total instantaneous values of S and T. If we recall the Eulerian form of the total derivative (e.g. $dS/dt = \partial S/\partial t + u\partial S/\partial x + v\partial S/\partial y + w\partial S/\partial z$) the problem is clear. We must use Eulerian equations for velocity and hence for S and T also. Terms such as $u\partial S/\partial x$ arise; they are called advective terms because they represent changes caused by the motion of the fluid. Here u, v and w are the total instantaneous velocity components. We cannot solve for them in practice as discussed in Chapter 7; therefore we cannot use the equations 10.1 and 10.2 directly. (As with velocity, the fact that boundary and initial conditions for S and T would never be known sufficiently well to solve for instantaneous values is a further difficulty.) The problem arises from the presence of the advective terms involving cross-products of velocity components and the scalar quantity together with the inability to calculate the velocity.

As we did in Section 7.2, we adopt Reynolds' approach of splitting the total quantities into mean and fluctuating parts: $S = \bar{S} + S'$, $u = \bar{u} + u'$, etc., and take the average of the equation. (As the procedure is very similar to that used in Section 7.2 we will give less detail here.) Recalling that averages of terms involving a single fluctuating quantity vanish, we get from equation (10.1).

$$\frac{\partial \bar{S}}{\partial t} + \bar{u}\frac{\partial \bar{S}}{\partial x} + \bar{v}\frac{\partial \bar{S}}{\partial y} + \bar{w}\frac{\partial \bar{S}}{\partial z} + \overline{u'\frac{\partial S'}{\partial x}} + \overline{v'\frac{\partial S'}{\partial y}} + \overline{w'\frac{\partial S'}{\partial z}} = \kappa_s \nabla^2 \bar{S}. \qquad (10.3)$$

The first four terms may be written as $d\bar{S}/dt$ for the total derivative following the *mean* flow. The next three represent the effects of turbulence on the salinity field. The diffusivity term looks the same as before except that $\bar{S}$ replaces S.

10.31   *Reynolds fluxes and eddy diffusivity*

Using the continuity equation for the fluctuating velocity ($\nabla \cdot \mathbf{V}' = 0$) we can rewrite the turbulent terms by adding $S' \nabla \cdot \mathbf{V}'$ (which $= 0$) to the last three terms on the left-hand side of equation (10.3) which become

$$\frac{\partial \overline{(u'S')}}{\partial x} + \frac{\partial \overline{(v'S')}}{\partial y} + \frac{\partial \overline{(w'S')}}{\partial z}.$$

By analogy with the molecular case we suppose that the turbulent fluxes $\overline{u'S'}$, $\overline{v'S'}$, $\overline{w'S'}$ (also called the Reynolds fluxes because they arise when the Reynolds approach is used) are related to the mean gradients in a similar fashion. (As in Section 7.21 we use the simplest analogy, leaving more complex formulations to more advanced texts.) The analogy gives

$$\overline{u'S'} = -K_{Sx}\frac{\partial \bar{S}}{\partial x}; \quad \overline{v'S'} = -K_{Sy}\frac{\partial \bar{S}}{\partial y}; \quad \overline{w'S'} = -K_{Sz}\frac{\partial \bar{S}}{\partial z} \quad (10.4)$$

where K_{Sx}, K_{Sy} and K_{Sz} are kinematic *eddy diffusivities* (units $m^2 \, s^{-1}$).

Now the turbulent mixing is dominated by the turbulent flow field so it is common to assume that the eddy diffusivity is the same for all scalars (unlike the molecular values). Because of the static stability, K_{Sz} will be much smaller than K_{Sx}, K_{Sy} but these two should be similar. Thus we replace K_{Sz} by K_z, vertical eddy diffusivity, and the other two by K_H, the horizontal eddy diffusivity. The ranges of values for K_z and K_H are similar to those for A_z and A_H respectively (the eddy viscosities) because they are properties of the turbulent flow field. In a particular case, K_z and A_z or K_H and A_H may not have the same values but they probably have the same order of magnitude. Finally, neglecting the variations of the K's with space coordinates, neglecting the κ_S term compared with the turbulence terms and dropping the overbar for simplicity, equation (10.3) becomes

$$\frac{dS}{dt} = K_H\left(\frac{\partial^2 S}{\partial x^2} + \frac{\partial^2 S}{\partial y^2}\right) + K_z\frac{\partial^2 S}{\partial z^2}. \quad (10.5)$$

Here S is now the average salinity. Equation (10.5) looks quite similar to equation (10.1) for the instantaneous salinity.

In the same manner, an equation for the average temperature may be obtained. Overbars are omitted but all quantities are now averages including Q_T

$$\frac{dT}{dt} = K_H\left(\frac{\partial^2 T}{\partial x^2} + \frac{\partial^2 T}{\partial y^2}\right) + K_z\frac{\partial^2 T}{\partial z^2} + Q_T. \quad (10.6)$$

However, we have made bold assumptions in inserting the eddy diffusivities. All the cautions in Chapter 7 with regard to the eddy viscosities apply. Any results which depend strongly on the assumption of the eddy diffusivity (or viscosity) representation must be viewed with suspicion until confirmed by observations. Use of $K_z = A_z$ in the atmospheric surface layer is an example of a case for which the approach does work well.

10.4   Thermoclines and the thermohaline circulation

Let us consider the steady-state case and ignore Q_T because we are interested in the main thermocline. Then equation (10.6) becomes

$$\mathbf{V}_H \!\cdot\! \nabla_H T + w\frac{\partial T}{\partial z} = K_H \nabla_H^2 T + K_z \frac{\partial^2 T}{\partial z^2} \tag{10.7}$$

where $\mathbf{V}_H$ is the horizontal velocity ($\mathbf{i}u + \mathbf{j}v$) and ∇_H and ∇_H^2 are the horizontal gradient and Laplacian respectively. In words, equation (10.7) states that horizontal advection plus vertical advection equals horizontal plus vertical diffusion.

Our knowledge of the deep circulation is not sufficient to allow us to drop any of the terms as being small. Stommel's conceptual model suggests that both advective terms are needed in a thermohaline circulation theory and at least the vertical diffusion term. Lateral diffusion may well be important too. It is possible to try to balance pairs of these four terms while neglecting the other two to see what solutions are possible. At least five of the six possibilities have been tried and more than one of the possible balances can produce a reasonable looking thermocline structure.

The idea that vertical advection is balanced mainly by vertical diffusion with the other terms being fairly small has been considered a reasonable possibility for a long time. Assuming that this balance is correct we get

$$\frac{\partial^2 T}{\partial z^2} = \frac{w}{K_z}\frac{\partial T}{\partial z}. \tag{10.8}$$

Assuming w/K_z independent of z and $T = T_d$ for $z \ll -K_z/w$, then $T = T_d + (T_0 - T_d)\exp(wz/K_z)$ where T_0 is the temperature at $z = 0$ (taken to be at the bottom of the mixed layer). Adding a mixed layer on top and adjusting w/K_z one can produce a reasonable fit to observed vertical temperature profiles.

In the interior of the ocean we can use the geostrophic approximation (equation (8.10)). If we cross-differentiate to eliminate the pressure terms, ignoring the horizontal derivatives of α (the Boussinesq approximation), we get

$$\beta v + f\!\left(\frac{\partial u}{\partial x} + \frac{\partial v}{\partial y}\right) = 0.$$

Using the continuity equation gives

$$\beta v = f\!\left(\frac{\partial w}{\partial z}\right).$$

To have north–south flow, w must vary with z. As argued earlier, we must have north–south flow in the interior to keep the relative vorticity small. In the interior, w is thought to be upward. It could increase from zero at great depth to a maximum in the thermocline and decrease to zero at the base of the mixed layer (or to w_E if we include wind-driving). This vertical dependence of w would give the north–south flows needed to keep relative vorticity small. For our solution to equation (10.8) we require w/K_z to be constant. Thus K_z would have to be a maximum in the thermocline too. This behaviour for K_z seems contrary to the expectation that K_z would be lower in the region of strongest static stability. However, we do not know how K_z depends on height. Also there are likely to be internal waves in the thermocline and breaking of these waves could lead to sufficient mixing to make K_z larger there.

To keep things in perspective, remember that reasonable looking solutions have been found with $K_z = 0$. Clearly it is difficult to find satisfactory solutions even in the interior. To obtain a closed basin solution analytically (which has not yet been achieved) lateral boundary layers would have to be added.

Further discussion of thermocline and thermohaline circulation theories is left for more advanced texts, both because the mathematics rapidly becomes fairly complicated and because the observational data are so limited that it is difficult to tell whether or not the theories are realistic. Section 5.8 in G. Veronis' (1981) chapter on the "Dynamics of Large-scale Circulation" in *Evolution of Physical Oceanography* provides a recent review of the subject. Numerical modelling may eventually help in obtaining more complete analytical solutions by showing which terms must be retained and which may be neglected in the equations.

10.5   The mixed layer of the ocean

The top few tens of metres of the ocean are usually observed to be fairly well mixed, i.e. the temperature and salinity are fairly uniform. Below this region there is a thermocline (and perhaps a halocline) and hence a pycnocline region. The top layer is the oceanic planetary boundary layer where vertical friction effects are important. It is also called, as we did earlier, the Ekman layer. The dynamics governing the formation of this layer are of considerable interest. Convergences and divergences in the layer lead to circulations in the deeper water (Ekman pumping effect). This is also the region of (biological) primary productivity. The depth of the layer and mixing up of nutrients from below will be important factors in determining the productivity.

In addition, there are meteorological effects both for weather and climate. Much of the solar radiation, which is the ultimate source of energy for both atmospheric and oceanic motions (except tides), is first absorbed in the ocean's upper layer. A large part of the atmosphere's energy supply comes from heat exchange with this layer, mainly in the form of the latent heat of the water evaporated at the surface which is released when the water condenses higher in the atmosphere. For weather forecasting for a day or two ahead these energy inputs can be ignored. As the forecast period is increased the energy inputs become increasingly important. For an atmospheric model, the bottom boundary conditions require knowledge of the surface temperature (and the albedo which is reasonably well known for the ocean). From the surface temperature and humidity (which is determined by surface temperature) and the predicted air temperature, humidity and wind at the lowest level in the atmospheric model one can calculate surface friction, heat fluxes and back radiation although there is still some uncertainty in the coefficients used (e.g. the drag coefficient and equivalent coefficients for heat fluxes). To get the surface temperature, the evolution of the ocean mixed layer must be predicted too. For weather, or relatively short time-scales of a few days to perhaps a month, vertical transfers probably dominate in most places and horizontal advection and diffusion may be neglected. For climatic time-scales of a month to a few years the whole upper-layer circulation (based on present mean values) would need to be included because a sizeable part of the poleward heat transport required to maintain the global heat balance is provided by the ocean circulation in the top few hundred metres. For decades to centuries the whole ocean circulation must be included—a formidable task as we have shown.

Here we shall consider the short time-scale problem only. Without going into the mathematical details we shall try to indicate the processes which are included and the approach being taken in this continuing aspect of dynamical oceanography research. The conceptual model presented here comes from P. Niiler (1975). The temperature is assumed uniform within the mixed layer with a thin transition zone at the bottom where the temperature changes rapidly to the value in the thermocline below the mixed layer. (If salinity variations are important in determining density the effect may be included by defining an equivalent temperature.) The temperature structure below the mixed layer must be specified (presumably it can be obtained from observations with geographical and seasonal variations included) but hopefully it will be slowly varying on the time-scale for which predictions are required. The velocity is also assumed to be independent of depth throughout the bulk of the layer. There is a jump in the transition zone at the bottom to the value of zero in the non-turbulent thermocline region. (If a geostrophic flow is present in this region, it would have to be specified and the velocity considered would be the difference from the geostrophic value.) There is also a thin shear zone in the surface wave zone. The stress at the surface is specified.

With horizontal gradients assumed negligible, by continuity and $w = 0$ at the surface there is no mean vertical velocity and all the non-linear terms involving the mean velocity vanish. The equations for the horizontal velocity are

$$\frac{\partial u}{\partial t} - fv = -\frac{\partial \overline{(u'w')}}{\partial z} + F_x,$$

$$\frac{\partial v}{\partial t} + fu = -\frac{\partial \overline{(v'w')}}{\partial z} + F_y. \tag{10.10}$$

Primed quantities are fluctuations, unprimed are means. F_x and F_y are damping terms added to make inertial oscillations die out.

In the temperature equation the advective terms vanish because of the assumption of no horizontal gradients and the consequence that $w = 0$. The source term (solar radiation) is included in the specified surface heat flux since we are treating the layer as a whole and the temperature equation is

$$\frac{\partial T}{\partial t} + \frac{\partial \overline{(w'T')}}{\partial z} = 0. \tag{10.11}$$

Now in the layer u, v, T, F_x and F_y are independent of z (by assumption). Thus $\partial/\partial z$ of $\overline{u'w'}, \overline{v'w'}$ and $\overline{w'T'}$ must also be independent of z and the stresses and the heat flux are linear functions of z, going from the specified values at the surface to the values at the top of the transition zone required to bring fluid being mixed into the layer to the values of u, v and T within the layer. These expressions for $\overline{u'w'}, \overline{v'w'}$ and $\overline{w'T'}$ may be put into. equations (10.10) and (10.11). In this process an additional unknown, the layer depth h, is introduced. Thus we have three equations in four unknowns and therefore a closure problem as is usual in turbulence problems. The equation for conservation of turbulent kinetic energy is added and to close the system it is *assumed* that a fraction of the energy input by the wind coming from the upper shear layer is diffused downward and used to mix fluid up from below the mixed layer, the remainder being dissipated. If the surface is being cooled then the resulting convective instability can provide mixing energy; if it is being heated some of the mechanical mixing energy is used to overcome this stabilizing effect. The velocity "jump" at the bottom may also cause mixing. It causes layer deepening only if the layer is initially sufficiently thin and only for the first half-pendulum-day after a change in the wind. By that time the initial rapid deepening has thickened the layer and reduced the velocity jump; the Richardson Number (based on the velocity and temperature changes across the transition zone) becomes too large and the static stability prevents mixing from this source. If the initial layer depth is too large (greater than the depth to which it would have mixed if it were initially thin) the velocity jump at the bottom is never large enough to produce mixing. The fraction of turbulent energy from the surface zone which is assumed to diffuse down and cause mixing causes a slower continual deepening of the layer after the initial rapid deepening (if it occurs).

This model and similar models have had limited testing with real observations because data sufficiently detailed to test them fully are not available in the historical records. Once the free parameters in the model (the fractions of the three energy sources that are actually used for mixing and the damping terms) are established, weathership data can provide long time-series for further testing. What testing has been done suggests that the approach has promise. It seems to work fairly well during the heating season when the summer or seasonal thermocline is being formed. During the cooling season the continual slow erosion from the constant fraction of surface-generated turbulent energy, which is assumed to get to the bottom no matter how deep, makes the model mixed-layer too deep and the annual cycle does not close. In reality it is probable that when the layer gets sufficiently deep the surface-generated turbulent energy is all dissipated within the layer and none is left to deepen the layer further. Garwood (1977) gave a modified closure scheme which produced a closed annual cycle for idealized forcing. He also showed that including the diurnal heating cycle affects the result. Niiler and Krauss (1977) discussed the Niiler type model and how the parameters may change with depth.

Special observational programmes to obtain better field data were conducted in the late 1970s (e.g. Davies *et al.*, 1981; other accounts will appear in the literature in the 1980s). With further testing and modification a satisfactory model (or perhaps models depending on the application) will no doubt emerge. By "satisfactory" we mean that the model predicts what we want to know with acceptable accuracy—it need not reproduce all the details of the field data.

Many papers on this subject have appeared in the literature in the last few years and no doubt many more will be published. The reader interested in further information and applications of mixed-layer models to climate and ocean-circulation models should find the volume edited by Krauss (1977) and the article by Turner (1981) useful and then, as for any other subject area, would have to consult the recent literature.

11

Numerical Models

11.1 Introduction

Models such as those of Stommel and of Munk described in Chapter 9 were attempts to reproduce the general or climatological (long-term average) circulation and seemed to give qualitatively correct results. However, the actual magnitudes of the transports seemed too small. Stommel, in *The Gulf Stream* (1965), suggested that addition of thermohaline effects and perhaps some adjustment of the wind stress, because of its uncertainty, might give better agreement between Munk's model and the observations. At that time, the transport estimates were based on geostrophic calculations and so could be changed by varying the level of no motion which could also help to reduce the "discrepancy". Subsequently, more direct transport measurements in the Gulf Stream have shown that the difference between the northeast flow of the Gulf Stream and the wind-driven transport in the interior is even bigger than had been thought and probably cannot be explained by simple addition of thermohaline circulation, although it accounts for some of the difference. Of course, it had been known for a long time that if one looks at the Gulf Stream and similar currents over a short period of time they are much narrower and faster than the currents in the models. Nowadays, one can locate such features from surface-temperature measurements obtained by satellites because the surface-temperature field gives an indication of the location of the current. The currents are also known to meander, that is their position changes with time, and they follow quite a curved path at times. It was hoped that if these time-varying effects were averaged out then this average current would look like the currents in the models. As noted in Chapter 9, the various countercurrents in the western boundary region needed to be averaged with the Gulf Stream too.

Stommel also suggested another possible cause for the "discrepancy" which needed further examination. As noted, the actual "instantaneous" Gulf Stream is quite narrow and strong. One can look at the sizes of the terms in the equations of motion just as we did in Chapter 7 for the interior flow and as we did at the end of Chapter 8 for a strong current such as the Gulf Stream. We took the y-coordinate along the stream and the x-coordinate across the stream. When we examined the x-momentum equation, the Coriolis term fv still dominated because v is so large; it must be balanced by the pressure-gradient term $\alpha(\partial p/\partial x)$ because none of the other terms is large enough. Thus, the geostrophic balance still holds for this equation and this fact is the basis for using the geostrophic calculation to obtain the v or downstream velocity component as we noted before. In the y-momentum equation we expect friction to be important and estimate the eddy viscosity from the observed width, W, of the stream using the relationship from Munk's model $A_H = \beta W^3$ developed at the end of Section 9.14.1. If we examine the other terms in this equation using this estimate for A_H, rather than the maximum value as we did in Chapter 8, the non-linear terms are larger than the friction terms as noted in Chapter 9 and both effects are important (as the reader may easily verify). Thus to model the Gulf Stream as it actually occurs, non-linear terms must be included. Furthermore, if including these terms makes the stream stronger than it would be without them, a long-time average, while similar to the Munk model, would have a larger transport if any countercurrents also set up are not subtracted. Some attempts were made to actually solve the equations analytically with the non-linear terms included as noted in Chapter 9. However, only solutions restricted to very special cases could be obtained. It was the desire to include the non-linear terms in more adequate fashion that led to the early numerical models. Thus one tried to obtain solutions using approximate equations which could be solved numerically using a computer; hence the term *numerical modelling.*

There were other parallel developments which also led to numerical modelling attempts. Numerical or computer modelling of the atmospheric circulation had begun before these attempts at numerical models of the oceans were made. Nowadays, such models of the atmosphere are used to aid in operational weather forecasting as well as for research purposes to try to understand the atmospheric circulation better. Quite powerful computers are required for this purpose. The usefulness of such models for the atmospheric circulation naturally led to the idea of using them to model ocean circulation. Because such models, particularly the more detailed ones, require very powerful and fast computers the ocean modelling is often done in the same laboratories as the atmospheric modelling. The atmospheric modellers are also interested in the ocean and in modelling it because the ocean forms 70 % of the bottom boundary of the atmospheric models.

Another parallel development is numerical modelling of coastal regions, semi-enclosed seas and estuaries. This type of modelling has developed as an extension of numerical modelling of rivers. It is often done for engineering purposes as opposed to research purposes. Examples are for predicting the effects of construction or determining the best location for waste disposal. Prediction of tidal elevations and currents as an aid to navigation is another example. In addition, prediction of storm surges produced by the combined effects of wind and tide is also of considerable practical interest.

Since this introduction to dynamical oceanography is primarily concerned with the main features of the ocean circulation, neither coastal nor atmospheric modelling will be described, but the results of these types of models may be mentioned when they help us to understand the ocean models or when the results of such modelling provide some insight into the future potential of ocean models.

Two broad categories of models of the ocean circulation may be defined: *mechanistic* models and *simulation* models. In mechanistic models, the geometry is made as simple as possible and no terms are included in the model equations which are not essential to answering the question at hand. In such models one may try to examine the importance, for example, of the non-linear terms or of the effect of bottom topography on the circulation. In simulation models, one attempts to reproduce the circulation of an actual ocean for comparison with oceanographic observations. In such models the actual geometry of the ocean basins is included with all the possible driving effects and terms in the equations. Simulation models produce an enormous amount of data and it may be difficult to untangle which dynamical effects are most important, and a great deal of analysis of the output of the model is required. Both types of models are important in increasing our understanding of ocean circulation. Mechanistic models are easier to interpret and help to advance our understanding of particular aspects of ocean circulation dynamics. On the other hand, simulation models allow direct comparison with nature which is essential to prove that ocean models really do represent nature and that the mechanistic models which form the basis of the simulation models lead to an understanding of the dynamics of the real world.

A detailed discussion of numerical models is, of course, beyond the scope of this book. However, because many of the readers may be unfamiliar with the approaches taken, the next section will give a very general description of the more common approaches used. Then we will describe a number of ocean models and their results.

11.2 Numerical methods

The method most commonly used in numerical models, particularly in the simulation models, is that of *finite differences*. We have already made use of some finite-difference approximations, for example in calculating the horizontal divergence approximately in Chapter 4.

In such methods we do not attempt to find solutions which will allow us to give the velocity at a particular time at *any* position, but to find values on a grid of points. For example, a two-dimensional grid may be used if we are considering the case in which we either deal with variables vertically averaged over the water column or assume that there is no vertical variation in the flow, or assume that there is no variation in the cross-stream direction and consider only variations in the long-stream and vertical directions. We need a three-dimensional grid if we allow variations in all spatial dimensions. We shall also calculate values at discrete time $t_1, t_2, \ldots, t_n$, rather than continuously.

To calculate values at a new time from values at a previous time, using the equations of motion, we have to be able to calculate spatial gradients of quantities. For example, suppose we want to determine $\partial u/\partial x$ at the point x_j. We could approximate this value by $(u_j - u_{j-1})/\Delta x$, where Δx is the grid spacing in the x-direction, u_j is the value at x_j and u_{j-1} is the value at x_{j-1}. This is called a *backward difference*. We could use $(u_{j+1} - u_j)/\Delta x$, a *forward difference*. A Taylor series expansion around x_j gives

$$u_{j+1} - u_j = \frac{\partial u}{\partial x}\Delta x + \frac{1}{2}\frac{\partial^2 u}{\partial x^2}(\Delta x)^2 + \frac{1}{6}\frac{\partial^3 u}{\partial x^3}(\Delta x)^3 + \frac{1}{24}\frac{\partial^4 u}{\partial x^4}(\Delta x)^4 + O(\Delta x)^5,$$

$$u_j - u_{j-1} = \frac{\partial u}{\partial x}\Delta x - \frac{1}{2}\frac{\partial^2 u}{\partial x^2}(\Delta x)^2 + \frac{1}{6}\frac{\partial^3 u}{\partial x^3}(\Delta x)^3 - \frac{1}{24}\frac{\partial^4 u}{\partial x^4}(\Delta x)^4 + O(\Delta x)^5$$

where derivatives are evaluated at x_j, and $O(a)^n$ means that the term is a finite number times a^n, called "of order a^n". The largest term of the error in the gradient in either the forward or backward approximations of $\partial u/\partial x$ is $(\partial^2 u/\partial x^2)\,\Delta x/2$, that is proportional to Δx. Such schemes are said to be first-order accurate, because the error in the approximation depends on Δx to the first power. A more accurate approximation to $\partial u/\partial x$, obtained by adding the equations and dividing by $2\Delta x$, is

$$\frac{(u_{j+1} - u_{j-1})}{2\Delta x} = \frac{\partial u}{\partial x} + \frac{1}{6}\frac{\partial^3 u}{\partial x^3}(\Delta x)^2 + O(\Delta x)^4.$$

The largest error term is now $O(\Delta x)^2$ and this scheme, also called a *centred difference* since it used a value on either side of x_j, is called second-order accurate. The next higher term is $O(\Delta x)^4$ because the terms $O(\Delta x)^3$ cancel. Higher accuracy approximations can be obtained by using two points on either side of the point where we want an approximation to the derivative. Again using Taylor series expansions we have

$$\frac{(u_{j+2} - u_{j-2})}{4\Delta x} = \frac{\partial u}{\partial x} + \frac{4}{6}\frac{\partial^3 u}{\partial x^3}(\Delta x)^2 + O(\Delta x)^4.$$

Then
$$\frac{\partial u}{\partial x} = \frac{4}{3}\left(\frac{u_{j+1} - u_{j-1}}{2\Delta x}\right) - \frac{1}{3}\left(\frac{u_{j+2} - u_{j-2}}{4\Delta x}\right) + O(\Delta x)^4$$

is fourth-order accurate because now the $\partial^3 u/\partial x^3$ terms cancel.

One could continue this procedure but the computer programming would get more complicated and the treatment near boundaries becomes more complicated too. For example, if we use the centred fourth-order scheme for

$\partial u/\partial x$ the first two points near a north–south boundary must be given special treatment.

Likewise, higher derivatives such as $\partial^2 u/\partial x^2$ can be approximated. A second-order accurate approximation can be obtained by taking

$$(u_{j+1} - u_j) - (u_j - u_{j-1}) = u_{j+1} + u_{j-1} - 2u_j = \frac{\partial^2 u}{\partial x^2}(\Delta x)^2 + \mathrm{O}(\Delta x)^4.$$

Higher order approximations can be obtained by using more points.

Given that we can calculate first- and second-order *spatial* derivatives we can then approximate the *temporal* derivatives in the equations to follow the time evolution in our model. The simplest approach is to use a forward difference approximation, using values at the past time step to calculate the spatial gradients, that is $(u_{n+1} - u_n)/\Delta t = $ a function of spatial derivatives at time $t = n\Delta t$, where Δt is the time step. (Here the subscripts indicate the time at which we evaluate u at each point. For simplicity we have omitted a second subscript indicating the spatial location.) Such a scheme is first order accurate in Δt. A scheme which is second-order accurate (time errors proportional to $(\Delta t)^2$) is the leap-frog scheme $(u_{n+1} - u_{n-1})/2\Delta t = $ function of derivatives at time $t = n\Delta t$. This scheme is more accurate but involves storing values in the computer at three time levels rather than two.

These schemes where the values at new times are calculated from values at previous times are called *explicit*. One can proceed from grid point to grid point calculating values for the next time step. It is also possible to devise schemes in which the values at the new time step depend on spatial gradients of values at the new time step. One then writes down equations for each grid point and must solve the whole set of linear equations simultaneously. Such methods are called *implicit*. Because for a large grid, a large system of equations must be solved, one may require more high-speed storage area in one's computer for an implicit scheme than for an explicit one.

There are many ways in which the equations of motion and continuity can be formulated in finite difference form. The question arises as to whether these various formulations may be distinguished on grounds of relative merit. Broadly speaking, the assessment of a particular scheme is carried out by comparing the solutions obtained using a finite difference representation of the linearized equations of motion and continuity with known simple analytical solutions. In this manner there emerge a number of criteria crucial to the ability of the numerical model to simulate oceanographic phenomena correctly. Thus, for example, there exists the problem of linear computational instability. In the case of straightforward explicit schemes there is an upper limit on the time step which is determined by the distance step and the maximum wave propagation speed C_{max} in the model according to the relation $\Delta t \leqslant \Delta x/C_{max}$. If this time step is exceeded there results an explosive growth of small errors inevitably present in the numerical operations. Implicit methods seem to have better computational stability and it may be possible to use larger time steps than those given by the condition above. However, if one makes the time step too large it may be that the implicit solution obtained, while it is stable, is not a true solution.

Further problems can occur in that the amplitude and speed of a wave in the numerical model can differ significantly from that in nature, and further that spurious oscillations may be introduced which are entirely of computational origin. Instances of the latter have been encountered, for example, in schemes using centred differences in time. In addition, non-linear terms such as the advective accelerations may lead to non-linear instability, e.g. a rapid spurious increase of energy at small scales which is unrelated to any real physical phenomenon. In practice, various techniques have been developed over the years to deal with these numerical problems. It is important to emphasize, however, that a particular model is limited to a certain range of motions within the sea which it will simulate satisfactorily.

Further discussion of these aspects of numerical modelling is beyond the scope of this book. Here we have tried to present an outline of the approach used; the reader desiring further information should consult the literature. The symposium proceedings on Numerical Modelling (Reid, 1975) referred to in the Further Reading list should provide a starting point.

11.3   General approach to numerical modelling of ocean circulations

The equations of motion, continuity and, where applicable, heat and salt conservation are put in finite difference form. All terms, including the non-linear ones, can be incorporated simultaneously, a real advantage over analytical methods. Some suitable grid in two or three dimensions is chosen. Boundary conditions, for example wind stress, temperature and salinity at the sea surface, and temperature, salinity and velocity on lateral boundaries, are chosen. Initial conditions at time $t = 0$ must also be chosen. Often a state of rest, that is all velocities initially zero, is assumed. The temperature and salinity values may be taken to be uniform in the interior but it is more usual to prescribe a depth distribution approximating that in the real ocean. The calculation then proceeds by stepping forward in time, one step at a time, and the process is spoken of as time "integration".

As the spatial resolution is increased (i.e. the separation between grid points is reduced) the time step must be decreased to maintain computational stability, at least for the explicit techniques that are generally used. Thus, for a two-dimensional grid, doubling the resolution probably requires an eightfold increase in the computation time; in three dimensions, if the number of vertical levels is also doubled the amount of work goes up sixteenfold. Thus it is easy to see why a desire to improve the resolution of a model may be limited by the speed and storage capacity of the computer.

Sufficient friction is generally required for computational stability in these non-linear models, either built into the numerical scheme or included in the equations being used. Nature works this way too, of course; as the motion becomes stronger, frictional effects increase until there is a balance between the rate at which energy is being put into the system and the rate at which it is being lost through friction. Because energy sources are finite the motion remains bounded although it may be quite strong and have large variations in both time and space.

As discussed in Chapter 7 for momentum and in Chapter 10 for heat and salt, the system of equations is not closed. Some additional equations must be incorporated to relate the frictional effects to the calculated large-scale velocity field, and the diffusive effects to the gradients of the temperature and salinity fields. These effects are produced in some turbulence-like fashion by motions on scales too small to be resolved by the grid. These are called sub-grid scale phenomena. To close the system the effects of these scales are usually parameterized in a simple way through the use of eddy viscosities and diffusivities. Typically, constant values are used (with different values in the horizontal and vertical, of course).

The eddy viscosity and the grid spacing or the resolution of the model are related to some extent. As noted, friction must be kept at a reasonable level and at the same time velocity differences between adjacent grid points must not become too large. Thus if we use larger grid spacing the maximum velocity gradient is limited. Because the frictional stress is taken as equal to the eddy viscosity times the velocity gradient (e.g. equation (7.4)), then to maintain friction approximately constant the eddy viscosity must be increased as the grid spacing is increased. Because the resolution is limited by the speed of available computers the eddy viscosities used are often larger than inferred on the basis of our limited observations and may prevent the non-linear terms from playing a proper role.

In some atmospheric models, variable eddy viscosity has been adopted. It is made proportional to the root mean square rate of strain in a calculated flow (the rate of strain is derived from the calculated shears). Physically, this variable viscosity with friction being larger where the shears or velocity gradients are larger seems a better representation than constant eddy viscosity and it seems to improve the atmospheric models. As yet this approach has had little use in oceanic models. (O'Brien's model, described later, is one exception.) Also, the variable eddy viscosity seems to damp scales of motion near the grid scale more strongly relative to larger scales than constant eddy viscosity. Thus, computational stability can be maintained with lower, more realistic, average viscosity values. The use of variable viscosity may be helpful in allowing non-linear effects to be more realistic with the resolution capabilities of present computers.

As noted above, one of the boundary conditions to be imposed is the forcing by the wind stress. As discussed in Chapter 9, the procedure for calculating the wind stress from the observed wind velocities has very recently become reasonably well established, particularly for stronger winds. In the models to be described in the next section the imposed wind stress is usually based on computations made by Hellerman (1967, 1968). These computations are based on climatological wind data and have been carefully done in an attempt to take into account the variations in wind speed and direction at a given location. The drag-coefficient formulation that Hellerman used in these computations was the smoothed version of the step function with values for C_D of about 0.8 $\times 10^{-3}$ at low wind speed rising rapidly but smoothly in the vicinity of 7 m s^{-1} to 2.4×10^{-3}. It was known at the time (about 1965) that Hellerman did the calculation that the drag coefficient does not have this rapid transition, but Hellerman was not aware of the more recent results. Thus this calculation may have overestimated somewhat the contribution of strong winds and also perhaps underestimated the contributions in regions of light winds, that is in the regions in the middle of gyres where the stress changes sign and the curl of the wind stress is a maximum, although even here, climatologically, winds less than 6-8 m s^{-1} are unimportant. The spatial resolution of Hellerman's stress values is also limited because of the way in which the climatological data are compiled as described in Section 9.9. Because of this limited resolution the stress gradients or the curl of the wind stress may be underestimated. It is difficult to know how serious these errors are but it is not likely that the true values of the maximum of the curl are more than about 50% larger than Hellerman's calculations. Some more detailed calculations for the regions of maximum curl in the North Atlantic, which have better spatial resolution and use more up-to-date drag coefficient values, give similar values for the maximum curl to those obtained using calculations similar to those of Hellerman (Saunders, 1976). However, these calculations produce a long-time average and do not take into account seasonal variations in the position of maximum curl. As noted in Chapter 9, such a calculation may give a lower value than if we calculated a maximum curl on a daily, weekly or monthly basis and then averaged these maximum values together regardless of the latitude of the maximum curl. It is the latter value of the maximum curl which is probably most appropriate for calculating the maximum transport due to the wind using the simplified Sverdrup equation (9.21). As noted in Chapter 9, the latest estimates for the wind-driven transport in the interior in the Atlantic are quite similar to those of Munk.

11.4   Descriptions of some models of individual oceans

In this section we shall outline the features and main results of a few numerical simulation models to show the sort of results which can be achieved and also some of the limitations. We start with a fairly simple model and proceed to somewhat more detailed models.

11.41   O'Brien's two-dimensional wind-driven model of the North Pacific (1971)

This is a rather simple vertically averaged model similar to the classic Munk analytical model described in Chapter 9, except that it uses more realistic geometry and includes non-linear effects. It has the following features:

(1) Density is taken to be uniform, that is, the model is barotropic; the velocities are independent of depth, reducing the problem to two dimensions.
(2) It is driven by the wind using the climatological average wind stress calculated by Hellerman.
(3) The equations are in spherical polar form, instead of using the beta plane approximation as Munk did.
(4) The "grid" is 2° in latitude and in longitude.
(5) Bottom topography is included above 2000 m, the assumed depth of the ocean.
(6) Lateral friction is used with a variable eddy viscosity as in some atmospheric models as described in Section 11.3.

This model is started from a state of rest. As the integration proceeds, flow begins to occur and to build up into a series of gyres. This process is spoken of as "spinning up". After some time a fairly steady circulation is apparent but there are fluctuations superimposed upon it. The transport pattern after 72 days of integration is shown in Fig. 11.1. Figure 11.1(a) shows the detailed structure at the western boundary and Fig. 11.1(b) the broad-scale pattern over the whole ocean basin. Note that the sign convention for the stream function in this figure is opposite to that which we adopted in Section 9.7. This difference is present in all the figures in this chapter, so that the modellers whose work is reported here are all consistent but contrary to the usual fluid mechanics convention. In this chapter, a gyre with a positive ψ value at its centre indicates a clockwise gyre. There are arrows in some of the figures but in any case the reader should know the directions of the main circulations by now and be prepared for the lack of consistency in the literature.

The model seems to produce known features of the circulation reasonably well, although there are differences in detail. The Kuroshio leaves the coast at 35°N and has a transport of about 60 Sv in reasonable agreement with observations. Earlier estimates from geostrophic calculations suggested a transport of 65 Sv but more recent estimates are somewhat higher, perhaps 80 to 90 Sv. Munk's calculations gave a transport of 38 Sv and the current did not leave the coast until about 45°N. The numerical model thus seems to give better

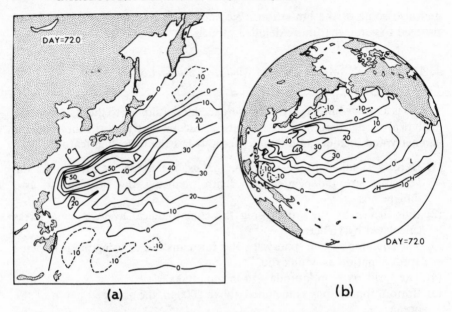

(a)   **(b)**

FIG. 11.1   *Mass transport stream function* (ψ) *computed for a wind-driven model of uniform density* (*North Pacific*): (*a*) *the circulation pattern at the western boundary in units of* 10^6 *tonnes* $s^{-1} \simeq$ *units of sverdrups* ($10^6\ m^3\ s^{-1}$), (*b*) *the pattern over the entire basin.* (*From O'Brien, 1971.*) (*See note in text on the signs for* ψ.)

results. The departure from the coast at 35°N is partially an effect of the topography. (Topographic effects may be overemphasized in barotropic models as we shall see. However, in this case, the bottom is taken to be flat for real depths greater than 2000 m so the problem is probably not serious in this model.)

The greater transport, compared with Munk's linear model, is probably due to enhancement of the flow by non-linear effects (the advective acceleration terms). The early numerical models mentioned previously which were used to investigate the effect of the non-linear terms showed quite clearly that the flow could be enhanced by these effects by comparison with the results obtained when non-linear terms were omitted. Thus, as Stommel suggested, non-linear or inertial enhancement can lead to stronger circulation. Much of the additional transport is a local recirculation, that is, the additional flow in the western boundary current is mainly returned in a fairly strong countercurrent close to the boundary current. O'Brien's result (Fig. 11.1) shows this behaviour with about one-third of the Kuroshio transport associated with local recirculation. The southward flow in the interior for the main gyre is about 40 Sv, just what one would calculate using the Sverdrup equation as must be the case for a model with a flat bottom, provided that the non-linear and bottom friction effects are unimportant in the interior as we expect to be the case.

As we shall see, other numerical models of the North Pacific of similar resolution, with constant eddy viscosity and with the same wind stress, give transports more like that of Munk's study. It appears that the variable eddy viscosity allows lower overall viscosity and hence more important non-linear effects, and leads to enhancement of the transport of the Kuroshio to more realistic values.

While O'Brien's model seems to produce results fairly consistent with our limited knowledge of the actual circulation, the maximum transport values produced are still somewhat lower than the most recent estimates although there is uncertainty both in these transport estimates and in the wind stress which O'Brien used in his calculations. However, the assumption of constant density leaves out any thermohaline circulation completely. This separation has always been necessary in analytical work because the whole system cannot be treated at once. In addition to the non-linear nature of the equations describing the velocity field, there is a fundamental non-linearity—the advective terms in the equations for salt and heat conservation as discussed in Chapter 10. It seems most unlikely that the flows produced by the two driving forces do not interact, which provides, as we shall see, another possible enhancement mechanism, in addition to the simple linear addition of the thermohaline flow to the wind-driven flow that was originally suggested by Stommel. O'Brien's model shows the effect of inertial enhancement in a simulation model and how Munk's approach works when treated more realistically.

11.42   *Cox's model of the Indian Ocean* (1970)

This model is an attempt to see if the current reversals in the Indian Ocean that occur as the monsoon winds change can be reproduced in a numerical model. The model is three-dimensional with up to seven levels in the vertical depending on the depth. It incorporates the following features:

(1) Wind driving is seasonal, based on 3-month averages over each of the four seasons using the Hellerman wind stress. The stress is made to vary smoothly from season to season.

(2) Thermohaline driving is included by imposing the surface temperatures and salinities. While surface fluxes of heat and salt would be preferable, the surface salt and temperature values are imposed because the fluxes are not nearly as well known. Temperatures are imposed with smooth variation from winter to summer values. The salinity field is held constant because the seasonal variation of the surface salinity appears to be small.

(3) The "rigid lid" approximation is imposed as is quite common in ocean

models (O'Brien's is an exception). This approximation is achieved by requiring that the vertical velocity at the surface be zero. The effects of real variations of sea surface elevations appear as the equivalent pressure distribution on the rigid, level, upper surface of the model. The purpose of this approximation is to allow longer time steps to be used; with a rigid upper surface, surface gravity waves which have rapid propagation speeds are not allowed. Rossby waves in which the variation of the Coriolis force with latitude provides the restoring force (see Chapter 12) are still permitted but are modified somewhat by the presence of a rigid lid instead of a free surface. It is believed that this modification produces no important errors. It may modify the initial "spin up" processes but probably produces no serious errors in the final quasi-steady state.

(4) The model is started from a state of rest, the initial salinity and temperature values are taken to be horizontally uniform but with a vertical distribution based on observed data. On open boundaries, salinity and temperature are specified at all depths using observed data. The barotropic (i.e. depth independent) and baroclinic (depth varying) flows are calculated separately subject to no net vertically integrated flow through the open boundary to the south.

(5) The friction and diffusion of heat and salt are parameterized with constant eddy viscosity and diffusivity. A_z, eddy viscosity for vertical friction, $= 10^{-2}$ m² s⁻¹ for the top layer (50 m) and 10^{-4} m² s⁻¹ for the remaining layers. A_H, eddy viscosity for horizontal friction, is large at first, 2×10^5 m² s⁻¹, but is finally reduced to 5×10^3 m² s⁻¹ as the grid is refined (as described below). Vertical eddy diffusivity is taken to be constant at 10^{-4} m² s⁻¹ throughout; horizontal eddy diffusivity is initially 10^4 m² s⁻¹ and finally 5×10^3 m² s⁻¹.

(6) The density field is not allowed to be statically unstable. Whenever such a state is predicted in a new time step it is assumed that vertical mixing occurs immediately to produce a neutral density structure with heat and salt conserved. This mixing hypothesis is commonly used to avoid the problem of static instability in numerical models of the ocean (and of the atmosphere).

(7) Because of the long response time of the density field in the interior (of the order of 200 years) the calculations are done in three stages. First, bottom topography is ignored and computations are carried out for a 4° grid. The integration proceeds for 130 years of model time. (It takes 0.2 hour of computer time per year of model ocean time.) In the next stage the grid is reduced to 2° and the integration proceeds using the values from the first stage as a starting point. Bottom topography is included in this stage and the integration proceeds from 130 to 185 years in the model (at 1.7 hours of computer time per year of model time). In the third and final stage which proceeds using the second-stage result as a starting point the grid is reduced

to 1°, the surface layer is split into two layers and integration from 185 to 192 years is done (at 22 h/year). Thus the total computation requires about 270 hours of computer time on a Univac 1108. Refinement to a 0.5° grid would require 200 to 300 hours of computer time per year of model time. Faster computers have now become available but at the time when Cox did his work, resolution better than the 1° grid of his final stage was clearly impractical.

The general features of the solution for the final year of the computation agree fairly well with the rather sparse observations. A Somali Current (western boundary) appears in the model and shows the proper seasonal variations. Figure 11.2 shows the longshore component of the velocity in vertical sections normal to the Somali coast. The Somali Current is well developed in the northern hemisphere summer solution (southwest monsoon) and absent in the winter (northeast monsoon). One cannot compare the calculated transport in the Somali Current with an observed value because of insufficient field data. This example illustrates the serious and typical problem of insufficient verification data.

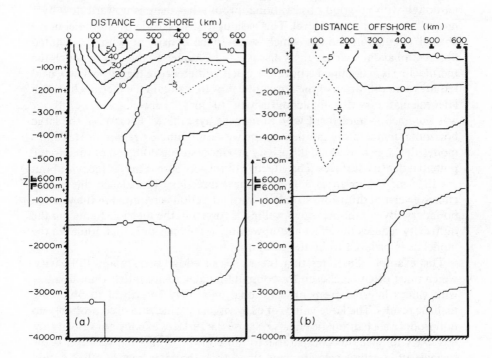

FIG. 11.2 *Cross-sections showing the long-shore velocity component* $(cm\,s^{-1})$ *near the Somali coast in an Indian Ocean model: (a) during the southwest monsoon (May–September), (b) during the northeast monsoon (November–March). (From Cox, 1970.)*

Although the transports and general features of the density field look reasonable, the computed velocities in the Somali Current are too small and the horizontal density gradients are not as large as in nature. These discrepancies are probably due to using eddy viscosity and diffusivity values that are larger than they should be, causing the boundary current to be "smeared out". To see if these effects are serious in the sense that the solution is not correct to the first order (that is it is not a smoothed version of what one would obtain with finer resolution but something quite different because, for example, of suppression of non-linear terms) the calculation would have to be continued at finer resolution. With the newer, faster computers that have now come into use, such calculations would be feasible.

While the seasonal response seems to have been correctly reproduced here, the question of whether or not models with large eddy viscosities and diffusivities produce results correct to first order is far from simple. For example, in a model of the North Atlantic done by Holland which uses the same value for the horizontal eddy diffusivity for heat as the Cox model, the vertical circulation is quite different from what we expect. The model produces *downwelling* in the vicinity of the main thermocline, but it is generally believed (although with no good observational proof) that there is upward flow over most of the ocean at this level. This assumption was made by Stommel in his abyssal circulation models which seem to be consistent with the limited observational data, although as discussed in Chapter 10 the way in which the thermocline is maintained is not clear. Such a difference in the vertical flow field requires a modification of the horizontal flow to maintain the thermocline. In a later calculation with a smaller diffusivity (10^3 m^{2}s^{-1}) upwelling in the interior was obtained in agreement with Stommel's hypotheses. Because of the large horizontal temperature gradient in the western boundary region in Holland's model, the large horizontal diffusivity produces strong diffusion normal to the potential density surfaces. The effective diffusivity across the surfaces is about 5×10^{-3} m^2 s^{-1} in an essentially vertical direction. To balance this strong effective vertical diffusion, strong upward advection is required in the western boundary. By continuity, downwelling is forced in the interior. Reducing the diffusivity reduces the effect and upwelling in the interior is then found in the model as is believed to be the case in nature.

This example illustrates that the values of eddy viscosity and diffusivity, which must often be chosen for computational stability rather than to agree with values inferred from observational data, may be critical to obtaining realistic results. The large values of eddy viscosity which are used probably do not produce such dramatic effects because the surfaces of constant velocity are less likely to have tilts comparable to those of the isothermal surfaces. If the surfaces of constant velocity were tilted then the same sort of effect would occur and the effective vertical diffusion of momentum might be stronger than that in reality, leading to greater coupling of the layers in the vertical than is realistic. To answer such questions requires running the models with finer resolution and lower viscosity and diffusivity to see if the results are different from those from the coarser resolution models.

11.43 *Holland and Hirschman's model of the Atlantic Ocean* (1972)

The computational scheme used in this model is similar to that of the Cox (1970) model of the Indian Ocean. However, unlike the Cox model it is what is called a *diagnostic* model rather than a *predictive* or *prognostic* model. The density field is prescribed, that is it is based on observed data and is held fixed rather than predicted as part of the calculation. The North Atlantic is chosen because it is the area of the world ocean with the best observed density data, and also the best knowledge of the circulation, although our knowledge is still very limited, even here. The great advantage of the diagnostic model is that it takes only a small fraction of the computer time of a prognostic model. The model ocean reaches its state of statistical equilibrium in about a month instead of several hundred years. Thus, more parameters can be varied and finer resolution can be used than in a prognostic model.

The features incorporated in the model are:

(1) Horizontal resolution of $1°$ in latitude and longitude and 14 levels in the vertical.
(2) The basin extends from $11.5°S$ to $50.5°N$, thus including the equatorial region to test the model there.
(3) Climatological density data are used, smoothed to the $1°$ grid.
(4) Climatological wind-driving, based on the Hellerman wind stress.
(5) Vertical eddy viscosity is taken to be $10^{-4}\,m^2\,s^{-1}$; some tests have indicated that this value is not critical in a diagnostic calculation.
(6) The horizontal eddy viscosity is $4 \times 10^4\,m^2\,s^{-1}$. (Some calculations were attempted with the smaller value of $10^4\,m^2\,s^{-1}$ but eddies with size comparable to the grid spacing began to appear. To prevent accumulation of energy in eddies of the smallest size that can be resolved, which could lead to non-linear computational instability, the larger value of eddy viscosity is required.)

This model seems to produce the main features of the circulation rather well. However, the surface currents of the equatorial region and the region south of the equator appear rather more broken up than observations suggest. This characteristic of the model is perhaps due to insufficient resolution and to inadequate density data. Transport through the Florida Straits is much smaller than observations indicate because of the limited resolution. The Gulf Stream gyre, both the surface circulation and the transport (maximum value 81 Sv), looks reasonably correct although the Gulf Stream is broadened and the velocity values are low due to the large eddy viscosity. Recent observations based on direct measurements suggest maximum transports of 100 to 150 Sv so the calculated value is still below what the observations indicate although it is over twice that of the classical linear theory of Munk.

The calculated Gulf Stream transport is much larger than the Sverdrup transport due to the wind (calculated from the simplified Sverdrup equation (9.21)). This transport would be about 25 Sv, similar to but somewhat smaller than the value obtained by Munk. It is, of course, based on the Hellerman wind stress and may be somewhat lower than it should be, but even doubling the value, which is probably the outer limit, would not produce a balance between the Sverdrup transport and the Gulf Stream transport.

The reason for the enhancement of the Gulf Stream in this model has been quite clearly demonstrated by doing computations with simplified versions of the model. Figure 11.3 shows the transport stream function for three cases. Figure 11.3(a) shows the case in which the density is taken to be constant, all other features of the model being unchanged. Thus, this is the barotropic case but with the real bottom topography within the resolution limits of the model. The Gulf Stream transport for this case is only 14 Sv. The bottom topography has reduced the flow below what one would expect using the simplified Sverdrup equation (9.21). As discussed in Chapter 9, when the current extends to the bottom and the bottom is not level, an additional term has to be added to the simplified Sverdrup equation. As is to be expected, topographic steering of the currents is very important and is probably much stronger than in reality. In Fig. 11.3(b) the transport pattern is shown for the case in which the observed density field is used but the bottom is assumed to be level at a depth of 1273 m. The transport pattern is quite smooth and much like the results of the classical Munk theory except for a little distortion because of the more realistic geometry. This result could have been anticipated since the model is now essentially the same as the Munk model except for the more realistic coastlines. The transport is virtually the same as that calculated using the simplified Sverdrup equation for the interior transport as expected. Figure 11.3(c) shows the transport stream function for the case in which all the features of the model are included. The transport pattern has become quite complicated and, in addition, the maximum transport in the Gulf Stream has increased to 81 Sv as noted earlier. This enchancement of the Gulf Stream is produced by what has been called the joint effect of baroclinicity and bottom topography. Baroclinicity and thermohaline driving are required to produce deep currents that interact with the bottom topography through the term involving the deep current and the bottom slopes in the case of a non-level bottom. A purely barotropic case is not correct because topographic effects reach to the surface instead of modifying the deep flow, which then interacts with the upper layer. The interaction of the deep currents with the sloping bottom produces pressure torques which can accelerate the flow and put vorticity into the flow just as the wind-stress curl does. In the model, this effect can be larger than the wind-stress

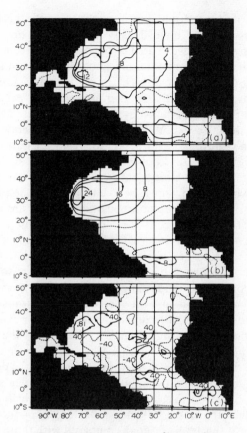

FIG. 11.3   *Transport stream functions (in sverdrups) for the Atlantic Ocean from diagnostic calculations based on the observed density field: (a) for uniform density, (b) for the observed density field but uniform depth of 1273 m, (c) realistic case of observed density field and bottom topography, allowing bottom pressure torques. (From Holland and Hirschman, 1972.)*

curl in some regions, particularly in the Western Atlantic. This effect has also been demonstrated in a mechanistic model with idealized geometry by Holland (1973), and in a diagnostic model of the North Atlantic done by Sarkisian and Ivanov (1971). There is an analogue of this effect in the atmospheric flow, where the presence of mountains can, through pressure torques, add vorticity to the flow.

Unfortunately it is not certain how important this effect is in the real ocean. The possible errors in the input-density data prevent the diagnostic calculations from being conclusive. These errors arise both from instrumentation limits and from the difficulty of obtaining good values of a time-varying field. It seems that there is enough evidence for the possible importance of this mechanism to justify a carefully planned observational experiment.

The transport pattern produced by the full model is rather complicated, much more so than we expect from the classical pictures of Chapter 9. However, we don't know, in fact, whether the real transport pattern is like the one of the model or not. Our knowledge of the large-scale flow of the ocean below the surface is based almost entirely on indirect evidence either from trying to guess what the flow must be like to produce the observed property distributions, or from geostrophic calculations which are always uncertain because of lack of knowledge of the level of no motion. The only direct observations of currents that we have for large regions are of the surface drift. The surface-drift values are based on ship's navigation records and any individual observations may have quite a lot of error, so a great deal of averaging and consequent smoothing is done in the process of producing charts of these currents. In spite of the complicated transport pattern, the results of the Holland and Hirschman model are not inconsistent with our limited observational knowledge. Figure 11.4(a) shows the surface-pressure distribution (as p/ρ_0) for the full model. The model is a rigid-lid one, so this surface-pressure distribution if divided by g would be the equivalent surface elevation. Although it is slightly smoother it shows strong similarities to the dynamic height based on a 1000-m reference level, which is shown in Fig. 11.4(b). Although in the model 1000 m is not a level of no motion the dynamic height variations at this level are small compared with those of the surface, so the model pattern of equivalent surface elevation is very similar to that of the dynamic height computation. It is worth noting that the dynamic topography is not as smooth and simple as the classical stream function pictures that we showed in Chapter 9. Except near the equator the flow is nearly geostrophic and so is more or less along the contours and the Gulf Stream is quite evident in both Figs. 11.4(a) and 11.4(b). Finally, Fig. 11.4(c) shows the surface current calculated from the model and except for the equatorial region it is fairly similar to the currents shown on Pilot Charts based on ship's records.

The large eddy viscosity (near the upper end of the likely range) poses a problem in a model such as this. Non-linear effects are undoubtedly suppressed. In addition, the resolution is insufficient to allow for the presence of mesoscale eddies, now known to exist in the ocean, which may be important to the dynamics of the large-scale flow. It would be interesting to know if either inertial enhancement or the effects of mesoscale eddies or some combinations are large enough to give the additional enhancement of the Gulf Stream needed to match the observations.

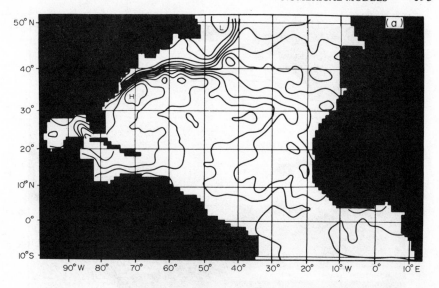

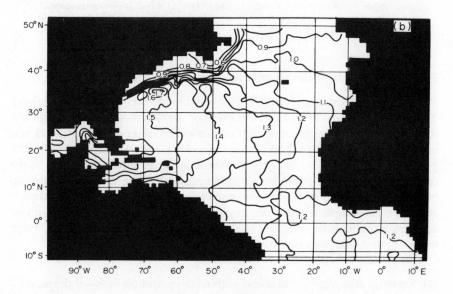

FIG. 11.4 *For the diagnostic model of the Atlantic Ocean:* (a) *the surface pressure* (p/ρ_0) *calculated from the results for Fig.* 11.3(c) *after reaching the steady state (after 35 days), contour interval* 1 $m^2 s^{-2}$ (0.1 *dyn m*), (b) *the dynamic topography based on a 1000 m level of no motion, contour interval* 1 $m^2 s^{-2}$ (*continued*).

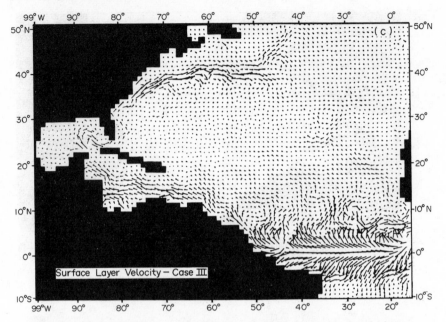

FIG. 11.4   *(continued) Diagnostic model of the Atlantic Ocean: (c) horizontal velocity vectors for the surface layer; the lengths of the vectors in the Gulf Stream and equatorial regions are limited to a maximum of 0.75 m s^{-1} in order not to obscure the smaller scale features, the largest velocity being 1.1 m s^{-1} in the Gulf Stream. (From Holland and Hirschman, 1972.)*

11.5   Two models of the circulation of the world ocean

The first, by Bryan and Cox (1972), is a constant density or barotropic model. The second, by Cox (1975), is an extension of the first model to the diagnostic baroclinic case (temperature and salinity based on observed mean values and held fixed) and finally to a short period (2.3 years of model ocean time) of a fully predictive baroclinic case using the final state of the diagnostic model for initial conditions. It was not possible to run the predictive case to final equilibrium because this calculation would have required several centuries of ocean time at 10 hours of computer time per year of ocean time. However, the initial adjustment in boundary currents to removing the imposed density field, which takes about a year, can be examined.

The models have the following features:

(1) Realistic topography and coasts within the resolution limits of the models.
(2) In the baroclinic case, up to nine levels depending on the depth.
(3) Horizontal eddy viscosity of 8×10^4 m^2 s^{-1}; with the resolution that can be used in this model, smaller values of eddy viscosity lead to small-scale computational noise.

(4) Vertical eddy viscosity of $10^{-4}\,\mathrm{m^2\,s^{-1}}$.

(5) Wind stress based on the Hellerman computation with extension to the polar regions.

(6) Resolution of $2°$ from $62°$S to $62°$N. Separate $2°$ spherical grids are used for both higher latitude regions with considerable overlap between these grids and the lower latitude grid. Each grid is integrated separately for several time steps and the boundary conditions are updated in the overlap region. This approach is required because, on a full spherical grid, the grid spacing becomes small near the poles requiring unreasonably small time steps everywhere.

(7) No conditions on open boundaries (fluid boundaries within the ocean's interior) are required as all lateral boundary openings are included in a world ocean model. Such boundary conditions are always somewhat arbitrary due to lack of adequate observations and are a problem when attempting to model a part of the world ocean.

The transport pattern for the barotropic case is shown in Fig. 11.5(a). Topographic steering effects on the current appear to be much stronger than they should be; western boundary currents such as the Kuroshio and Gulf Stream are too weak. The Antarctic Circumpolar Current is particularly weak (about 21 Sv). In a separate calculation (with the different horizontal eddy viscosity of $4 \times 10^4\,\mathrm{m^2\,s^{-1}}$) the barotropic model was run with a level bottom and the Kuroshio and the Gulf Stream then had transports similar to those calculated by Munk, as can be seen in Fig. 11.5(b). As can also be seen in this figure the Antarctic Circumpolar Current is very large (greater than 600 Sv); it was still increasing at the end of the calculation.

In the North Pacific, the level bottom case is almost equivalent to O'Brien's North Pacific model shown in Fig. 11.1, except for the difference in eddy viscosity. The difference in the strengths of the Kuroshio Current in these two models provides evidence for our earlier suggestion that inertial enhancement produces the larger transport in the O'Brien model.

The transport pattern from the baroclinic diagnostic model by Cox is shown in Fig. 11.6(a). The results appear to be much more realistic with more reasonable transport values for the western boundary currents, although they are still smaller than observed values and the currents are broader and weaker because of the limited resolution and large eddy viscosity. The Antarctic Circumpolar Current appears more reasonable too, with a transport value of 184×10^6 tonnes $\mathrm{s}^{-1} \simeq 180$ Sv, near the upper end of the range of values suggested by observations. Again, the inclusion of baroclinicity and bottom topography has enhanced the western boundary currents to levels beyond what is expected from the simplified Sverdrup equation (9.21). The enhancement (to about 60 Sv) is not as large in the Gulf Stream as found in the Holland and Hirschman's model, perhaps because of the larger viscosity and lower

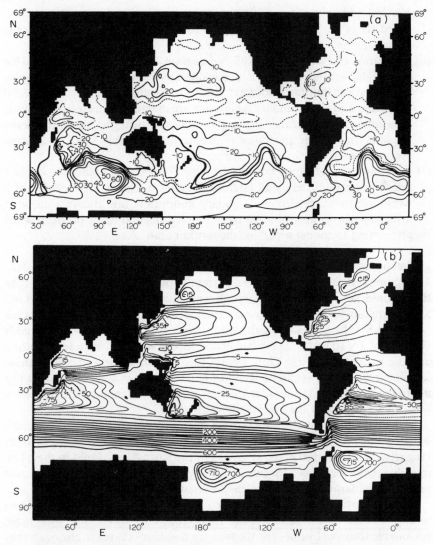

FIG. 11.5   *World ocean circulation: (a) horizontal mass transport stream function for uniform density with realistic topography (positive values indicate clockwise flow), Antarctic Circumpolar Current* $= 22 \times 10^6$ *tonnes* $s^{-1} \simeq 21\, Sv$. *(From Cox, 1975.) (b) Transport streamlines for uniform density and depth showing much larger Antarctic Circumpolar Current* ($\sim 600\, Sv$). *(From Bryan and Cox, 1972.)*

resolution. Many comparisons may be made between the model and observed features of the ocean; good agreement is obtained in some regions and not in others. The model is only a first step toward a model representative of the world ocean, but it is an important one.

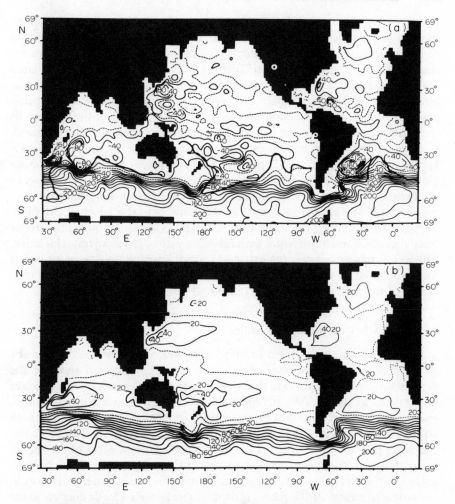

FIG. 11.6   *World ocean circulation: (a) pattern of mass transport for the diagnostic
case based on the observed density field, (b) pattern of mass transport for the predictive
case based on a 2.3-year numerical integration of a three-dimensional model using
observed density and the mass transport of Fig. 11.6(a) as initial conditions. (From Cox,
1975.)*

   The transport pattern for the predictive model is shown in Fig. 11.6(b). There
is a smoothing out of various details, probably due in part to inaccurate density
values; in addition the large eddy viscosity and diffusivity which must be used
probably lead to oversmoothing and to reduction in the transport values of the
Gulf Stream and Kuroshio. The Antarctic Circumpolar Current shows little
change; at the end of the run it is 186×10^6 tonnes s^{-1}.

One motivation for developing world ocean models is that they can then be used in joint atmosphere–ocean models to study climate and its possible changes. Development of such models is continuing with mixed-layer models, such as those described in Chapter 10, being included to improve the calculated sea surface temperature which is an essential boundary condition for the atmospheric models. Seasonal variation of the wind forcing is also being explored; it appears to be important in transferring heat from the summer to the winter hemisphere in the tropics, thus helping to moderate the climate. A brief review with many references is given by Haney (1979). Bryan and Lewis (1979) and Meehl *et al.* (1982) present results of recent attempts to produce global ocean models and, among other things, compare the heat transport of the oceans to estimates based on the differences between the total transport required and the atmospheric part of the transport. The effect of the time-dependent motions on the mean flow in these models still requires the use of eddy viscosities and diffusivities with variation with depth sometimes included. These closure schemes are somewhat *ad hoc* but hopefully can be improved as results from field observations and eddy-resolving numerical models (discussed in the next section) are incorporated into the large-scale models.

11.6   Models with mesoscale eddies

Starting in the 1960s and continuing with much increased activity in the 1970s, a considerable amount of evidence has been collected showing the existence of *mesoscale eddies* in the ocean. These eddies have characteristic sizes of the order of 100 to 500 km, time scales of one to a few months and kinetic energies between 10 and 100 times that of the mean flow in the interior. In strong currents, such as the Gulf Stream, the kinetic energy of the eddies and meanders is of the same order as the mean kinetic energy. Most of the energy associated with the mean flow is the potential energy of the tilted isobaric surfaces; this energy is of order a thousand times the kinetic energy. However, the potential energy of the eddies is comparable to their kinetic energy, so the total energy of the mean flow is larger than the total energy of the eddies even if they are found everywhere. How prevalent these eddies are and their importance to the mean flow is not well established as yet. The atmospheric analogues of the oceanic eddies are the storm systems in mid-latitudes. On a weather map they look like a series of large eddies. These eddies can be considered to be geostrophic turbulence which behaves like two-dimensional turbulence. In the atmosphere it is known that the eddies gain energy from the available mean potential energy and transfer kinetic energy and momentum to larger scales and to the mean flow. They are important in determining the strength of the westerlies and the jet stream, for example. If one tries to parameterize the effects of these eddies by an eddy viscosity in the usual simple way, the value is negative because they accelerate the flow in contrast to the more familiar smaller scale three-dimensional turbulence which acts to retard the flow. If the oceanic eddies are fairly common and have similar dynamical behaviour to their atmospheric counterparts they may be important in the dynamics of the mean flow.

The first evidence for these eddies that was taken seriously was obtained by Swallow when he first used his floats in an attempt in 1959–60 to observe the expected very slow flow in deep water. Indeed, more recent but very limited observations suggest that if he had gone further into the interior he might have found the quiet ocean at great depth which he expected—there do seem to be some quiet regions. In any case, much to his surprise he found that his floats went off in various directions at speeds at least 10 times greater than expected, so that he was unable to follow them for very long. Once one accepts the idea that eddies are a feature of the ocean, one can find evidence in the historical records. While most of the early cruises had station spacing too large to show the eddies clearly, there were some detailed observations. In these the eddies can be seen in the density field (and in the temperature field as the dominant factor determining density in the open sea at low and mid-latitudes) because the flow around the eddies is nearly geostrophic. (As in the western boundary currents and atmospheric weather systems, the Coriolis force associated with the along-stream component dominates the cross-stream momentum equation.) In many ways it is curious that oceanographers did not consider the possible importance of eddies and look for them sooner. Everyone is well aware of the meteorological analogue—the weather may easily make a pleasant climate unbearable. Oceanographers continued to try to find and understand the ocean climate (the mean state) perhaps rather longer than they should have. Eddy effects were misinterpreted as internal wave noise or observation errors or were just ignored. Of course, hindsight is better than foresight. Also, until about 1960 it was not possible to make extensive direct measurements of the currents in the open ocean.

Intensive observational efforts to look for eddies and to determine their characteristics were made in the 1970s, e.g. the U.S.S.R. POLYGON experiment (Brekhovskikh *et al.*, 1971), the U.S.A.–U.K. MODE (Mid-Ocean Dynamics Experiment, the MODE Group, 1978) and the international POLYMODE experiment. (Some early results and references for this experiment may be found in the MODE Group article which also reviews the theoretical work related to these experiments.) These experiments were conducted in the North Atlantic but evidence of eddy features has been found in many other parts of the world ocean, mainly from their signature in the temperature field using closely spaced XBT (expendable bathythermograph = temperature/depth recorder) temperature profile observations, although there are direct current observations also. References to such observations are given in the review by McWilliams (1979); this review also gives references to the theoretical work on these eddies. Efforts have also been made to get a picture of the overall distribution of eddy activity. Emery (1982) gives results for the North Atlantic and North Pacific of dynamic height variability, an indication of eddy kinetic energy, and estimates of eddy potential energy based on temperature variations at 300 m depth. He also gives results of several other investigations for comparison. Note that in these overviews, the energy of the fluctuations includes other motions than closed mesoscale eddies and closed circulations such as the Gulf Stream rings produced when large meanders pinch off—see *Descriptive Physical Oceanography*, Pickard and Emery (1982) for a description of this process. The energy of meanders is included and for some types of calculations some of the energy is not from the mesoscale frequency range, e.g. internal waves will make some contribution to isotherm displacements. Also there is quite a lot of noise in some of the observations used, e.g. ship's drift data used to estimate mean and fluctuating kinetic energy of the surface flow.

Based on observations and on numerical models which contain mesoscale eddies (to be described shortly) the strongest eddy activity is found in the regions of strong currents but there is evidence of some eddy activity virtually everywhere. Since eddy activity is high in strong current regions they are likely source regions. In mid-ocean regions, variations in the wind stress may contribute to the eddy activity but this contribution is thought to be a small fraction of the total.

Up to now it has not been possible to use fine enough resolution in a simulation model to allow for the presence of mesoscale eddies. However, mechanistic models with simplified geometry of sufficient resolution have been examined and have shown that the mesoscale eddies may produce important effects on the general circulation. For example, they can extract energy from the potential energy of the mean flow of the upper-layer wind-driven circulation, transfer it to eddies and mean circulation in the deep layer of the ocean, and in this way increase the total transport of the system.

The instability mechanism which allows these eddies to grow with their energy supply coming directly from the mean potential energy field is called *baroclinic instability*. Baroclinic flow with vertical shear is required for this mechanism to be possible. It is different from dynamic instability (Chap. 7) in which eddies may be generated by shear in the mean flow and gain their energy from the mean flow kinetic energy. In contrast to baroclinic instability, this shear instability is called *barotropic*; as there is no vertical shear, horizontal shear is required for disturbances to grow. Both of these instability mechanisms are important in the atmosphere; their relative importance in the ocean has not been established but as eddies exist both types of instability are likely sources. Both types are found in the numerical models which generate mesoscale eddies.

Rotation, while it does not directly affect stability, plays an indirect role. Because of the importance of the Coriolis term much larger isobar slopes are required than in the non-rotating case so that the pressure gradient can balance most of the Coriolis term. The larger isobaric slopes lead to a much larger mean potential energy which is a much bigger possible energy source for instability than that available in the non-rotating case.

Theoretical work on mesoscale eddies and their possible role in the general circulation of the oceans has been extensive in the last decade and no doubt will continue. Much of this effort has involved mechanistic numerical models with sufficient resolution to include mesoscale eddies explicitly. Such models have played an important role in designing the observational programmes mentioned earlier in this section. Work on such models has proceeded along two lines. One is with what are termed *process models* which are attempts to understand the dynamics and evolution of the eddies themselves. The geometry in these models is usually quite simple, e.g. an isolated region with open boundaries in mid-ocean or an open-ended channel. Topography may be included to explore its effects. Many references to this type of work may be found in the reviews cited before (McWilliams, 1979; the MODE Group, 1978); more recent work may be found in the literature. The other approach is to examine the role of eddies in the general circulation in what are termed EGCMs (Eddy-resolving General Circulation Models). The first model of this type was by Holland and Lin (1975). References to more recent eddy-resolving models may be found in Schmitz and Holland (1982). In the present models, the geometry is simple (square or rectangular) and the forcing is simple (sinusoidal east–west wind stress and approximately linear variation with latitude of surface heating in the few cases in which thermal forcing is included). The bottom is usually taken to be flat but a simple continental slope has been included in some cases or small-scale bottom roughness added in others. The model domains are small relative to the actual oceans because of computer speed limitations with fine resolution. The largest horizontal size seems to be that reported by Schmitz and Holland which is 4000 km by 4000 km and with the wind forcing chosen to give a double gyre pattern in the wind-driven upper layer. Free-slip boundary conditions must be used as otherwise eddies do not occur. This limitation is due to the present restriction on resolution; friction cannot be made small enough to maintain numerical stability and get eddies too with no-slip boundary conditions.

Friction associated with constant eddy viscosity has the form $A_H \nabla_H^2 u$ and $A_H \nabla_H^2 v$ in the horizontal momentum equations. As noted before in this chapter and in Chapter 7, it is a somewhat *ad hoc* approach and an interim measure. In some models this usual formulation is replaced with $A_4 \nabla^4 u$ and $A_4 \nabla^4 v$. Because higher-order derivatives are used, this type of friction acts on the small scales much more strongly than on the larger scales so that frictional effects on the mesoscale eddies and the large-scale flow can be much smaller while still maintaining computational stability. This type of friction is called *biharmonic friction* while the constant eddy viscosity friction is termed *Laplacian*. If an equation for the stream function is derived (e.g. Munk's equation, Chapter 9) then biharmonic friction takes the form $A_4 \nabla^6 \psi$ ($= A_4 [\partial^6 \psi / \partial x^6 + 3 \partial^6 \psi / \partial x^4 \partial y^2 + 3 \partial^6 \psi / \partial x^2 \partial y^4 + \partial^6 \psi / \partial y^6]$). Friction involving vertical shear, when included, is taken to be of the usual eddy viscosity form (e.g. $A_z \partial^2 u / \partial z^2$, etc.). In some models, bottom friction is included to explore its role in the overall balance; indications from EGCMs are that it may be very important, contrary to past opinions.

Models using the momentum equations and the continuity equation (and equations for heat and salt conservation along with an equation of state when necessary) to solve for u, v, w, etc., are termed *primitive equation* models. With fine resolution these require long calculations which limits the size of the domain, the resolution and the parameter variations which can be explored. To allow faster calculations, a factor of 10 or more, the quasi-geostrophic approximation has been used in EGCM models (and process models). Even in eddies and strong currents, the flow is nearly geostrophic. Thus if we can obtain a solution for the geostrophic part of the flow it should be the same as the total flow within a few percent. To zero order, the horizontal momentum equations (6.2) are

$$f_0 v_0 = \alpha \frac{\partial p_0}{\partial x}; \qquad f_0 u_0 = -\alpha \frac{\partial p_0}{\partial y}$$

where f_0 is the constant value of f in the β-plane approximation ($f = f_0 + \beta y$). Taking $\alpha =$ constant within individual layers in a multilayer representation of the ocean, $-\alpha p_0 / f_0$ is a stream function for the geostrophic part of the velocity (u_0, v_0). By cross-differentiating the momentum equations we can obtain equations for this stream function (like the Munk equation of Chapter 9). In these, terms which are of first order in the Rossby and Ekman numbers are retained and terms of second order in the Rossby number are neglected. These stream-function equations are forms of the vorticity equation and can be solved more rapidly than the primitive equations. They allow larger domains and more parameter variations. A few comparisons of primitive and quasi-geostrophic equation models show quite good agreement. Primitive equation models show that the quasi-geostrophic approximation is not good in limited regions—parts of the strong currents of the model analogues of currents such as the Gulf Stream. Thus quasi-geostrophic models, after they have been used to find the parameters that give the most "oceanic" results, should probably be checked with primitive equation models.

Clearly EGCMs are at an early stage of their development. The details of the results should not be taken too seriously at present but the general features probably have relevance. Holland (1978) summarizes the general features on which the following is based in part. Suppose, as indicated by scaling arguments of the type used in Chapter 7 and at the end of Chapter 8, that frictional effects in the ocean are small for the large-scale flow and the eddies, except for Ekman layers at the top and bottom. Wind and thermohaline driving then accelerates the upper layers until critical values of horizontal and/or vertical shear are reached. Eddies then form due to baroclinic or barotropic instability (which type of instability dominates varies with the model). In the real oceanic flows, both instabilities are possible and with present data one cannot tell which may dominate. The dominance may vary regionally as it does in the models. Kinetic energy is transferred downward to both the eddies and to the mean gyres in the deep ocean. This transfer can occur only if eddies are generated in the upper ocean, although once eddies are present some of this downward transfer is directly from kinetic energy of the mean upper layer flow to kinetic energy of the mean lower layer flow. The bottom Ekman layer, i.e. bottom friction, is the main cause of energy loss. The eddies limit the flow in the upper layer but then cause deep mean circulations and eddies. While the deep currents are weaker, because of their greater depths the transports are comparable to those of the upper layers and the analogue Gulf Stream transport can be double that of the part in the upper layer. The extra transport in the deep layer and the extra transport in the upper layer associated with inertial enhancement recirculate in the boundary current regions (western boundaries) or in the region where the Gulf Stream analogue penetrates the ocean as an eastern jet. Probably because of the geometry, the real Gulf Stream enters the ocean as a northeastward jet; recent analyses of field data, e.g. Worthington (1976), also indicate large recirculations near the Gulf Stream.

Schmitz and Holland (1982) give comparisons between the results of some of the runs in Holland (1978) and extensions to larger domains and field observations mainly from MODE and POLYMODE. Comparisons of energy levels and spectra (distributions of energy with frequency) are reasonable although the results must be geometrically scaled because the eastward jet in the model does not penetrate as far into the ocean as the real Gulf Stream. As the authors note, the comparison is preliminary both because of the simplicity of the model and because of the sparseness of the field observations (in spite of a decade of effort). The comparison was made to guide both the development of the models and the conduct of future field observation programmes and is a very valuable contribution.

Extension of such models to more layers and realistic geometry, i.e. an eddy resolving simulation model of the North Atlantic, is in progress. With such simulations and "tuning" of the models to match field observations we should see considerable progress in ocean general circulation models in the next decade. Once the EGCMs are well developed, parameterization schemes for the effects of the mesoscale eddies based on them can be developed for coarser resolution world ocean models. The EGCMs in mechanistic form indicate that the heat transport associated with eddies behaves approximately as eddy diffusivity models predict although this result needs verification in more detailed models. Negative viscosity effects at the eastward end of the Gulf

Stream are found so momentum parameterizations may need more complicated treatment.

11.7   Comments on the numerical model solutions

Many models use the Hellerman wind-stress calculation although it is not the best calculation that could be done, as discussed earlier in this chapter. Nevertheless, using the same stress values for different models is useful for comparison and the inaccuracy is probably not too critical at the present stage of modelling. However, one cannot make very good quantitative arguments about the accuracy of the Sverdrup equation (9.21) using this computation although recent assessments for the Atlantic suggest that Hellerman's 1967 values are not too bad. He is presently redoing his stress computation with more extensive data and in a form which will allow the drag coefficient's variation with wind speed and stability to be updated as more results become available.

Some modelling to test the sensitivity of the results to variations in the stress input would be most helpful in determining how well this forcing function needs to be known. In diagnostic models this stress function seems to be almost immaterial; the information is in the imposed density field that the wind is partially responsible for setting up. For example, in the model of Holland and Hirschman, putting the wind stress to zero only lowered the Gulf Stream transport by 5%.

In principle, the advantage of the numerical modelling approach is that all terms in the equation of motion can be included and the topography and coastline can be realistic. At present, at least in the simulation models, the friction seems too large to allow the non-linear terms to play a realistic role. Exploration of possible parameter ranges has also been limited by the speed of available computers. If the mesoscale eddies are as important in the dynamics of the ocean as they are in the atmosphere, then they present a serious resolution problem. It is known in atmospheric models that 250 km resolution gives rather better results than 500 km resolution. Then all scales that make important contributions to the total kinetic energy, including the storms or mesoscale eddies, are fairly well resolved. The scale of these eddies seems to be proportional to the internal Rossby radius of deformation, $\lambda_i = [g(\Delta\rho/\rho) D]^{1/2}/f$ (see Section 9.14.2), where $\Delta\rho$ is the density difference between the two main layers of the fluid, D is the thickness of the layer, and f is the Coriolis parameter. For the atmosphere, λ_i is of the order of 1000 km. For the ocean, λ_i is of the order of 100 km because both $\Delta\rho/\rho$ and D are smaller. Thus, it appears that a resolution of order 25 km is required in an ocean model to resolve all energetic scales. To model the world ocean with such resolution is impossible at present and for the near future at least. However, modelling of limited regions is being done (although the region cannot be too limited because the results may then depend mainly on the poorly known conditions on open boundaries). All these problems are related to the limitations of the size and speed of available computers; however, the new faster machines that have now become available and new numerical techniques that are being developed should help to overcome them.

As noted at the beginning of this chapter, there appears to be a discrepancy between the observations of the Gulf Stream and Kuroshio transports and the predictions of the linear wind-driven theory presented in Chapter 9. Simple linear addition of the thermohaline circulation discussed in Chapter 10 probably cannot resolve the "discrepancy" but plays a role. While the numerical models discussed in this chapter do not provide the final answer, they have suggested a number of mechanisms which might explain the "discrepancy". The three possibilities are: inertial enhancement, bottom topography–baroclinicity effects and mesoscale eddies. All of these mechanisms enhance the Gulf Stream or Kuroshio but are cancelled by recirculation adjacent to these currents so that most of the interior has a nearly Sverdrup balance.

Finally we offer a cautionary note about this "discrepancy". Our observations are still very limited and have considerable uncertainties. Some recent tests of the simplified Sverdrup relation (equation (9.21)) in the interior, with the wind-stress curl calculated with more up-to-date drag coefficients and better spatial resolution, seem to give good results as discussed in Chapter 9. The flow was calculated by the geostrophic method and assumed to be baroclinic. (In the interior, if the bottom is level the generalized Sverdrup relation tells us that the net north–south transport is given by curl τ_η. Any additional thermohaline flow must have zero net transport over the whole depth—the poleward deep flow is exactly balanced by the equatorward flow in the upper layer. Of course the bottom of the ocean is not level and indeed is quite rough in places (e.g. the Mid-Atlantic Ridge) so the baroclinicity–bottom topography effect may play a role. In this case, separation into interior and western boundary regions may be difficult.) The calculated geostrophic transport agreed very well with the transport calculated from the wind-stress curl. When extrapolated for the full width of the ocean, the sections being taken well within the interior to avoid lateral boundary effects, the calculated transports also agreed very well with the transports in the Florida Straits which are 30–35 Sv; it is only much further downstream that the very large Gulf Stream transports (in excess of 100 Sv) have been observed. This test of the Sverdrup relation was done where it ought to be done, in the interior, as was the fairly successful test in the eastern equatorial Pacific described in Chapter 9. Some direct check on the geostrophic calculation is really needed also, of course, but the test described above is much better than trying to compare the interior Sverdrup transport to the Gulf Stream transport which may not be sufficiently well observed. With the possibilities of enhancement due to inertial effects, mesoscale eddies, baroclinicity–bottom topography interactions and perhaps other effects, the western boundary region is rather more extensive than the region of the northeastward-flowing Gulf Stream or Kuroshio. The observed transports for these currents (directly observed or geostrophically calculated with some direct current observations to fix the level of no motion between hydrographic stations) may be based on observations which do not extend far enough offshore to detect all of the counter-flows which *must* be subtracted before comparison with the interior transport calculated from the wind-stress curl using the Sverdrup relation (9.21). There is evidence for a strong *sub-surface* countercurrent offshore of the Gulf Stream. This and other counter-flows have not been well observed. Thus the "discrepancy" between the net western boundary region transport and the Sverdrup transport may not be as large as thought previously.

12

Waves

12.1 Introduction

The word *waves* usually brings to mind a picture of undulations on the surface of the sea or a lake, often with some semblance of regularity, and usually progressing from a region of formation to a coast where they are generally dissipated as surf or may, in part, be reflected. Less evident are the related movements of the water below the surface, waves which are entirely beneath the air/water surface (internal waves) and a variety of waves which are usually not evident by visual observations. The main classes of waves and their causes are:

(1) *ripples, wind waves* and *swell*—due to the effects of the wind on the air/water interface;

(2) *internal waves*—which may occur when vertical density variations are present—various causes, e.g. current shear, surface disturbances;

(3) *tsunamis*—generated by seismic disturbances of the sea bottom or shore;

(4) *gyroscopic-gravity waves*—(surface and internal) of sufficiently long period that the Coriolis effect is important—various causes, e.g. wind stress changes, atmospheric pressure changes;

(5) *Rossby* or *planetary waves*—large-scale and long period, evident as time varying currents—various causes, e.g. time variations in wind stress and perhaps baroclinic and/or barotropic instability (mentioned in Chapter 11);

(6) *tides*—due to fluctuating gravitational forces of the moon and sun.

In this chapter we shall discuss some of the characteristics of the first three classes above, describe the fourth and fifth classes briefly and leave the discussion of tides to Chapter 13. Our material is mainly descriptive and presents, for the non-specialist, the terminology used by physical oceanographers and some of the results of wave studies. For physical oceanographers, a more extensive and detailed treatment such as that by LeBlond and Mysak (1978) is needed.

As mentioned previously, sound (acoustic) waves are also possible in the ocean and have many applications. These include the determination of depth (echo sounders) and of sub-bottom structure, detecting objects in the body of the ocean (e.g. fish, submarines), transmitting information (e.g. from sub-surface instruments or drifters), determining ship speed relative to the bottom and current speeds using the Doppler effect from sound reflected from objects in the sea, determining the wind speed from the ambient noise and, more recently, determining the distribution of water properties from the time of travel of sound pulses (acoustic tomography). We shall not discuss sound waves explicitly but note that much of the terminology and of the physics presented here would apply to them. More information on ocean acoustics may be found in the text by Urick (1975) or those by Officer (1958) or Tolstoy and Clay (1966) listed in the Bibliography, although the reader is warned that the latter two texts are rather detailed and mathematical.

The classical approach to the study of waves is to consider the fluid dynamics of ideal waves which, in side view, have a sinusoidal shape (i.e. sine or cosine curves) and to progress to other regular surface shapes. This approach gives a great deal of information about the relations between the surface shape, the progress of the waves and the motions of the water below the surface. The least satisfactory feature of this approach is that the ideal regular waves thus studied bear only a limited resemblance to real waves observed at sea which are characterized by their irregularity in form and period.

A more recent and pragmatic approach is to start from observations of the shape of the irregular sea surface, regard it as a composite of a wide range of possible ideal components and carry out a spectral analysis to determine the characteristics of the spectrum of components. For example, if we have observations of the surface elevation, η, at a point for a period of time, we can consider this record to be the sum of sine or cosine waves of different amplitudes, phases and frequencies. Spectral analysis then consists of finding the amplitudes and phases as a function of frequency. A plot of amplitude squared (proportional to the wave energy) *versus* frequency is called a *wave energy spectrum*. Figure 12.1 shows the range of frequencies and periods observed for the ocean and gives a rough idea of the spectrum of energy associated. Over most of the frequency range, the spectrum is continuous; only at tidal frequencies is there a significant line spectrum, i.e. energy peaks at specific frequencies. For comparison, the potential energy (associated with the mass distribution) of the large-scale mean circulation of the world ocean is ~ 3 times the total energy of all the time-dependent motions together, while the kinetic energy of the mean circulation is only of 0(0.001) times the potential energy. For the time-dependent motions, the potential energy is approximately equal to the kinetic energy.

From the spectrum and the classical theory for each component we should be able, in principle, to calculate the total effect of the wave field by summing over all the components using the appropriate amplitudes and phases. To get a complete picture, the direction of travel should be considered as well and a

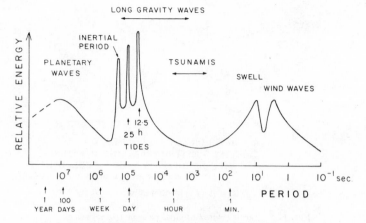

FIG. 12.1 *Schematic energy spectrum of oceanic variability, showing approximate relative energy levels. The area under the curve is roughly proportional to the contribution to the total energy of the time-dependent motions. Amplitudes for various parts of the spectrum reflect order of magnitude estimates and are most uncertain for the planetary waves whose average energy levels are least well established.*

spectrum including the direction of travel information is called a *directional spectrum*. This *statistical approach* is required in applications of our knowledge of waves, e.g. to determine the total effects of waves on ships and engineering structures near the sea surface.

Here we will start with the classical approach in which the wave shape is assumed to be a sine or cosine wave.

12.2 Some general characteristics of waves

Assuming that the waves on the sea surface are simple cosine waves (in vertical section) some terms which we shall use are illustrated in Fig. 12.2. The quantity H, called the *height* of a wave (the vertical distance from trough to crest), is twice what the physicist calls the *amplitude A* (the maximum displacement) of the vertical oscillatory motion of the surface above or below the mean water level. For all types of waves, the *speed* at which a particular part of the wave passes a fixed point is $C = \Lambda/T$ where Λ is the *wavelength* (the distance from crest to crest or trough to trough) and T is the *period* (the time between two successive crests or troughs passing a fixed point). The height H is basically independent of C, Λ or T but is limited by *breaking* which occurs when the non-linear terms become important. (The symbol C comes from the alternative word "celerity" used in the older literature.)

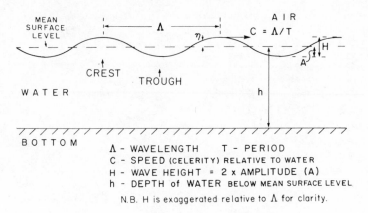

FIG. 12.2 *Terms related to ideal (sine or cosine) waves.*

To simplify the notation, the *radian frequency* ($\omega = 2\pi/T$) and the *radian wave number* ($k = 2\pi/\Lambda$) are often used. The vector wavenumber **k** has magnitude k and points in the direction of propagation of the waves, e.g. perpendicular to the crests of surface gravity waves. Then $C = \Lambda/T = \omega/k$.

For convenience in referring to them, it is common to classify surface waves according to their periods as in Table 12.1. In all these surface waves, gravity is the primary restoring force, allowing oscillations to occur. (If some water is lifted up and allowed to fall back under the action of gravity its inertia will cause it to overshoot the equilibrium position; pressure forces will then push it back up and oscillations will ensue.) Ripples of wavelength less than about 5 cm are also affected by surface tension; these waves are of very small amplitude and will not be discussed here, although it is considered that they may play a role in determining the drag of wind on water. For surface waves of periods of several hours or more, the Coriolis force must also be included in the analysis as will be discussed briefly near the end of this chapter. Tides will be discussed in Chapter 13.

TABLE 12.1. *Waves classified by period*

Period	Wavelength	Name
0–0.2 s	centimetres	ripples
0.2–9 s	to about 130 m	wind waves
9–15 s	hundreds of metres	swell
15–30 s	many hundreds of metres	long swell or forerunners
0.5 min–hours	to thousands of kilometres	long period waves including tsunamis
12.5, 25 h, etc.	thousands of kilometres	tides

Wind waves are the locally generated waves. As they have a fairly wide range of directions of propagation the sea surface is quite irregular. *Swell* is the term for waves which have been generated elsewhere, it travels in one direction and is much more regular. Also, as we shall show, the longer waves travel faster than shorter ones and so at some distance from the source area, at any one time, the swell has a narrow range of frequencies which also makes it more regular than wind waves. The ranges of periods of wind waves and swell actually overlap considerably—wind waves may have periods of up to 15 seconds or so if the wind speed is very large, while swell with periods of only a few seconds is possible.

12.3 Small amplitude waves

12.31 *Pure waves (single frequency)*

The word "small" here is used in a comparative manner and refers to the "relative height" or *steepness*, H/Λ. For the simple (linear) theory to be correct within a few percent, this ratio should be less than about $1/20$ and in many cases for real waves it is $1/50$ or less. For example, for swell of $\Lambda = 200$ m, the small amplitude theory would apply up to about $H = 10$ m which would, in fact, be very high for swell of that wavelength. (For clarity in Fig. 12.2 and other figures we have exaggerated the wave height relative to the wavelength.) Here we shall consider the first-order (linear) theory, i.e. we neglect terms which are of order H/Λ (or higher powers, e.g. H^2/Λ^2) times the terms retained. (The extra terms arise from the non-linear terms in the equation of motion but the non-linear terms remain small compared with the other terms, i.e. are of higher order in H/Λ.)

For a progressive sine or cosine wave, the vertical displacement η of the free surface from the mean level is given in terms of time t and displacement x (for a wave travelling in the x-direction) by

$$\eta = A \cos\left[2\pi\left(\frac{x}{\Lambda} - \frac{t}{T}\right)\right]. \tag{12.1}$$

Using radian wavenumber and frequency, equation (12.1) can be written more compactly as

$$\eta = A \cos(kx - \omega t). \tag{12.1'}$$

The argument $(kx - \omega t)$ of the cosine function is termed the *phase* of the wave; it goes from 0 to 2π as one goes from one crest to the next (distance Λ) at a fixed time or through one period (T) at a fixed point. Also, since C is the speed at which a point of fixed phase travels it is more completely termed the *phase speed*.

For such waves it can be shown that the phase speed

$$C = \left[\frac{g\Lambda}{2\pi}\tanh\frac{2\pi h}{\Lambda}\right]^{1/2} = \left[\frac{g}{k}\tanh kh\right]^{1/2} \tag{12.2}$$

where g = acceleration due to gravity, "tanh" is the hyperbolic tangent and h is the water depth. For $\Lambda < 2h$, $\tanh 2\pi h/\Lambda = \tanh kh = 1$ within 0.5%, while for $\Lambda > 20h$, $\tanh 2\pi h/\Lambda = \tanh kh = kh$ within 3%, so that for most applications the expression for C may be simplified as follows:

(1) for $\Lambda < 2h$, called SHORT or DEEP-WATER waves ($\because h > \Lambda/2$),

$$\text{then } C_s = (g\Lambda/2\pi)^{1/2} = (g/k)^{1/2}, \tag{12.2'a}$$

(2) for $\Lambda > 20h$, called LONG or SHALLOW-WATER waves ($\because h < \Lambda/20$),

$$\text{then } C_l = (gh)^{1/2}. \tag{12.2'b}$$

Figure 12.3 illustrates these nomenclatures. .

The relationship between ω and k, or equivalently between wavelength and period, is termed the *dispersion relation* and is derived from the equation of motion (see LeBlond and Mysak, 1978, section 11). It is

$$\omega^2 = gk\tanh kh. \tag{12.3}$$

Equation (12.3) together with $C^2 = \omega^2/k^2$ are used to obtain equation (12.2) relating the phase speed and wavelength.

Figure 12.4 shows plots of equation (12.2) as speed C against water depth h for a selection of wavelengths fom 10 m to 1 km. The left-hand (straight) line is the plot of $C_l = (gh)^{1/2}$ (long-wave speed). Then the line for $\Lambda = 200$ m (for example) shows that the speed follows the long-wave line up to about 10 m water depth ($h = \Lambda/20$) where it commences to curve to the right on the figure eventually reaching its constant value of $C_s = 17.7$ m s^{-1} at about 100 m water depth ($h = \Lambda/2$). The zone in the figure to the right of the dashed line is where the short-wave speed approximation holds, and the intermediate zone between the long-wave speed line and the dashed line is where the full expression of equation (12.2) must be used to calculate the speed. In practice, the long-wave

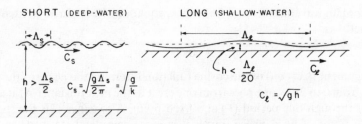

FIG. 12.3 *Properties of short and long waves.*

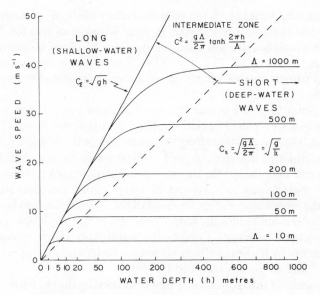

FIG. 12.4   *Wave speed versus water depth for various wavelengths.*

and short-wave approximations find most use; the intermediate zone applies chiefly in the study of the surf zone near the beach.

If we introduce the values for the constants in the two expressions (12.2′) for the wave speeds we obtain the expressions in Table 12.2 in which are also included a few numerical values for wave properties. The values for the short waves give an idea of the properties of wind waves and swell, while the first two

TABLE 12.2.   *Short and long-wave formulae and sample values*

SHORT (deep-water) WAVES	LONG (shallow-water) WAVES
$C_s = (g\Lambda_s/2\pi)^{1/2} = (g/k)^{1/2} = g/\omega$	$C_l = (gh)^{1/2}$
$C_s = 1.56T = 1.25\,(\Lambda_s)^{1/2}$	$= 3.13(h)^{1/2}$
$\Lambda_s = 1.56T^2$	$\Lambda_l = 3.13(h)^{1/2}T$

(With Λ and h in m, T in s, C in m s^{-1}, $g = 9.8$ m s^{-2})

Examples

T	=	5	15 s		$h =$	5 20 4000 m	
		Wind-wave	swell			(tsunami)	
	C_s	=	7.8	23 m s^{-1}		$C_l =$	7 14 200 m s^{-1}
or	C_s	=	28	84 km h^{-1}		or $C_l =$ 25 50 710 km h^{-1}	
and	Λ_s	=	39	350 m		and for the tsunami:	

if $\quad \Lambda_l = 200$ km
then $\quad h/\Lambda_l = 1/50$
and $\quad T = 17$ min

214   INTRODUCTORY DYNAMICAL OCEANOGRAPHY

examples for long waves show the retarding effect of shoaling water on such waves. The last example, for $h = 4000$ m, may seem out-of-line for "shallow" water, but it is included to emphasize that the term "shallow" is only relative to the wavelength (see Figs. 12.3 and 12.4). The example is typical for the (quite long) tsunami waves generated by undersea seismic disturbances (see Section 12.8).

Another point to notice is that the speed of short waves depends on their wavelength and so on their period, i.e. they are *dispersive* waves. This term refers to separation in position along their direction of travel, not to separation in direction, although this also occurs. For short waves, the speed of the longer waves is greater than that of the shorter ones. Therefore, if a number of waves of different wavelengths (a spectrum of wavelengths) are generated simultaneously, the longer ones will move ahead of the shorter ones and be observed first at a distant point (hence the term "forerunners" for the longer period, i.e. longer wavelength, waves generated by the wind). Also, shorter waves tend to lose their energy by frictional effects somewhat faster and die out sooner than longer ones, and so do not travel so far.

A consequence of this dispersion is that by observing the swell for a few days at one location it is often possible to determine how far away was the storm which generated the swell (see Section 12.33).

12.32   *Groups of waves; group speed; dispersion*

Real ocean waves, even swell from a distant storm, are not pure sine waves but are a sum of sine waves with a range of wavelengths, corresponding periods and amplitudes. To see that groups of waves may not travel at the phase speed, consider the simplest possible case of two cosine waves, η_1 and η_2, of the same amplitude (taken as unity here) but of slightly different wavenumber and frequency, proceeding simultaneously over the same ocean area, i.e.

$$\eta_1 = \cos(k_1 x - \omega_1 t) \quad \text{and} \quad \eta_2 = \cos(k_2 x - \omega_2 t)$$

with $\quad k_1 = k + \Delta k, k_2 = k - \Delta k, \text{ and } \omega_1 = \omega + \Delta\omega, \quad \omega_2 = \omega - \Delta\omega$

where $\quad \Delta k = (k_1 - k_2)/2 \ll k = (k_1 + k_2)/2,$

and $\quad \Delta\omega = (\omega_1 - \omega_2)/2 \ll \omega = (\omega_1 + \omega_2)/2.$

Then $\quad \eta_1 = \cos[(kx - \omega t) + (\Delta kx - \Delta\omega t)],$
$\quad\quad\quad \eta_2 = \cos[(kx - \omega t) - (\Delta kx - \Delta\omega t)].$

Using $\quad \cos(A \pm B) = \cos A \cos B \mp \sin A \sin B,$

then $\quad \eta = \eta_1 + \eta_2 = 2\cos(kx - \omega t)\cos(\Delta kx - \Delta\omega t).$ (12.4)

Equation (12.4) describes a higher frequency wave $\cos(kx - \omega t)$ of wavelength $\Lambda = 2\pi/k$ whose amplitude is modulated by a lower frequency term $\cos(\Delta kx - \Delta\omega t)$. Figure 12.5 shows what η looks like as a function of x at a

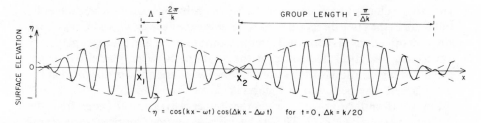

FIG. 12.5   *Surface elevations for a "group" made up of two pure cosine waves.*

fixed time $t = 0$ for $\Delta k/k = 1/20$, and presents one complete cycle of the envelope $\cos(\Delta kx - \Delta\omega t)$ whose amplitude goes from zero through the sum of the amplitudes of η_1 and η_2 back to zero in a "group length" $= \pi/\Delta k$.

In this simple, idealized case the pattern would repeat to infinity in both directions. One can produce more realistic examples where η is non-zero over a finite range of x but the results are basically the same as for the simple example above; more realistic wave groups than those of Fig. 12.5 may be seen in Fig. 12.10.

Now at a fixed time, as we go from one zero to the next of the envelope, (Δkx) changes by π so the "group length" $= \pi/\Delta k$; likewise at a fixed point $(\Delta\omega t)$ changes by π as the successive envelope zeros pass by, so the "group period" $= \pi/\Delta\omega$.

The *group speed*, C_g, at which the group travels is the group length/group period, i.e. $C_g = \Delta\omega/\Delta k$. In the limit as $\Delta\omega$ and $\Delta k \to 0$, the group speed $C_g = d\omega/dk$ which can be evaluated from the dispersion relation (equation (12.3)) to give

$$C_g = \frac{C}{2}\left[1 + \frac{2kh}{\sinh 2kh}\right] \qquad (12.5)$$

where sinh is the hyperbolic sine. For long waves, $kh \ll 1$ and $\sinh 2kh \to 2kh$ and $C_g = C_l$; for short waves, $kh \gg 1$, $\sinh 2kh \gg 2kh$ and $C_g = C_s/2$.

Wave energy travels at the group speed. One can see this intuitively if one considers that (in Fig. 12.5) initially at the position $x = x_2$ there is a zero wave amplitude and therefore zero wave energy, while at $x = x_1$ there is maximum amplitude and maximum energy. As time increases, the group proceeds to the right and the maximum energy region of the group will reach $x = x_2$ after an interval corresponding to the travel time from x_1 to x_2 at the group speed. So energy travels at the group speed. Since coded fluctuations of energy, e.g. variations in the amplitude or frequency of the modulation, may carry information, the group speed is also the speed of propagation of information by the waves. The *group velocity*, $\mathbf{C}_g$, where the vector indicates the direction of travel, is actually a more important characteristic than the phase velocity, $\mathbf{C}$.

The group velocity indicates where the waves go. In the case of surface waves C_g and C are in the same direction but for other types of waves, C_g and C may be in different directions (as described later for some internal and for most planetary waves). An analytical discussion of the relations between phase and group speed and justification for the statement that wave energy is propagated at the group speed may be found in Lamb (1932, sections 236, 237).

For short surface gravity waves, the longer waves travel faster than the shorter ones and $C_g < C$. This behaviour is termed "normal dispersion" (because it was studied first). As the individual waves in the group travel at different speeds, the shape of the group changes as it travels and it becomes more spread out with longer waves leading. If one observes the waves while travelling with the group, the individual waves appear near the back of the group, travel forward through it, and disappear at the front. One can often see this happen when observing waves which are in the deep water regime.

For very short waves, when surface tension dominates (i.e. when $\Lambda \lesssim 0.5$ cm), the shorter waves travel faster than the longer ones and $C_g > C$. This situation is termed "anomalous dispersion". Again the group spreads out as it travels but with the shorter waves leading. If one travels with the group, the individual waves appear near the front, fall behind and disappear at the back.

12.33   Estimating the distance to a wave generation region

Let us now return to determining the distance to a storm which has produced the swell which we see at a wave-observing station. The distance will be $d = C_g(t - t_0)$ where t is the time of arrival and t_0 is the time of generation. From equations (12.2) and (12.3), for short waves $\omega C_s = g$ and $C_g = C_s/2 = g/2\omega$, so that $\omega = g(t - t_0)/2d$ at the observing station. Therefore the observed frequency increases linearly with time and from a plot of ω versus t one can determine the generation time t_0 from the intercept $t = t_0$ when $\omega = 0$, and the distance d from the slope which equals $(g/2d)$. The reader should note that in practice the wave records will provide information about more complex groups than the simple ones of Fig. 12.5, and the calculation is, in fact, a little more complicated than stated above, since one must track the peak frequency of the swell spectrum from the generation area in the presence of waves from other areas and local waves.

Snodgrass et al. (1966) used several stations across the Pacific and tracked swell from storms in the Antarctic all the way to Alaska to demonstrate the application of the method. Although there was considerable loss of amplitude, the results also showed that the wave decay from non-linear and/or viscous effects was very slow.

The observations at a single wave station give the distance to the generation area but not the direction. However, if observations of swell from the same storm are available from two separate wave stations, then the intersection of the two radial distances from the stations will indicate the location of generation.

12.34 *Orbital motion of water particles and wave pressure*

For the case of a surface wave with elevation $\eta = A \cos (kx - \omega t)$, the water velocity components at depth z for deep water are

$$u = \omega A \exp (kz) \cos (kx - \omega t)$$
$$w = \omega A \exp (kz) \sin (kx - \omega t)$$

and the pressure in the water is p (*hydrostatic*) $+ p$ (*wave*) where p (hydrostatic) $= -\rho g z$ as usual and the component due to the wave elevation is

$$p_w = \rho g \eta \exp (kz) = \rho g A \exp (kz) \cos (kx - \omega t).$$

Derivations of these relations may be found in Neumann and Pierson (1966), Wiegel (1964) or LeBlond and Mysak (1978). Figure 12.6(a) shows a wave profile with the surface velocities indicated by arrows at various points and the wave pressure indicated below them. At the first crest (P), u is a maximum and in the direction of wave travel, while p_w is a maximum in the column below the

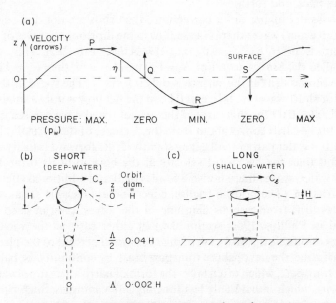

FIG. 12.6   (a) *Velocity and wave pressure for surface gravity waves; orbits of water particle motion for* (b) *short (deep-water) waves,* (c) *long (shallow-water) waves (for which the water depth is exaggerated by about five times relative to wavelength for clarity).*

crest. At the first zero crossing (Q), w is a maximum (upward), u and p_w are zero; at the trough (R), $w = 0$ and u and p_w are at their minimal values (u being opposite to the direction of wave travel); at the second zero crossing (S), u and p_w are zero again and w is a minimum (downward).

Note that it is only the shape of the wave which moves forward continuously at the speed C_s or C_i; the water particles do not travel across the ocean but circulate in *orbits*, circular for short (deep-water) waves and elliptical for long (shallow-water) waves. These orbits decrease in size with increase in depth (Fig. 12.6(b,c)). For short waves, the diameter of the (circular) orbit is $D_z = H \exp(kz) = H \exp(2\pi z/\Lambda)$ where H is the wave height at the surface and z is the level (numerically negative with $z = 0$ at the average surface level as usual). For example, at $z = -\Lambda$, the orbit diameter will be only 0.002 of that at the surface (Fig. 12.6(b)).

For long (shallow-water) waves, the orbits are elliptical at the surface (Fig. 12.6(c)) where the vertical dimension of the orbit is H just as for the short waves. The horizontal dimension is $(H \Lambda)/(2\pi h)$ and is considerably larger than H since $\Lambda/h \gg 1$; however, since $H/h \lesssim 0.8$ at breaking, the horizontal dimension is always small compared with Λ, i.e. it is $\lesssim \Lambda/8$. The horizontal dimension decreases only slightly while the vertical dimension decreases linearly with increasing depth, until near the bottom (if it is flat) the motion will be simply back and forth.

If we consider higher order corrections, the orbits are not quite closed; for short (deep water) waves there is a net flow in the direction of travel of the wave of magnitude $(\pi^2 H^2/\Lambda^2) C_s \exp(4\pi z/\Lambda) = A^2 k^2 C_s \exp(2kz)$. This net transport is called the *Stokes drift*. For $\Lambda = 100$ m, $H = 3$ m, then $C_s = 12.5$ m s^{-1} while the Stokes drift at the surface is only 0.1 m s^{-1}. The speed of the orbital motion for short waves is $AkC_s \exp(kz)$, so the net flow is only a small fraction $Ak \exp(kz)$ ($\leqslant 0.09$ for this case) of the orbital speed. One can see intuitively how the Stokes drift comes about from the decrease of the horizontal velocity with depth; a water particle will have a slightly larger forward velocity at the top of its orbit than the backward velocity at the bottom of its orbit. For long (shallow water) waves there is also a Stokes drift and additional effects due to bottom friction. (The bottom friction effect and appreciable enhancement of the Stokes drift from viscous damping of the waves, even in deep water, is discussed in Phillips, 1966, section 3.4.) In either case, as the wave reaches breaking conditions the Stokes drift may be several percent of the phase speed. In the surf zone, the net onshore transport must, by continuity, be balanced by offshore transport, which often takes the form of narrow jets to seaward, called *rip currents*, which individually last for only a few minutes. Such rip currents are often too fast to swim against toward the shore—the best tactic if caught in one is to swim either way parallel to the shore to get out of the usually narrow outward rip current.

If the waves are approaching the shore at an angle, the net transport will have a component along the shore giving a longshore current. It will be significant over a long period in transporting sand, etc., along beaches after the material has been stirred up by the waves. The strongest effect, by far, is inside the breaker zone where the longshore current is increased by the momentum from the breaking waves until its speed is limited by friction.

12.35   *Wave energy and momentum*

Surface gravity waves have energy associated with them, kinetic energy of the water particle motions and gravitational potential energy associated with vertical displacement. Values for these may be found by averaging over one wavelength or one wave period. In this case, although it is not true for all waves, the average potential and kinetic energies are equal and the total energy per unit area of sea surface is $E = (\rho g A^2)/2 = (\rho g H^2)/8$ (joules m^{-2}). Likewise, waves have momentum associated with them. Averaged over a wave, the momentum per unit area $= E/C$. Thus energy and momentum are related by the phase speed. Both energy and momentum propagate at the group speed C_g.

R. S. Arthur (lecture notes) once estimated that the surface wave energy of the world oceans at any time is about 10^{18} joules and that the rate at which the energy reaches the west coast of the United States (wave power) is about 4×10^{10} J s^{-1} $= 4 \times 10^7$ kilowatts or about 25 times the power generated by a fairly large hydroelectric station. The power reaching the total shoreline of all the oceans was estimated at about 2×10^9 kW. If all this energy were converted to heat and distributed without loss uniformly through the oceans it would take about 90 000 years to raise their temperature by 1 C°, i.e. the rate of heat contribution to the oceans by dissipation of wave energy is negligible compared with the solar contribution (about 3×10^{12} kW). The rate will be higher in the surf zone where the wave energy is dissipated but even there it will be small and likely to be removed by circulation too quickly to be detected; measurements have not shown any significant temperature rise in the surf zone.

In recent years, much effort has been put into investigating the possibility of extracting energy from ocean waves. Some of the techniques show promise but it appears that physically extensive structures would be needed to obtain significant amounts of energy.

12.4   Finite amplitude effects

So far, except for the Stokes drift, we have limited our discussion to the small amplitude (linear) theory. For short (deep water) waves one can work out corrections for the terms neglected in the linear theory. The effect is to add higher harmonics to a wave of a given frequency, e.g. for η, terms involving $\cos 2 (kx - \omega t)$, $\cos 3 (kx - \omega t)$, etc. For example, to first order in Ak

$$\eta = A[\cos (kx - \omega t) + \tfrac{1}{2} Ak \cos 2 (kx - \omega t)]. \qquad (12.5)$$

The effect of adding these higher harmonics is to make the crests sharper and the troughs flatter, e.g. compare the second-order Stokes wave (Fig. 12.7(b)) with the simple cosine (linear) wave (Fig. 12.7(a)). This sharpening of the crests is easily observed with real waves. There is also a correction to the phase speed as

$$C_s = (g/k)^{1/2} \left(1 + \tfrac{1}{2} A^2 k^2\right), \tag{12.6}$$

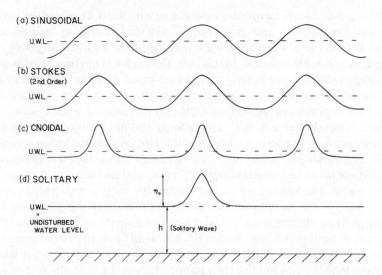

FIG. 12.7 *Shapes of (a) pure sine wave; finite amplitude waves, (b) Stokes (2nd order), (c) cnoidal and (d) solitary. (Wave heights are exaggerated relative to lengths by about seven times in order to show differences between shapes more clearly.)*

i.e. the larger amplitude waves travel a little faster than smaller amplitude ones. At breaking, $Ak \simeq 1/4$ (when $H/\Lambda \simeq 1/12$), so at most C_s is increased by 3% which is why the linear theory is very good for calculating wave propagation.

In fairly shallow water there will be variations of C_l with both wavelength and amplitude. It is possible to find solutions where the two types of dispersion balance to permit finite amplitude waves of permanent form, i.e. the shape does not change as they travel. The solutions are shown in Fig. 12.7 for (c) *cnoidal waves*—a periodic wave train, and (d) for the *solitary wave*, an isolated travelling disturbance. (The term "cnoidal" comes from the abbreviation "cn" used for one of the elliptic functions which is used in the solution of the higher-order equations.) For the solitary wave, $C_{sol} = (gh)^{1/2} (1 + \eta_0/2h)$ and $\eta_0/h \lesssim 0.7$ so that the phase and group speeds can be considerably larger than the value $(gh)^{1/2}$ from linear theory.

In very shallow water, as $H/\Lambda \ll 1$ there will be no dispersion of waves of different lengths but the larger amplitude waves will travel faster than smaller amplitude ones. The crest, where the local $(gh)^{1/2}$ is largest, travels faster than the trough, the forward face steepens, and a "shock wave" may form. Tidal bores in river estuaries, described in Section 13.62, are an example of this phenomenon as is the formation of surf when waves approach a beach over a shoaling bottom.

12.5   Refraction and breaking in shallow water; diffraction

12.51   *Refraction*

Small amplitude long (shallow-water) waves all travel at the same speed in water of a given depth h but where the bottom depth is changing their direction of travel may change. More generally, as waves move into shallow water their period remains constant but C decreases and therefore Λ decreases. As an example, Table 12.3 shows the decrease of speed and wavelength for waves of period 8 seconds on entering shoaling water.

TABLE 12.3 *Decrease of speed and wavelength in shoaling water for waves of period 8 s and length 100 m in deep water*

h =	50+	10	5	2	m
C =	12.5	8.9	6.6	4.3	$m\,s^{-1}$
Λ =	100	71	53	35	m

Hence, if a series of parallel-crested waves approaches at an angle to a straight shoreline (Fig. 12.8(a)) over a smooth sea bottom which shoals gradually, they progressively change direction as the end of the wave nearer to the shore (P in Fig. 12.8(a)) slows down earlier than that farther away (Q in Fig. 12.8(a)). As a result, the waves become more parallel to the shore by the time that they pile up as surf. The change in direction (e.g. PP', QQ') associated with the change of speed is called *refraction*. The same phenomenon occurs abruptly to light waves travelling from air to water but gradually, as for water waves in this case, for light coming from the sun at an angle to the vertical and entering the gradually increasing density of the atmosphere or to sound waves passing through the thermocline at an angle to the horizontal.

If the sea bottom does not have a uniform slope along the full length of the shore, the refraction may be more complicated. Two simple examples are where there is an underwater ridge running out at right angles to the shore or where there is an underwater valley. The refraction pattern for waves coming straight towards the shore would then be as in Fig. 12.8(b). In this figure are shown not

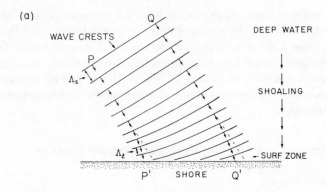

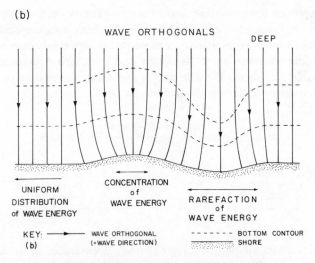

FIG. 12.8   (a) *Refraction of waves approaching a beach over a smoothly shoaling bottom,* (b) *wave refraction on approaching an underwater ridge (left) and a valley (right). Dashed lines represent depth contours, full lines represent wave orthogonals (directions of travel of waves), in (b).*

the wave crests but the wave *orthogonals* which are perpendiculars to the wave crests, i.e. parallels to $\mathbf{k}$, $\mathbf{C}$ and $\mathbf{C}_g$, which indicate the direction of travel of the waves, as do the arrows in Fig. 12.8(a).

Refraction of waves round a headland, for instance, occurs if the water deepens gradually to seaward from the land but not if the water is of relatively uniform depth off the headland. Waves are often observed to be refracted round islands and one can sometimes see an interference pattern set up where the waves which are refracted around the two sides of a small island meet behind it.

12.52   *Wave breaking*

As the waves move inshore and slow down, not only does the wavelength decrease but also the wave height changes. In a steady state, the energy flux (the product of the wave energy per square metre times the group speed C_g) is constant as the waves move inshore (until they break), with a value averaged between two successive crests of $(\rho g H^2 C_g)/8$ joules per metre width of crest per second. If the flux is not constant, the divergence of flux, $\partial(E C_g)/\partial x$, would cause the energy level to change locally, i.e. $\partial E/\partial t$ would not vanish.

If the waves are initially long, e.g. tsunamis, $C_g = C_l$, both speeds decrease, and H increases. However, if the waves are initially short, at first C_g increases, reaching a maximum value of 1.2 times the deep-water value when $h/\Lambda \simeq 0.19$, where Λ is the local value not the deep-water value ($h/\Lambda_d = 0.16$). In this zone, H decreases to a minimum of about 90 % of the deep-water value when C_g is a maximum. H/Λ is nearly constant at first but begins to increase before $h/\Lambda \simeq 0.19$ because Λ decreases faster than H. At $h/\Lambda = 0.19$, the steepness H/Λ is about 10 % greater than the deep-water value. As the waves move further inshore, C_g decreases and H must increase. However, the decrease in Λ dominates the H/Λ changes. In the example given in Table 12.3 when $h = 2$ m, Λ is about 35 % of the deep-water value but the increase in H is only about 25 %. For initially long waves, Λ decreases proportionally to $C_l = (gh)^{1/2}$ while H increases proportionally to $1/(C_l)^{1/2}$. There is a limit to how much H/Λ may increase. Theory puts this limit at $H/\Lambda = 1/7$ but in practice it is rare for waves to get steeper than $H/\Lambda \simeq 1/12$. When the wave steepness approaches this limit the waves tend to change from a symmetrical sine shape (Fig. 12.9 (a)) to a more peaked shape (Fig. 12.9(b)) and finally, as H/Λ reaches about 1/12, the waves become unstable and break. Breaking may also occur before $H/\Lambda = 1/12$ if the water becomes too shallow; breaking will occur when H/h reaches about 0.8 regardless of the value of H/Λ. In the shallow water where the ratio

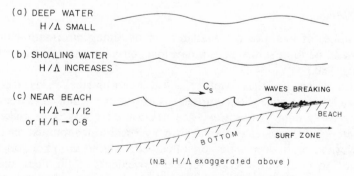

(a) DEEP WATER
H / Λ SMALL

(b) SHOALING WATER
H / Λ INCREASES

(c) NEAR BEACH
H / Λ → 1 / 12
or H / h → 0·8

C_s

WAVES BREAKING

BEACH

BOTTOM   SURF ZONE

(N.B. H / Λ exaggerated above)

FIG. 12.9   *Shape of waves:* (a) *in deep water,* (b) *in shoaling water,* (c) *close to beach.*

H/Λ is approaching about 1/12, or H/h is approaching about 0.8, the waves tend to behave as individuals (solitary waves) rather than as a successive train, and they become progressively more unsymmetrical in side view until they break as surf (Fig. 12.9(c)). Usually, rather shallow water must be reached; breaking does not occur until the wave height is comparable to $\sim 0.8\,h$ or until $H/\Lambda \simeq 1/12$, whichever occurs first.

To double the short-wave steepness requires $h/\Lambda_s = 0.047$, i.e. 4.7 m depth for a wave of 100 m deep-water wavelength. If the initial height is 4 m ($H_s/\Lambda_s = 0.04$), which is quite large for 8-second waves, the waves will break at about 5 m water depth because $H/h \simeq 0.8$ just before $H/\Lambda \simeq 1/12$. If the initial height is somewhat larger, $H/\Lambda \simeq 1/12$ will be reached before $H/h \simeq 0.8$. Thus only waves which are generated nearby and are fairly steep initially will break because of the H/Λ limit; waves from some distance will break because the water becomes too shallow ($H/h \simeq 0.8$).

12.53   *Diffraction*

Diffraction can also occur with water waves, as for light or sound waves. For example, if ocean waves arrive at a harbour entrance this will act as a wave source for the harbour area itself. If the entrance is wide (compared with the wavelength) much of the wave energy will continue into the harbour in the original direction of wave travel but, near the sides of the opening, some wave energy will be diffracted into the geometrical "shadow" area behind the harbour walls. When the gap in a sea wall or reef is narrow compared with the wavelength of the waves, then it will approximate a point source and most of the wave energy will spread over the harbour area as diffracted waves in the form of circular arcs centred on the harbour opening.

12.6   Measurement of waves

12.61   *Methods*

A number of methods are available for obtaining information on wave heights and periods, some approximate and some accurate. (For a review see Stewart, 1980.)

The simplest way of all is simply to look at the sea and make a visual estimate. It takes a lot of practice with comparisons with instrumental measurements to obtain reliable data in this way. The next simplest way is to make visual observations of the water surface against a vertical scale mounted on a pier in shallow water or, in deep water, on a float with a large, deep horizontal plate "damper" to limit short-period vertical movements of the float and scale. Because the period of water waves is relatively short, only a few seconds, such visual measurements are limited to estimates of the height of only a proportion of the waves. Because of the variety of heights usually present in most real wave conditions it is common to quote the mean height of the highest one-third of the waves (called the *significant wave height*, H_s) as a descriptive characteristic. For many purposes this value is more useful than the value for single waves, e.g. the single highest wave. The latter is most important for its possible destructive effects on ships or drilling platforms.

A third method is to use a fixed pressure-sensor mounted below the depth of the deepest troughs. The hydrostatic pressure below surface waves varies periodically as the depth of water from the surface to the sensor varies (see Section 12.34) and so a continuous record of pressure against time will give information on surface shape. This method is only practical for short (deep-water) waves when the sensor can be mounted in a fixed position at a relatively small distance below the troughs because the pressure variations decrease rapidly with depth increase (Section 12.34), so that the deeper the sensor the smaller are the pressure variations available to be measured. For long (shallow-water) waves, the decrease in pressure variations with depth is slow, so a sensor mounted on the bottom will record almost the full pressure variations due to the variations of the height of the water surface. Therefore bottom-mounted pressure sensors can be used effectively near shore. They can also be used even in the deep ocean (thousands of metres) to measure tsunamis or tides because these are long (shallow-water) waves.

Information about the details of the surface height variations when waves are present has been obtained with electrical devices mounted to penetrate the sea surface. One method is to mount on a vertical rod, at intervals of a few centimetres, a series of pairs of wires with a small gap between the wires of each pair. Then the wire pairs which are immersed in the conducting sea water will be short-circuited and by recording continuously against time the number of pairs from the bottom which are short-circuited a record of sea surface level is obtained. An alternative method is to use a vertical, bare resistance wire and then a record against time of the value of the resistance from the top of the wire to the sea will yield a record of the height variations of the sea surface. A third alternative is to mount a thin, insulated wire vertically through the sea surface and to measure the electrical capacitance between the wire as one electrode and the sea as the other. Of course, there are technical difficulties to be overcome. For the first method, drops of water may remain between the wire pairs when the sea surface falls below them, or in the other two methods a thin film of sea water may be left behind briefly as the water level falls, so indicating too great a water height. (Remember that we would be dealing with fluctuations of water level of only seconds duration.) The use of a hydrophobic wire covering for the capacitance wire system helps to minimize this source of error.

A method developed by the Institute of Oceanographic Sciences in England makes it possible to measure waves at sea from a ship. The water pressure is measured at a low point on the hull, normally below even the troughs of the waves, to give the wave height relative to the ship. At the same time, the vertical motion of the ship is measured with a vertical accelerometer (integrating twice to show vertical height variations, i.e. $\int a \, dt \rightarrow w$ and then $\int w \, dt \rightarrow$ height, and added algebraically to the pressure record). Wave profiles may also be obtained from the records of a vertical accelerometer mounted in a buoy floating at the surface of the sea; this method is now used regularly and extensively.

All of the above methods have a major failing—like current meters they provide information at one point only. If real surface waves were a sum of pure sine waves travelling in a single direction the problem would not be serious, but even a few minutes observation will show that the real sea surface is usually quite irregular, small waves superimposed on larger waves superimposed on swell, and the crests of the waves are generally quite short along their length (only a few wavelengths at most) so that a real water surface varies with both x and y and with time. To obtain more complete information, stereo-photographs of the sea surface may be made. However, the analysis of such photographs is a very laborious process and this method has not been used much.

Another method to obtain information on the spatial structure has been to fly over the sea at as nearly constant an altitude as possible (measuring and correcting for variations) and record the shape of the sea surface with a narrow-beam radar altimeter or, more recently, with a pulsed laser altimeter (to look at only a very small area of the sea surface at a time). This method only yields the surface elevation along the flight path but flights can be made in several directions over an area and a statistical picture built up of the sea surface shape. Finally, wave measurements are being made from satellites. Using radar altimeters, mean wave heights to about ± 1 m or 25 % of the actual height are possible while measurements of microwave scattering give statistical information over areas both for waves and for surface wind speeds and directions. (For papers on satellite techniques see Gower, 1981.)

To test some aspects of wave theory it is desirable to obtain information on the direction of propagation of waves. In principle this can be obtained by mounting a number of wave recorders in a geometrical pattern and examining the phase relationships between the records. The results are limited by the number of measuring points and the analysis is complicated. In principle, this directional information can also be obtained from the stereo-photographs and it can be obtained from aircraft laser altimeter records and from the satellite instrumentation records.

12.62   Real waves

The records from such instruments and procedures make it clear that the sea surface rarely has the ideal sine shape as in Figs. 12.2 and 12.3 except for swell, but is more likely to look like Fig. 12.10 because there are usually a large

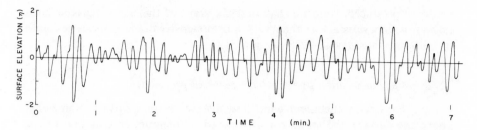

FIG. 12.10 *Character of real wave train (profile) to compare with ideal sine waves of Figs. 12.2 and 12.3. (Note that the vertical scale is much exaggerated as before.)*

number of wave components present simultaneously. The only practical way to deal with this situation is the statistical one to be discussed in the next section in which the spectrum of wave energy is related to wave period (e.g. as in Fig. 12.11).

12.7   The generation of wind waves; practical wave analysis and forecasting

12.71   *General*

Wind waves are started by the wind blowing for some hours *duration* over a sea surface many kilometres long called a *fetch*. The fitful gusts of wind generate a choppy and irregular *sea*. These oscillations of the surface, once set up, continue to run across the sea far beyond the direct influence of the wind. Under these conditions they are called *swell*. Swell consists of uniform wave trains with a broad sideways extent of the individual crests. Because it is comparatively uniform we can numerically describe the height and period of a wave at the beginning of the swell zone (i.e. at the end of the fetch) and during its subsequent progress. The swell *decays* for a long distance while its wavelength increases and wave height decreases. As the swell enters shallow water it *feels bottom* and a rejuvenation takes place. The wave speed and length decrease and the height increases, but the period remains constant. The swell finally peaks up into *waves*, breaks, and is dissipated as *surf*.

The above paragraph gives a brief sketch of the generation and dissipation of wind waves. Clearly the wind is responsible for the generation of most óf the surface waves which are almost always present. How are we to get more quantitative information on what sort of waves will be present under given conditions of fetch, duration and wind speed? Later we shall describe an empirical approach based on observations of the waves themselves under a variety of conditions, which has to be used for practical purposes (e.g. for prediction) because a quantitatively accurate theory based on physical laws has yet to be produced. Here we shall describe some of the steps that have been taken toward a satisfactory theory. (A more extensive discussion can be found in LeBlond and Mysak, 1978, section 51.)

12.72   *Wave generation by the wind—physical processes*

First, consider a turbulent wind flowing over a solid surface. Somewhat above the surface, the stress (or downward momentum flux) is due to the turbulent Reynolds stresses ($-\overline{\rho u'w'}$ and $-\overline{\rho v'w'}$ derived in Chapter 7). As the surface is approached, the amplitudes of the turbulent velocity fluctuations are suppressed by the boundary condition of no flow parallel or perpendicular to the surface. At the same time, the mean shears ($\partial u/\partial z$ and $\partial v/\partial z$) increase. If the surface is smooth (e.g. a sheet of glass) the final stress transfer at the surface is by molecular viscous shear stresses ($\mu\partial u/\partial z$ and $\mu\partial v/\partial z$ in the x- and y-directions respectively). If the surface is rough (e.g. a sand beach) the bumps on the surface will cause the air flow to separate from them leaving stagnant regions behind them. There will be a positive pressure difference from the upwind to the downwind sides of the "roughness" elements. Part, and perhaps most, of the final stress transfer will be due to these pressure differences or what is termed "form drag". As the pressure is a normal stress (force per unit area perpendicular rather than parallel to the surface) one may also say that the final momentum transfer (flux) is mainly by normal rather than shear stresses for a rough surface.

Now surface waves are nearly irrotational (i.e. have almost zero vorticity). Irrotational motions are produced by normal stresses while rotational motions are produced by shear stresses. Thus, most of the wave generation must be due to normal stresses (pressure). The facts that waves are nearly irrotational and nearly linear (the advective acceleration terms are of higher order in H/Λ than the local acceleration terms) are important in explaining why waves outside active generation areas (where wave breaking is important) decay very slowly. Thus, swell is not likely to produce turbulence and hence turbulent friction to damp it. Further, for exactly irrotational motion the molecular viscous terms vanish identically. As mentioned earlier, swell generated near Antarctica has been traced some 14 000 km across the Pacific Ocean to Alaska, albeit with considerable loss of amplitude over this very long distance. The decrease in amplitude will be due in part to spreading of the energy because of differences in direction of travel as well as to viscous losses.

One of the fairly early theories for wave development, once the waves already existed, was that they grew by form drag with flow separation at their crests. This theory was proposed by Jeffreys in the mid-1920s. Tests with flows over solid models showed that the effect was too small to explain observed rates of wave growth. A problem with these tests, not clearly recognized at the time, is that results for solid surfaces cannot be applied directly to a moving fluid surface such as that of the ocean. Another problem, not recognized for a long time, is that a condition for flow separation is that wave breaking must occur, which may be of fundamental importance as we shall see.

Renewed attempts to solve the problem were made in the late 1950s and early 1960s. First, Phillips suggested a means for getting waves started on an initially undisturbed surface. He pointed out that as the air flow is turbulent, not only are there velocity fluctuations but pressure fluctuations as well. These pressure fluctuations may start wave motion; they lead to a growth of wave energy proportional to the time, t. (There is some observational evidence to support Phillips' theory of initial growth.) Once the waves exist they may modify the air flow so that the growth rate becomes proportional to the wave amplitude (or energy) and hence exponential in time. Assuming that waves of *small* amplitude have formed, the growth process may be calculated using linearized stability theory as was done by Miles. However, his calculated growth rate, though exponential, was soon found to be much smaller than observations indicated. The observations themselves show considerable variation but there is little doubt that there is a real discrepancy.

Further ideas have been suggested in attempts to overcome this disagreement between observations and theory, e.g. that the momentum input to the waves is largely due to the very short waves and ripples and that these in turn become very steep and perhaps break (a very non-linear process) and transfer at least some of their momentum and energy to the larger waves. Again, while this mechanism may play a role, it does not resolve the observational–theoretical discrepancy either.

Recently, numerical calculations have been made showing the air flow over water waves in more detail. The rate of growth due to pressure depends quite strongly on the wave steepness, H/Λ. These calculations also show that about one-half of the total momentum transfer is through the pressure field (normal stresses). These results are supported by observational data as well as by the argument that, to explain observed growth rates, much of the total stress must go into the waves by normal stresses to produce nearly irrotational wave motions. The shear stress is also greatest near the crest, giving some support to the idea of initial input to short waves which then transfer their momentum to longer waves. However, even these results do not lead to growth rates large enough to explain the observed values.

Other very recent observations have suggested that the momentum transfer is much increased when wave breaking occurs. Thus Jeffreys' original argument may be reasonably correct *if* breaking is taken into account (which Jeffreys did not do explicitly). Non-linear transfer from shorter to longer waves no doubt plays a role too. A quantitatively correct wave-generation theory remains to be established but the recent results suggest further research which may lead to it.

Other approaches are being taken too. If one observes the growth of the wave spectrum and calculates the non-linear transfer and viscous dissipation, the input function can be calculated by differencing (Hasselman *et al.*, 1973). Attempts at the direct measurement of the input have been and continue to be made. Such measurements are difficult as the reader may appreciate if he considers how to make observations in a breaking-wave field while trying to keep instruments *at* the moving surface. Of course, such observations do not "explain" the wave-generation process but they do give the theoretician something to try to reproduce. (See LeBlond and Mysak, 1978, sections 39 and 51.)

Non-linear effects in surface waves are weak but they are not negligible. In a fully-developed sea (one for which fetch and duration are not limiting, and in which further growth no longer occurs because momentum/energy loss by wave breaking balances the input from the wind) there are components in the spectrum whose phase speeds are greater than the wind speed a few metres above the surface. Indeed, the peak values of the spectrum have this characteristic. If one translates to coordinates moving at the phase speed of the waves near the spectral peak then the air flow is contrary to them and it is very difficult to imagine how the wind can enhance them. However, this argument may not be correct. In reality, one always deals with groups of waves. In coordinates moving at the group speed of the waves near the spectral peak, the air flow is still such that it may do work on the waves and enhance them. (In Phillips' original study of the initiation of waves he considered the phase speed and got a growth rate proportional to t^2, but because the waves travel in groups at the group speed they are always falling behind or getting out of phase with the pressure fluctuations, and the growth rate is really proportional to t.) The calculations on non-linear transfers also indicate that the longer waves in the spectrum gain energy by non-linear transfer from shorter waves.

Considering that much of the momentum input at the surface goes, at least initially, into the waves one may wonder about using the total stress as a forcing function for large-scale ocean circulations. What if the waves radiate away and take their momentum somewhere else or even to a distant shore? It seems that this possibility is not serious. The wave field develops quite rapidly; in the developed stage the momentum input to the waves is transferred through wave breaking to the current quite quickly (with transfer to longer waves a possible intermediate stage). Only in rather small regions of rather strong winds is significant radiation of momentum out of the area likely; even then, on a global scale most of the momentum goes into the currents locally. Thus, although much of the momentum input from the wind passes through the wave field, the net input to the wind-driven circulation is probably not affected significantly.

Finally, a comment should be made on the constancy or near constancy of the drag coefficient. For a solid rough surface, the drag coefficient (defined in Chapter 9) is constant (provided that increased wind speed does not change the geometry, e.g. a hay field becomes flatter and "smoother" as the wind becomes stronger, but a field of boulders does not). The drag coefficient over the ocean is constant on average to speeds of ~ 10 m s^{-1} and then slowly increases as wind speed increases (Section 9.9) Thus, to the air flow the sea presents a nearly constant roughness. (An observer riding in a small ship would hardly agree with the air flow on this point!) Perhaps the apparent roughness is due to the short waves which develop quickly; perhaps the effect(s) of flow separation occurring over longer distances with longer waves at higher wind speeds gives an equal effect. Further investigations are needed. There is some evidence that the drag coefficient is higher when the wave steepness is greater, when the waves are growing rapidly after an increase in wind speed and when the wind direction reverses. Again further investigations are needed.

12.73   *Wave generation by the wind—empirical relations*

Although there is still uncertainty in our knowledge of many details of the actual mechanisms of wave generation at the sea surface by the wind, many observational data on related wind and wave properties have been accumulated and graphical relations assembled. Some features of one of the sets of relations (Pierson, Neumann and James, 1955) will be described to illustrate their character. (Because of the importance of wave effects on marine structures, much work has been done recently to develop techniques for predicting wave characteristics but much of it is proprietary and not readily available.)

The wind factors are *wind speed, fetch* (the linear distance over which the wind is blowing over the sea) and the *duration* (the time for which the wind has been blowing over the fetch). Wave properties are the *significant wave height* H_s (the average height of the one-third highest waves) and the range of *wave periods* or *frequencies* in the wave spectrum.

Figure 12.11 shows three plots of the square of the wave height (H^2) against frequency and period for a system developed on the sea by winds of 10, 15 and 20 m s^{-1} respectively. One feature of these curves is that the quantity H^2 which is related to wave energy, increases very much more rapidly than wind speed; another feature is that the spectrum of wave energy as a function of frequency is peaked and that the peak occurs at lower frequencies (longer periods) at higher wind speeds. These curves are for a *fully developed sea*, i.e. when the wind has been blowing for long enough and with a sufficient fetch for the steady state to be established with the energy spectrum at a maximum for that speed. The numerical information on which such spectra are based is obtained from measured wave records at sea at various wind speeds.

From such energy spectrum curves are developed *co-cumulative spectrum curves* such as those in Fig. 12.12 (full lines for wind speeds of 10, 15 and 20 m s^{-1}). The ordinate for any point on each curve is proportional to the total cumulative wave energy from infinite frequency (zero period) to the frequency represented by the point on the curve. (Note that the frequency scale increases to the right but the period scale increases to the left.) In these

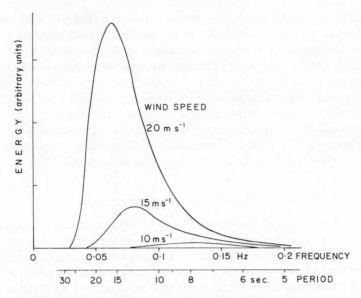

FIG. 12.11   *Idealized spectra of wind-wave energy versus frequency and period for three wind speeds for a fully-developed sea.*

plots the ordinate (energy scale) has been arranged so that it is linear in significant wave height while the abscissa is linear in period from zero to 20 seconds and then compressed for longer periods.

To illustrate the information available in this plot we will take the 15 m s^{-1} wind-speed curve (full line) as an example. Reading from the right, the curve indicates that the cumulative wave energy (and the significant wave height H_s) increases with period slowly at first, then more rapidly, then more slowly and finally levels out (expressed as H_s at about $H_s = 6.3$ m) a little before 20 s period. The fact that there is a maximum value indicates that for any given wind speed there is a maximum total wave energy possible (wave breaking causing the limit). This maximum is seen to increase with increased wind speed. The steepest part of the curve (at about 12 s period for the 15 m s^{-1} wind speed) corresponds to the peak of the H^2 spectrum curve of Fig. 12.11.

In Fig. 12.12, in addition to the three sample wind speed curves (full lines) there are two sets of cross lines, dashed lines for 10, 20, 30 and 40 hours duration (assuming unlimited fetch) and dotted lines for 100, 500, 1000 and 1500 km fetch (assuming unlimited duration). These lines indicate the progressive development of wave height and period with these parameters. For instance, the 20-h duration line intersects the 15 m s^{-1} wind speed line at 5.8 m significant wave height and 14 s period. The former indicates that this value of H_s is reached after 20 h and the latter indicates that the majority of the wave

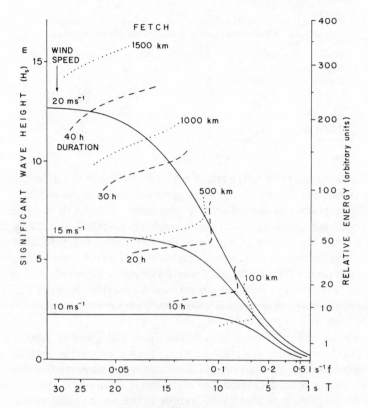

FETCH

FIG. 12.12   *Co-cumulative wave spectra as significant wave height* (*H_s*) *and wave energy against frequency* (*f*) *and period* (*T*) *for three wind speeds (full lines), four fetches (dotted lines) and four durations (dashed lines).* (*Adapted from Pierson, Neumann and James,* 1955)

energy will be present at periods of 14 s *or less.* Alternatively, in terms of fetch the (dotted) 500-km line indicates that at the end of this distance, for a steady wind of 15 m s^{-1}, the value of H_s will be 6.2 m and most of the wave energy will be at periods of 17 s or less.

The levelling off of the wind-speed lines indicates that the sea is fully developed, i.e. that the rate of input of energy and momentum by the wind is balanced by the rate of loss by wave breaking. In principle, this fully developed state requires infinite duration and fetch but in practice it is considered to be effectively reached in a finite duration and fetch as shown in Table 12.4.

The wave periods in Table 12.4 are defined such that 5 % of the total energy will be at periods greater than T_U and 3 % at periods less than T_L, i.e. 92 % will be between these two periods. For a non-fully developed sea, either the duration

TABLE 12.4 *Effective duration and fetch for a fully developed sea with corresponding significant wave height and range of wave periods*

Wind speed	$(m s^{-1})$	5	10	15	20	25
Duration	(h)	2.3	9.5	22	40	64
Fetch	(km)	20	130	480	1200	2400
H_s	(m)	0.4	2.2	6.2	13	22
Range of	T_U (s)	6	11	16	21	26
periods	T_L (s)	1	3	5	6	8

or the fetch may be the limiting factor; which should be used will be determined by the conditions for which the wave calculations are to be made.

Statistical studies indicate that for a long series of waves the average height of all waves, the significant wave height (highest one-third), and the average of the one-tenth highest waves will be in the ratios 0.6:1.0:1.3. It must also be realized that there will be a range of heights in the highest one-third waves and it is to be expected that the longer the series of waves observed, the higher will be the highest ones. For instance, for 100 waves observed there is a 1/20 chance that the highest will be over $1.9 H_s$, while for 1000 waves there is a 1/20 chance that the highest will be over $2.2 H_s$.

The above examples have been drawn from the Pierson, Neumann and James (1955) procedure. It should be pointed out that other investigators have also analysed wave observations and developed wind/wave relationship graphs and calculation procedures and that all methods do not give identical results. The differences may be due to differences in the wave characteristics in the different regions from which the data were drawn, to differences in observing techniques or to differences in the treatment of data. However, the procedure described above gives an indication of how an empirical approach may be employed to obtain useful results even though the generation mechanism is not understood in detail.

Further information on the spectral shape and the dependence of H_s and the period on wind speed, fetch and duration may be found in Wiegel (1964), Hasselman *et al.* (1973), LeBlond and Mysak (1978). Forristal (1978) gives recent information on wave statistics.

There are numerous applications for wind/wave relationships. One of the needs which led to the development of such procedures (by Sverdrup and Munk (1947) in the first instance) was to forecast wave conditions, from forecast winds, for beach landings during World War II. The forecasting of wave conditions for other operations, such as for laying undersea cables and pipes, is a more recent application. The calculation of wave characteristics for regions for which wind data are available, but for which no wave data are available, can be important for ship hull design or for ship routing (because ship speeds are decreased by increased wave heights). For structures such as oil-drilling rigs at sea, an important prediction from wave statistics is the probable highest wave over a period of, say, 10 or 100 years to avoid going to the often considerable expense of building the structure stronger than it need be.

12.8   Tsunamis or seismic sea waves

Tsunami is the transliteration of a Japanese wave meaning "harbour wave" (as distinct from the ordinary tidal rise and fall), and it is now generally used to refer to long water-waves generated by sea bottom movements associated with earthquakes. The term "seismic sea wave" is also used. The older term "tidal wave" is incorrect and should not be used because the generating mechanism for tsunamis is quite different from that for the normal tides. It is known that virtually all tsunamis follow earthquakes occurring under the sea or near the shore, but not all earthquakes generate tsunamis. It is considered probable that only those earthquakes which involve a significant component of motion perpendicular to the bottom, i.e. a rise, fall or tilt, are likely to be generators, while those in which the motion is horizontal do not generate tsunamis.

Occasionally, tsunamis are generated by other earth movements, such as large landslides into the sea, or are possibly associated with marine volcanic activity. However, the effects are usually not widespread, whereas tsunamis generated by seismic activity on one side of the Pacific have caused devastation on the other side, as in the case of the tsunami generated by the 1960 earthquake in Chile which caused serious damage there and also in Japan nearly 20 000 km away.

Tsunamis behave in the open sea just like other surface waves but because of their long wavelength, of the order of 200 km, they behave as shallow-water waves even in the deep ocean because the ratio $h/\Lambda \sim 1/50$. Their calculated amplitude at the surface of the deep ocean is of the order of 1 m and so they are of no significance to ships there. It is only when they slow down and peak up near shore that they become dangerous. The effect observed is an abnormal rise and fall of sea level of up to several metres amplitude, and period typically of 15 min to an hour or so. Sometimes there is an initial rise of sea level, sometimes an initial fall, and the unusual oscillations may continue for hours, occasionally for a day or two. During the abnormal fall of sea level, ships may be left aground, tip on their side and be swamped on the succeeding rise. During an abnormal rise there will be little significant effect at a steep cliff shoreline, but where the shore is flat and only slightly above normal high-tide level, the sea may pour across flat lands and sweep away buildings or carry ships inland and strand them there. Refraction effects also play a significant role near shore and make certain ports particularly susceptible to damage, while others may be less so.

Tsunami amplitudes are measurable along a coast from tide gauge records, by special tsunami recording gauges, and from surveys of damage caused.

There have not yet been any reports of direct measurements of the amplitude of tsunamis in deep water but gauges for this purpose are in experimental use.

In the Pacific Ocean, where most tsunamis occur because of the large number of earthquake zones, there is a Tsunami Warning System in operation based near Honolulu, Hawaii, and with input from many countries around the Pacific. For this System, seismological observatories around the ocean provide information to a Centre at the Honolulu Magnetic and Seismological Observatory within about one half-hour of any earthquake occurring under or near the sea. From the earthquake epicentre information, if a tsunami is generated its time of arrival at any point around the Pacific can be calculated to an accuracy of a few minutes because the depth distribution for the Ocean is sufficiently well known for this purpose. Because it is not yet possible to predict whether or not a significant tsunami will be generated, the procedure is to alert observing stations along the coast on either side of the epicentre and they report to Hawaii immediately if any significant wave is observed. (Many of the stations are automatic and report to Honolulu without the necessity for a human observer to visit the gauge.) In the case of a significant wave being observed, all other countries in the System are warned of the possibility of a tsunami arriving so that they may take whatever precautions have been planned. If no significant waves are observed, the alert can be cancelled.

Descriptions of tsunamis and a discussion of their dynamics may be found in Murty (1977).

12.9   Internal waves

12.91   *Interfacial waves in a two-layer ocean*

Thus far we have considered so-called "surface" waves occurring at the air/water interface. Similar waves can occur at surfaces between different density layers within the sea, i.e. *internal waves*, because the density difference leads to a gravitational or hydrostatic pressure (caused by gravity) restoring force if fluid is displaced vertically. Particular surfaces are in the thermocline in oceanic waters, where the density difference is chiefly due to temperature difference, or at the halocline in coastal waters, where the density difference is mainly due to salinity difference. Of course, the water movements are not limited to the interface itself, but extend through the water above and below it. For surface waves, the density of air is so small compared with that of water (ratio about 1/800) that the former could be ignored and the air density did not appear in the formulae for wave speed. However, for internal waves the densities of the two water layers are nearly the same. Figure 12.13 illustrates the situation being considered with an upper layer of depth h_1 and density ρ_1 and a lower layer of depth h_2 and density ρ_2 with an internal wave of wavelength Λ_i at the interface between the layers. The dispersion relation can again be derived

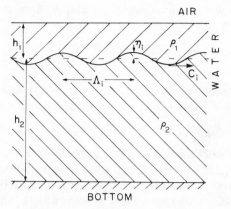

FIG. 12.13 *Internal boundary waves
at the interface between two layers of
water of different densities. (Note that
the wave height is exaggerated for
clarity.)*

from the equations of motion. With $(\rho_2 - \rho_1)/\rho_2 \ll 1$ as is the case in the ocean ($\Delta\rho/\rho$ rarely being greater than 3×10^{-3} in the open ocean; in extreme cases in coastal waters it may reach 20×10^{-3}), the dispersion relation, correct to $0(\Delta\rho/\rho)$ is

$$\left[\omega^2 - gk\tanh k(h_1 + h_2)\right]\left[\omega^2 - \frac{(\rho_2 - \rho_1)gk}{\rho_2\coth kh_2 + \rho_1\coth kh_1}\right] = 0 \quad (12.7)$$

where coth is the hyperbolic cotangent ($= 1/\tanh$). The first term is the dispersion relation for surface gravity waves in water of total depth $(h_1 + h_2)$. Because $\Delta\rho/\rho \ll 1$, the surface waves are essentially unaffected by the stratification. The second term is the dispersion relation for internal waves at the interface. (For a rigid upper surface it is exact rather than being correct only to $O(\Delta\rho/\rho)$.)

As with surface gravity waves we can consider special cases of short and long internal waves. For short (deep water) waves, i.e. Λ_i/h_1 and $\Lambda_i/h_2 < 2$, coth kh_1 and coth kh_2 are essentially equal to 1 and

$$\omega_i^2 = \left[\frac{\rho_2 - \rho_1}{\rho_2 + \rho_1}\right]gk; \quad C_i^2 = \frac{g\,\Delta\rho}{k(\rho_1 + \rho_2)}. \quad (12.8)$$

These waves are dispersive with the longer waves travelling faster than the shorter ones; also $C_{gi} = C_i/2$. Because $\Delta\rho/\rho \ll 1$, the phase speeds are much smaller than those of surface gravity waves; for $\Delta\rho/\rho = 2 \times 10^{-3}$,

$$C_i/C_s = [\Delta\rho/(\rho_1 + \rho_2)]^{1/2} \simeq 1/30. \quad (12.9)$$

For long (shallow-water) waves, i.e. Λ_i/h_1 and $\Lambda_i/h_2 > 20$, $\coth kh_1 \rightarrow 1/kh_1$ and $\coth kh_2 \rightarrow 1/kh_2$ and

$$\omega_i^2 = \frac{gk^2 \Delta\rho h_1 h_2}{\rho_2 h_1 + \rho_1 h_2} \simeq \left(\frac{gk^2 \Delta\rho}{\rho_2}\right)\left(\frac{h_1 h_2}{h_1 + h_2}\right). \qquad (12.10)$$

These waves are non-dispersive. Again the phase speed is much less than that of long surface waves $= [g(h_1 + h_2)]^{1/2}$; e.g. for $\Delta\rho/\rho = 2 \times 10^{-3}$ and $h_1 = h_2$,

$$C_i/C_l = [(\Delta\rho\, h_1 h_2)/\rho(h_1 + h_2)^2]^{1/2} \simeq 1/45. \qquad (12.11)$$

Finally one can consider the case of a thin upper layer ($h_1 < \Lambda_i/20$) over a deep lower layer ($h_2 > \Lambda_i/2$). This would be a good approximation in coastal regions near river inputs and for the pycnocline beneath the mixed layer in the open ocean. In this case, $\coth kh_1 \rightarrow 1/kh_1 \gg \coth kh_2 \simeq 1$ and the dispersion relation is

$$\omega_i^2 = \frac{gh_1 k^2 \Delta\rho}{\rho_1} \quad \text{with} \quad C_i^2 = \frac{\Delta\rho}{\rho_1} gh_1. \qquad (12.12)$$

These waves are non-dispersive and analogous to long surface waves in water of depth h_1 but with "reduced gravity" $g' = (\Delta\rho/\rho)g$ because of the small density difference. For example, for coastal waters with $S_1 = 0$ (fresh water), $S_2 = 30, T_1 = T_2 = 10°C$ and $h_1 = 5$ m, then $C_i = 1.1$ m s^{-1}, while for the open sea with $S_1 = S_2 = 35$, $T_1 = 25°C$, $T_2 = 20°C$ and $h_1 = 50$ m, then $C_i = 0.8$ m s^{-1}. Again, speeds are much less than those for surface waves, a typical feature of internal waves. At the same time, the small density difference between layers (23.4 kg m^{-3} for the coastal case and 1.4 kg m^{-3} for the open ocean case above, compared with about 1024 kg m^{-3} for the ocean water) permits the height of the internal waves to be quite large even for a small energy content.

The analogy with surface waves holds for the water speeds and orbital motions too. For short interfacial waves, the orbits are circular and they, the speeds and wave pressure decay exponentially away from the interface (as $\exp(-kd)$ where d is the vertical distance from the interface). For long interfacial waves the orbits are elliptical, the horizontal dimension of the orbit, horizontal speed and wave pressure are nearly independent of z. The vertical speed and vertical dimension of the orbit decrease linearly away from the interface. They go to zero for a solid level boundary. For an upper free surface they go to $\Delta\rho/\rho_1$ of the values at the interface.

If the upper layer is thin ($h_1 \lesssim \Lambda_i/20$), there will also be a wave system at the sea surface associated with the internal waves (and independent of any wind waves at the surface) but the amplitude of this surface system is approximately $(\Delta\rho/\rho)\eta_i$ and is normally too small to see and difficult to measure because wind waves are usually present. However, the presence of the internal waves can often be detected visually by secondary effects if the upper layer is not very thick. As the internal waves travel along, the upper layer alternately gets thicker and thinner. Thus there are travelling convergences and divergences in this layer. The convergences may cause bands of irregular ripples on the sea surface above them (perhaps by compressing short surface waves making them steeper and more visible), while there is smoother water over the divergences. The ripples are positioned just behind the crests of the internal waves. In other circumstances, generally when the upper layer is somewhat thicker, the convergence in the upper layer brings together organic material on the surface of the sea and thus changes its surface tension, tending to suppress any ripples which would otherwise be formed by a light wind at the surface. The result is a smooth band over the convergent region with wind ripples elsewhere. Thus, if a pattern of alternate bands of smooth and rippled water is seen on the sea surface, it is quite probable that there is an internal wave train below the surface. Another characteristic which may reveal the presence of internal waves is light-coloured silt in the upper layer, as from a river. If the upper layer is thin (a few metres) then over the troughs of the internal waves there will be a thicker layer of silty water which looks lighter in colour than the thinner layer over the crests of the internal waves (which generally have clear, dark water below).

In the upper layers of coastal waters, internal wave periods of a few minutes and amplitudes of several metres are often observed; in ocean waters where the density differences are smaller, periods of up to 12 hours and amplitudes of 10 to 300 m or more have been recorded (e.g. Fig. 13.11).

12.92   *Internal waves when the density varies continuously with depth*

The non-specialist may find the material of this subsection somewhat heavy going but it is included because of the significance of internal waves to the ocean structure and because some of the characteristics, e.g. the group velocity being directed *parallel* to wave crests in some situations, are very different from one's usual ideas about waves.

The previous discussion has been of the simple two-layer situation in which there is assumed to be a layer of water of uniform lower density over a layer of uniform higher density with a sharp interface, leading to a simple internal wave system. This model is quite realistic for coastal regions where there is a significant river runoff giving rise to an upper layer of low salinity over a deep layer of much higher salinity, with a steep gradient of salinity (i.e. density) between them. It is also a good approximation for internal waves at the base of the oceanic upper mixed layer. With continuous stratification, i.e. continuous variation of density with depth, internal waves still occur and, indeed, are much more interesting since unlike interfacial waves (for which the interface is generally almost horizontal) they do not always propagate horizontally. This fact is of considerable significance because it provides a means for propagating energy from the surface to the bottom of the ocean and vice versa and so influencing ocean movements and structure over the full depth of the ocean.

To illustrate what happens we consider the simplest possible situation of an ocean with a constant Brunt–Väisälä frequency given by $N^2 \simeq -(g/\rho)(\partial \sigma_t/\partial z)$. (See Chapter 5 for how to calculate N^2 exactly in the deep ocean when this approximate expression is not valid.) The Boussinesq approximation is used, that is we ignore density variations except in calculating $\partial \sigma_t/\partial z$ so ρ is taken to be a mean value. Now we must allow for non-horizontal propagation and treat $\mathbf{k}$ as a vector. The horizontal part of the propagation is taken to be in the x-direction; for a single wave we can always take the x-axis in the direction of horizontal propagation so the results to be described hold for any horizontal direction. Figure 12.14 shows the geometry for a wave propagating in a partly vertical direction.

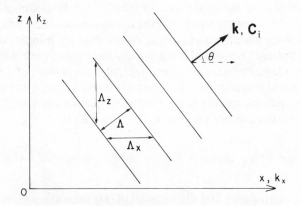

Fig. 12.14   *Geometry for an internal wave propagating at angle θ to the horizontal.*

The wave number $\mathbf{k}$ has components $k_x = 2\pi/\Lambda_x$ and $k_z = 2\pi/\Lambda_z$ and $k_x^2 + k_z^2 = k^2$ where $k = 2\pi/\Lambda = |\mathbf{k}|$. The dispersion relation may be derived from the equations of motion (e.g. LeBlond and Mysak, 1978, section 8) and is $\omega = Nk_x/k = N\cos\theta$ where θ is the angle between the x-axis and $\mathbf{k}$. Since $\cos\theta \leqslant 1$, $\omega \leqslant N$ and N is the upper limit in frequency for an internal wave in the continuously stratified case. (For interfacial waves, $\partial \sigma_t/\partial z \to \infty$ at the interface and there is no frequency limit. In practice, of course, the interface is not perfectly sharp and, as we shall discuss later, $\omega \leqslant N_{\max}$ in a pycnocline.)

The phase velocity, $\mathbf{C}_i$, is in the direction of $\mathbf{k}$ (by definition) and has magnitude $C_i = \omega/k = N\cos\theta/k$. For fixed N and θ, ω is a fixed constant too but there is, for a unbounded ocean, an infinite range of wavelengths! For fixed θ the waves are clearly dispersive since C_i depends on k and increases as wavelength increases (as k decreases). For fixed k (and wavelength), ω and C_i vary with θ (direction of propagation). Using $k^2 = k_x^2 + k_y^2$, we can rewrite the dispersion relation as $k_x^2(N^2 - \omega^2)/\omega^2 = k_z^2$. In the k_x, k_z plane for fixed N and ω these are straight lines of slope $\pm[(N^2 - \omega^2)/\omega^2]^{1/2}$. Figure 12.15(a) shows such a plot for several values of ω.

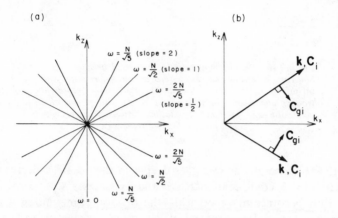

FIG. 12.15   (a) *Lines of* $\omega = constant$ *in* k_x, k_z *space, (b) relations between directions of* $\mathbf{k}$, $\mathbf{C}_i$ *and* $\mathbf{C}_{gi}$.

Now the group velocity must be treated as a vector and $\mathbf{C}_{gi} = \partial\omega/\partial\mathbf{k}$ is the gradient of ω in $\mathbf{k}$ space; it is perpendicular to lines of $\omega = $ constant and towards increasing ω, that is perpendicular to $\mathbf{k}$ (and to $\mathbf{C}_i$!) and towards the k_x axis; $|\mathbf{C}_{gi}| = Nk_z/k^2$. Although the waves are normally dispersive in the sense that long waves travel faster than short ones for a fixed θ, $C_{gi} \lesseqqgtr C_i$ for $\theta \lesseqqgtr \pi/4$. When $\mathbf{C}_i$ has a downward component, $\mathbf{C}_{gi}$ has an upward component and vice versa. As $\omega \to 0$, $C_i \to 0$, $\mathbf{C}_{gi}$ is nearly horizontal and C_{gi} is largest in magnitude for a given (fixed) value of k. As $\omega \to N$, $C_{gi} \to 0$, $\mathbf{C}_i$ is nearly horizontal, $\mathbf{C}_{gi}$ is nearly vertical and C_i is largest in magnitude for a given value of k. This behaviour may be seen in Fig. 12.16(a, b). The beams of waves are parallel to $\mathbf{C}_{gi}$ and become more vertical as ω/N increases. The crests propagate *across* the beam as shown in Fig. 12.16(c). Also the particle motions are perpendicular to $\mathbf{k}$ and $\mathbf{C}_i$ and the waves are termed *transverse*. Note that in a real situation corresponding to the sketch of Fig. 12.16(a, b), k would be fixed by the size of the wavemaker at the centre and thus θ decreases as ω/N increases but the beam angle increases since $\mathbf{C}_g$ is perpendicular to $\mathbf{k}$.

If N is variable but changes only slightly over one wavelength, then we can use the $N = $ constant theory to predict how internal waves will refract with changes in N just as we used the theory with $h = $ constant to predict how surface waves would be refracted in the presence of slow depth changes

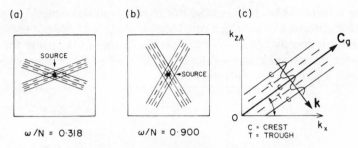

FIG. 12.16   *Phase configurations of internal gravity waves. Source is at centre and rays spread out in X-formation with (a) $\omega/N = 0.318$, (b) $\omega/N = 0.900$. Full lines represent troughs and dashed lines represent crests with propagation across the rays; (c) shows the relations between group velocity ($\mathbf{C}_g$) and phase velocity direction ($\mathbf{k}$).*

(Section 12.51). Generally as one goes deeper in the ocean N decreases. For fixed ω, with $\omega = N \cos\theta$, $\cos\theta$ increases and θ decreases; $\mathbf{C}_g$ is perpendicular to $\mathbf{k}$ and thus becomes more vertical—the rays or beams along which the energy propagates become more vertical. When a ray reaches the surface (or the base of the mixed layer) or the bottom it is reflected with the angle of reflection (*relative to the vertical*) being equal to the angle of incidence. Figure 12.17 shows such a situation where the internal waves of tidal frequency are being generated by the surface tidal wave as it goes over the continental shelf. The results shown in Fig. 12.17 may be readily shown in laboratory models but attempts to find such patterns in field observations have not been particularly successful.

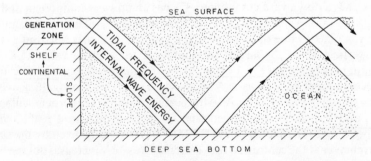

FIG. 12.17   *Internal wave generated by surface tide over continental shelf and being refracted in passage to the bottom, then reflected back to the surface, reflected again, etc. Stippled areas are "shadow zones" with no tidal internal wave energy. Note that the vertical scale is exaggerated as usual for vertical section plots and the rays are, in fact, nearly horizontal for tidal frequencies which are small compared with N.*

When the bottom is sloping, wave energy approaching the shore may be reflected *backwards* from the bottom instead of with a forward component. Backward reflection occurs when the bottom slope is greater than the ray slope (relative to the horizontal). Figure 12.18 shows a "transmitted" ray (1) and a "reflected" ray (2) with $N =$ constant for simplicity. (In considering this figure, it is important to remember that the rays (shown as lines) represent the *direction* of travel only. The waves themselves may be extensive (along their crests) and occupy a large proportion of the volume of the ocean represented by the section shown.) For the "transmitted" ray, the angle of reflection equals the angle of incidence about a *vertical* axis at the sea surface or bottom whereas for the "reflected" ray the angles of incidence and reflection are equal about a *horizontal* axis. This reflection process is quite different from that for light where the angle of reflection equals the angle of incidence about a *normal to the reflecting surface*. This approach of using ray theory with N varying slightly over a wavelength is applicable to internal waves which have wavelengths much less than the water depth. The case when the wavelength is comparable to the depth will be considered next.

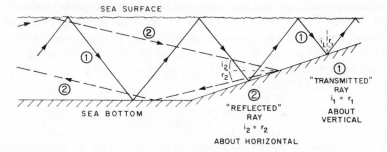

FIG. 12.18   *Internal wave energy reaching sloping bottom with ray 1 being "transmitted" toward shallow water while ray 2 is "reflected" from the bottom back to deep water. (See text for details.)*

12.93   *Normal modes for internal waves*

Consider the case with top and bottom boundaries. At the bottom, taken to be level, $w_{-h} = 0$ is required. At the top, with a free surface, a long surface wave is permitted. This is often termed the barotropic mode. For a long surface wave the amplitudes of the horizontal velocity components and the wave pressure are independent of z, isobaric surfaces are parallel over the whole depth and the term barotropic is appropriate. The surface slope is associated with this mode whereas for geostrophic currents the deep isobaric slope is connected with the barotropic flow and deviations from the deep isobaric slope are connected with the baroclinic flow as discussed in Chapter 8. Because the density variations are small, the barotropic (surface) wave mode is essentially unaffected.

In addition to the surface mode, internal modes are allowed; the amplitudes of the velocity components and wave pressure vary with z and the internal waves are appropriately termed baroclinic modes. The vertical velocity at the free surface associated with the internal waves has amplitude $0(\Delta\rho/\rho)$ times the maximum amplitude in the water column. Since this free surface amplitude is small it is a good approximation to take $w = 0$ at the free surface. Taking $w = 0$ is equivalent to saying that the surface is rigid or solid as far as the internal waves are concerned, simplifying the analysis. The surface is essentially rigid for the internal waves because the stratification is weak, that is the overall density variations are small. With $N = $ constant, the vertical velocity is obtained from the equations of motion and takes the form

$$w = A \sin\left[k_z(z + h)\right] \sin\left(k_x x - \omega t\right). \qquad (12.13)$$

To have $w = 0$ at $z = 0$ requires $\sin k_z h = 0$ or $k_z h = n\pi$, where $n = 1, 2, 3$, etc. Thus instead of having any value of k_z for fixed ω and N, only discrete values are allowed because of the boundaries. Note that the phase now is $(k_x x - \omega t)$ and involves only k_x so that the crests and energy propagate horizontally. To satisfy boundary conditions, two waves are required with equal upward and downward components of propagation which cancel leaving only horizontal propagation. (The top and bottom boundaries thus act as a waveguide.) Using the dispersion relation in the form $k_z = [(N^2 - \omega^2)/\omega^2]^{1/2} k_x$ and $k_z h = n\pi$ allows us to relate ω and k_x as

$$\omega^2 = \frac{N^2}{[1 + (n\pi/k_x h)^2]}. \qquad (12.14)$$

Here ω increases as k_x increases; shorter waves have higher frequency. Since we only see horizontal propagation now, with wavenumber k_x, C_i and C_{gi} are taken to be horizontal. The phase speed $C_i = \omega/k_x$ and decreases as k_x increases (short waves are slower than long waves) and $C_{gi} = \partial\omega/\partial k_x$ is less than C_i. The horizontal velocity component can be obtained from ω using the continuity equation (as $\partial u/\partial x + \partial w/\partial z = 0$ because $v = 0$ since motion is only allowed in the x, z plane by our assumption of propagation in the x-direction). Then

$$u = \frac{A k_z}{k_x} \cos\left[k_z(z + h)\right] \cos\left(k_x x - \omega t\right). \qquad (12.15)$$

The variations with depth of the amplitudes of w and u are shown in Fig. 12.19 for the case $N = $ constant. Note that the mode number n equals the number of zero crossings for u and the number of extrema (maxima and minima) for w. If N is not constant with depth, then the vertical structure of u and w will be modified. For a given $N(z)$ the modal structure can be calculated, numerically if necessary.

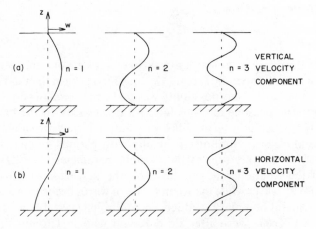

FIG. 12.19   *Variations with depth of amplitudes of (a)
vertical (w) and (b) horizontal (u) velocity components
for the first three normal modes of internal waves with
N = constant with depth.*

With N = constant, $\omega \leqslant N$ was required for internal waves to exist. With variable N the requirement is that $\omega \leqslant N_{\max}$ for internal waves to exist. The velocity amplitudes have modal structure in regions $\omega \leqslant N(z)$ and decay exponentially in regions where $\omega \geqslant N(z)$. If ω is near $N_{\max}$ in a pycnocline region then the waves will be "trapped" in a narrow range of depths. The ultimate pycnocline is an abrupt jump in density at an interface which leads to the interfacial waves described earlier in this section.

12.94   *Causes of internal waves*

Knowledge of the causes of these internal waves is incomplete. Possible causes are instability of flow where there is a strong vertical shear of velocity (e.g. strong tidal currents through passes or over bottom irregularities), inverted barometer effects associated with moving low atmospheric pressure systems and related short-period variations in wind stress and non-linear transfer from surface waves. Further discussions of possible generation and dissipation mechanisms for internal waves are given by LeBlond and Mysak (1978, section 53) and by Munk (1981) who also discusses the possible connections with small-scale mixing processes.

Because of the associated vertical oscillations of the water, internal waves can be a considerable nuisance when one is attempting to determine the steady-state distribution of water properties (e.g. see Defant, 1950). They are also probably a significant factor in promoting mixing between the upper and lower layers in the sea if they break to form "internal" surf.

12.10   Effects of rotation

12.10.1   *Introduction*

Consider the possible oscillatory (wavelike) motions of a thin layer of fluid under the action of gravitational and Coriolis forces when the fluid is taken to cover the whole earth. This possibility was first considered in connection with tidal theory. The possible motions fall into two classes. In the first class are gravity waves, which may be modified by rotation if they are sufficiently long. In the second class are motions, generally with periods greater than a day, associated with the variation of the Coriolis parameter ($f = 2\Omega \sin \phi$) with latitude. If the rotation rate goes to zero, the first class of waves become ordinary gravity waves while the second class of waves become steady currents. If a vertical lateral boundary is added, an additional special type of wave may exist, called a *Kelvin wave* after Lord Kelvin who first found the solution describing it.

The second class of waves are also called *planetary or Rossby waves* since Rossby was the first to investigate them on the β-plane (where $f = f_0 + \beta y$ is used with f_0 and β both taken to be constant). In the following, we shall discuss some of the basic characteristics of such waves.

12.10.2   *Modified gravity waves; Sverdrup and Poincaré waves*

For gravity waves with periods (T) approaching one-half pendulum day ($2\pi/f$), the Coriolis terms (fu, fv) become comparable in size to the local time-rate of change ($\partial u/\partial t$ or $\partial v/\partial t$). For example, if u, v vary as $\sin(2\pi/T)$, then the local time derivatives have amplitudes proportional to $(2\pi/T)u$ or $(2\pi/T)v$ which are equal to the Coriolis term for $T = 2\pi/f$. For much shorter periods (e.g. wind waves and swell and relatively short-period internal waves) the Coriolis terms may be ignored since $T \ll 2\pi/f$. For a horizontally infinite ocean with f = constant, it can be shown that the free modified gravity waves have periods $T < 2\pi/|f|$ for both surface and internal types of gravity wave. Such waves may be termed *gravitational–gyroscopic waves* to indicate that both gravity and rotation are important. (Sometimes the term *inertial* is used instead of gyroscopic; to avoid confusion with inertial oscillations (Section 8.2), *gyroscopic* should be used.) Inertial oscillations are the limiting case of gyroscopic waves at the "inertial" frequency $\omega = f$. Forced gravity waves, e.g. those produced by the tidal forcing with $T > 2\pi/|f|$ may occur. Free waves are the possible motions which may occur after some disturbance has occurred, e.g. a change in wind stress.

Let us take the ocean to be of uniform density (a valid model for a barotropic case as discussed in Section 7.4). The horizontal pressure gradients are then given by the surface elevation gradients (see Section 9.14.2). The equations for horizontal motion, neglecting the advective accelerations which we take to be small, are

$$\frac{\partial u}{\partial t} - fv = -g\frac{\partial \eta}{\partial x}$$

$$\frac{\partial v}{\partial t} + fu = -g\frac{\partial \eta}{\partial y}$$

(12.16)

while the equation of continuity with the time average depth, h, constant and $\eta \ll h$ is

$$\frac{\partial \eta}{\partial t} = -h\left(\frac{\partial u}{\partial x} + \frac{\partial v}{\partial y}\right).$$

(12.17)

Taking f = constant, a solution for a wave (which will be a long wave since T is large) travelling in the x-direction is

$$
\begin{aligned}
u &= [(g/h)/(1 - s^2)]^{1/2}\eta_0 \cos (kx - \omega t) \\
v &= s[g/h)/(1 - s^2)]^{1/2}\eta_0 \sin (kx - \omega t) \\
\eta &= \eta_0 \cos (kx - \omega t)
\end{aligned}
$$

(12.18)

where $s = fT/2\pi = f/\omega$, $\omega^2 = (ghk^2 + f^2)$, the phase and group speeds are

$$C = [gh/(1 - s^2)]^{1/2} \quad \text{and} \quad C_g = [gh(1 - s^2)]^{1/2} \text{ respectively.}$$

Note that $C \times C_g = gh$ and $C_g < (gh)^{1/2} < C$; rotation has made these waves dispersive. With the long waves travelling faster and $C_g < C$, the dispersion is normal. Clearly $s^2 < 1$ or $T < 2\pi/|f|$ (one-half pendulum day) for physically possible solutions. We see that the phase speed is increased by rotation and a horizontal component of the motion perpendicular to the direction of propagation is introduced. This latter behaviour should not be too surprising. As a particle moves forward it will tend to be deflected to the right (or left) by the Coriolis force and vice versa as it moves back. When the period is comparable to $2\pi/f$, the Coriolis effect will be important; when the period is much shorter it will not. Since these are long waves, $w \ll u$ or v; $w = \partial \eta/\partial t$ at the surface and decreases to zero at the bottom just as for ordinary long gravity waves.

These solutions are sometimes called *Sverdrup waves* because he used them in a paper published in 1926. Combinations of these waves which can propagate in a channel were discovered by Poincaré and were described in a publication in 1910. *Poincaré waves* are often required in studies of tidal propagation in semi-enclosed seas. Often the term Poincaré wave is taken to mean any long gravity wave for which rotational effects are important. Note also that a Poincaré wave requires $\omega > |f|$. If a Poincaré wave propagates poleward it cannot go past the latitude where $\omega = |f|$ and will be reflected there. Poincaré "waves" with $\omega < |f|$, (i.e. $T > 2\pi/|f|$) may be needed in forced problems to satisfy boundary conditions but will decay exponentially in space away from the boundary and do not propagate although they are still periodic in time.

Internal waves with periods comparable to $2\pi/|f|$ are also affected by rotation. For the case $N^2 \gg 4\Omega^2 \cos^2 \phi$ (in the ocean for depths < 3 km), $\omega = |f|$ is the lowest frequency, i.e. $T < 2\pi/|f|$ as noted above. Long internal wave solutions have the same form as the surface Sverdrup or Poincaré waves just discussed except that the h's in the formulae are replaced with an equivalent depth h_n, where $n = 1,2,3,\ldots$, etc., for the various modes. In general, $h_n \ll h$; for $N = $ constant, $h_n = N^2 h^2/gn^2 \pi^2$. With $N = 2 \times 10^{-3}$ rad s^{-1}, $h = 5$ km and $n = 1, h_1 \simeq 1$ m, so the internal modes are much slower than surface waves as usual. In the deep ocean, $N^2 \gg 4\Omega^2 \cos^2 \phi$ does not hold and the situation becomes more complicated. In particular, the horizontal component, $\Omega \cos \phi$, of the earth's angular velocity Ω and the Coriolis terms arising from it in the equation of motion (Chapter 6) cannot be neglected.

12.10.3   Kelvin waves

If a lateral (vertical) boundary exists, then the Kelvin wave solution to equations (12.16) and (12.17) is possible. It is, with the boundary parallel to the x-axis (i.e. east–west),

$$u = \eta(g/h)^{1/2}, \quad v = 0, \quad \eta = \eta_0 \exp(-fy/C) \cos(kx - \omega t) \quad (12.19)$$

with $C = (gh)^{1/2} = C_g$ and $h = $ constant. Kelvin waves propagate forward with the boundary on the right in the northern hemisphere (on the left in the southern hemisphere). The amplitude is greatest at the boundary and decays exponentially away from it (a characteristic which identifies it as a type of *boundary wave*). At each point at any time, the Coriolis force balances the pressure force due to the surface slope. Kelvin waves may also occur along the equator, where f changes sign, propagating from west to east, and over abrupt changes in bottom topography (where they are called "double" Kelvin waves because there is motion on both sides of the depth change).

Note that there is no restriction on the frequency of Kelvin waves except that it must be low enough that our assumption that we are dealing with long waves is correct. With $v = 0$, the x-momentum equation and continuity equation just give a long wave travelling at the usual speed $C = (gh)^{1/2}$ with no apparent effect of the rotation. The y-momentum equation is geostrophic so the rotation determines the fall off in amplitude away from the coast and the direction of propagation since with reverse propagation the amplitude would increase exponentially away from the boundary which is not physically realistic. The length scale of the amplitude decay is $C/|f| = (gh)^{1/2}/|f|$. With $h = 5$ km and $f = 10^{-4}$ s^{-1} (mid-latitudes), $C/|f| \simeq 2200$ km; for a shallow sea, $h \simeq 100$ m, $C/|f| \simeq 300$ km. The length scale here is similar to the Rossby radius of deformation, $\lambda = (g'D_0)^{1/2}/|f|$, introduced in Section 9.14.2. Thus $C/|f|$ is the external or barotropic Rossby radius of deformation (λ_e) while the radius introduced in Chapter 9 is one form of the internal or baroclinic Rossby radius of deformation (λ_i). For a two-layer fluid

$$\lambda_i = (g'(h_1 h_2)/(h_1 + h_2))^{1/2}/|f| \simeq (g'h_1)^{1/2}/|f| \text{ for } h_2 \gg h_1 = D_0$$

$$(12.20)$$

as in Chapter 9. The internal radius of deformation λ_i would be the scale length for the amplitude decay for an internal Kelvin wave whose form is the same as that given with h replaced by h_n; for $\Delta\rho/\rho \simeq 2 \times 10^{-3}$ and $h_1 \sim h_2$, $\lambda_i \sim 0.03$ $C/|f| = \lambda_e$. For continuous stratification, $\lambda_i = N h/|f|$.

Tides, to be discussed in the next chapter, generally have a Kelvin wave nature with Poincaré modes in some regions. When the Kelvin wave dominates, it propagates around basins with the coast on the right (northern hemisphere) with highest amplitudes at the coast. Examples are shown in Fig. 13.7 for the St. Lawrence estuary and in Fig. 13.8 for the North Atlantic.

12.10.4   Planetary or Rossby waves

These are waves of long period which are associated with the "β-effect", the variation of the Coriolis parameter f with latitude. Here we shall indicate some properties of the solutions on a β-plane, leaving more complete discussion for more advanced texts (e.g. LeBlond and Mysak, 1978, Chapter 3).

First we should consider how variations of f with latitude may lead to oscillatory motions. When discussing equatorial undercurrents in Chapter 9 we pointed out how an eastward current might oscillate if perturbed. More generally, suppose that we move northward a parcel of water whose initial relative vorticity is zero, with no influences such as friction or depth changes which would cause its potential vorticity to change. As f increases, the parcel will have negative relative vorticity when displaced (northward) and will circulate clockwise. Because of its variation, the Coriolis force will be maximal in size on the poleward part of the parcel and minimal on the equatorward part; the variation of f thus leads to a net southward force tending to produce southward displacement, i.e. a restoring force. If this force then pushes the parcel south of the latitude of zero relative vorticity (overshoots) the circulation becomes anticlockwise, and considering the f variation the parcel now suffers a net northward force, i.e. again a restoring force. Thus the variation of f provides a restoring force (in the horizontal plane) allowing oscillations to occur just as the effect of gravity does (in the vertical) for surface or internal waves. The flow will be nearly horizontal and for $T \gg f/2\pi$ will be essentially geostrophic (often termed "quasi-geostrophic") since the time derivatives (and advective accelerations) will be small compared with the Coriolis terms. Thus with sufficiently detailed observations, such flows (or waves) may be shown by geostrophic calculations.

From the equations of motion (e.g. LeBlond and Mysak, 1978, section 18) the dispersion relation for Rossby waves is

$$\omega_n = -\frac{\beta k_x}{k^2 + f^2/gh_n} \qquad (12.21)$$

where k_x is the x-component of the horizontal wave number $\mathbf{k}$ (and with ω_n positive, k_x is always negative), $k = 2\pi/\Lambda = (k_x^2 + k_y^2)^{1/2}$ and $\beta = \partial f/\partial y$. The barotropic (depth independent) mode is that with $n = 0$ and $h_0 = h$; $n = 1, 2,$ etc., are baroclinic (internal) modes with h_n being the equivalent depth as in the previous subsections. Note that as Λ decreases (k increases), ω decreases (period increases); the shorter waves are of longer period in contrast to surface gravity waves. The phase velocity has components

$$C_{nx} = -\frac{\beta k_x^2}{k^2(k^2 + f^2/gh_n)}, \quad C_{ny} = -\frac{\beta k_x k_y}{k^2(k^2 + f^2/gh_n)}. \qquad (12.22)$$

C_{nx} is always negative, that is the phase velocity always has a westward component; C_{ny} may be positive or negative (north or south); the group (energy propagation) velocity may, however, be in any direction and has components

$$C_{ngx} = \frac{\beta(k_x^2 - k_y^2 - f^2/gh_n)}{(k^2 + f^2/gh_n)^2}, \quad C_{ngy} = \frac{2\beta k_x k_y}{(k^2 + f^2/gh_n)^2}. \qquad (12.23)$$

Note also that $f^2/gh_n = 1/\lambda_e^2$ for $n = 0$ and equals $(n\pi/\lambda_i)^2$ for $n = 1, 2,$ etc. The term (f^2/gh_n) arises from including the horizontal divergence; for $\Lambda \leqslant \lambda_e, f^2/(gh) \leqslant 0.03\ k^2$ and may be neglected. For the open ocean, $h = 5$ km and $f = 10^{-4}\ \mathrm{s}^{-1}, \lambda_e \simeq 2200$ km; with $N \simeq 2 \times 10^{-3}\ \mathrm{rad\ s}^{-1}, \lambda_i \simeq 100$ km and the term $(n\pi/\lambda_i)^2$ is usually important.

To get some idea of the propagation speeds and periods, consider the barotropic mode for $\Lambda < \lambda_e$ and waves with $k_y = 0, -k_x = k$, i.e. phase propagation to the west. Then $\omega_0 = \beta/k, C_{0x} = -\beta/k^2, C_{0y} = 0 = C_{0gy}, C_{0gx} = \beta/k^2$ (opposite in sign to C_{0x}, i.e. to the east!). With $\beta = 2 \times 10^{-11}\ \mathrm{m}^{-1}\ \mathrm{s}^{-1}$ and $\Lambda = 1000$ km, $C_{0x} = -0.51\ \mathrm{m\ s}^{-1}$ while for $\Lambda = 300$ km, $C_{0x} = -0.05\ \mathrm{m\ s}^{-1}$. The corresponding periods are 23 days and 76 days. For phase propagation to the southwest, $k_x = k_y, k^2 = 2k_x^2, \omega_0 = (\beta/\sqrt{2}k), C_{0x} = C_{0y} = -\beta/2k^2, C_{0gx} = 0, C_{0gy} = \beta/k^2$ (to the north!).

For southwest phase propagation compared with westward phase propagation at a given k, the period is multiplied by 1.4 and the phase speed is multiplied by 0.7; the group speed stays the same but the direction of energy propagation changes by $90°$. For the baroclinic modes, the speeds will be lower and the periods longer because of the extra $f^2/(gh_n)$ terms.

The mesoscale "eddies" observed in the POLYGON and MODE experiments may be fitted by linear superpositions of Rossby waves (including internal types). However, the flow speeds in these eddies are of $O(10 \text{ cm s}^{-1})$ while the phase speeds are of $O(5 \text{ cm s}^{-1})$. In waves, the ratio of the flow speed to the phase speed is a measure of the relative importance of the non-linear terms. Although the eddies exhibit westward-phase propagation and can be fitted by Rossby waves with observations in a limited region, the dynamics cannot be linear.

Eddies may be generated by baroclinic or barotropic instability. If non-linear effects are important they tend to grow in size (behaving like two-dimensional turbulence). However, if they grow sufficiently they may become nearly linear (a superposition of Rossby waves) *before* they interact with the mean flow and will radiate their energy away (perhaps to be dissipated or reflected at boundaries). If they grow to the size where the dynamics are linear before interacting with the mean flow they may not be very important for the mean flow except perhaps as an energy sink. Since the eddies seem to start out rather small in the ocean, compared with those in the atmosphere, the eddies may not be very important for the mean flow except as a loss process in contrast to the atmospheric case. Their importance has not yet been established, as noted before in Chapter 11.

12.10.5 *Topographic effects*

More generally, it is not the f variations which matter but variations of f/h. When depth variations dominate (as often occurs near coasts) the possible Rossby waves are termed *topographic Rossby waves*. Indeed, in laboratory scale models, where rotation rate cannot be varied with position, the β-effect can be simulated by varying h.

Variations in topography may lead to *wave trapping* (concentration of wave energy in certain regions). (Variation of f may also lead to trapping, particularly near the equator.) Trapping may also occur with gravity waves for which the term *edge waves* is used or with Rossby-type waves for which the term is *shelf waves*. Kelvin waves may also be considered a special case; their very existence depends on the presence of a boundary (or on f changing sign at the equator or on an abrupt depth change). They are a boundary phenomenon in the sense that the oscillations are large near the boundary and decay away from it.

As an example, consider gravity waves approaching a shore at an angle (as shown in Fig. 12.8(a)) with the water depth decreasing but with a vertical boundary (cliff) so that the depth does not go to zero. By refraction, the wave crests will become more nearly parallel to the cliff (wave orthogonals more nearly perpendicular to the cliff). If the waves do not break they will be essentially totally reflected at the cliff at an angle equal to the angle of incidence. They will then travel outward, with refraction causing the crests to become perpendicular to the shore and then will turn inward to be reflected again. The

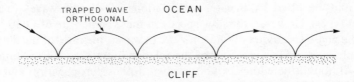

FIG. 12.20   *Refraction leading to wave trapping when reflection occurs before wave breaking. One wave orthogonal is shown.*

wave orthogonals will form a series of arcs as in Fig. 12.20. For planetary waves, f/h variations may produce similar behaviour.

Edge waves seem to play a role in the formation of series of cusps on sandy beaches and in the separation between rip currents. Shelf waves may give rise to marked oscillatory long-shore currents. Much fuller treatment of topographic waves and wave trapping may be found in LeBlond and Mysak (1978).

13

Tides

13.1 Introduction

The *tide* is the name given to the alternate rise and fall of sea level with an average period of 12.4 h (24.8 h in some places). Locally, the period varies by an hour or so either side of the average figure and the rise and fall sequence shows an almost infinite variety around the globe. Figure 13.1 shows samples of the

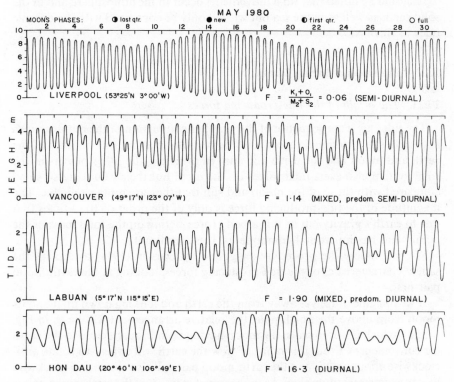

FIG. 13.1 *Tide curves for May 1980 (data from Admiralty Tide Tables) showing four types in terms of the "form ratio"* $F = (K_1 + O_1)/(M_2 + S_2)$ *of major diurnal to semi-diurnal constituents.*

four "forms" into which tides may be divided on the basis of the relative magnitudes of their main diurnal and semi-diurnal constituents, as will be described in Section 13.61. The rise and fall is the most obvious feature to most observers but fundamentally the prime phenomena are the horizontal tidal currents; the rise and fall near the coast is mainly a consequence of the convergence or divergence occurring there when the tidal currents flow toward or away from the shore.

The regular rise and fall and the tidal currents must have been observed by shore dwellers in pre-historic times; by the thirteenth century, serial observations led to empirical prediction techniques related to the moon's motions. However, it was only after Newton, about 1687, applied his Law of Gravitation to explain the basic physical cause of the tides that systematic methods for prediction could be developed. We will first present the basic dynamics to show how the present tidal theory and prediction techniques developed and will then describe some of the characteristics of tides and tidal currents in the oceans and near the coast.

It should be noted that tidal movements occur in the atmosphere and in the solid earth as well as in the sea but we will only be concerned with the oceanic tides in this text.

13.2   The tide-producing forces

13.21   *The origin of the tide-producing forces*

The tides are a consequence of the simultaneous action of the moon's, sun's and earth's gravitational forces, and the revolution about one another of the earth and moon and the earth and sun. In principle, the other planets in the solar system also exert tidal forces on the earth but their values are so small compared with those of the moon and sun that they are quite negligible. The magnitude of the tide-producing force is only of the order of 10^{-7} times that due to earth's gravity but as it is a body force, acting on the total mass of the ocean, and has horizontal components, it is significant.

We will follow the procedure suggested by Darwin (1911) to explain the salient characteristics of the tide-producing forces, looking at the earth/moon pair first.

If we imagine ourselves away from the earth and looking down on it and its moon from above the north pole, the relative arrangement will be as in Fig. 13.2(a) (which is not to scale, the moon having been brought close to the earth for convenience). From this point of view the earth will appear to rotate anti-clockwise about its axis and the earth/moon pair will also rotate anti-clockwise like an asymmetric dumbbell. As the centre of mass, Z, of the earth/moon pair is about one-quarter earth radius inside the earth, the rotation of the pair will be about an axis at the position Z and perpendicular to the paper. To understand

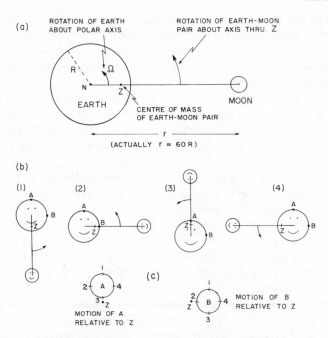

FIG. 13.2   (a) *Earth–moon pair rotates about a common centre of mass at Z*, (b) *successive positions of earth–moon pair with earth's axial rotation suppressed*, (c) *motions of two points, A and B, on earth's surface.*

the effect of the moon's gravitation plus the earth/moon rotation, we will assume for the present that the earth is not rotating on its axis. Figure 13.2(b) shows the character of the motion now in four stages (i) to (iv) of one complete rotation of the moon about the axis through Z. The motion of the earth now is an unusual one in that its *orientation* remains fixed (the face being drawn to emphasize this characteristic) but every part of the earth rotates in a circle of radius equal to the distance from the earth centre to the axis Z. (The easiest way to observe such a motion is to place your hand flat on a table, fingers outstretched, and to move it so that one point, e.g. the tip of the thumb, rotates in a horizontal circle of about 10 cm diameter, keeping your forearm pointing in the same direction all the time. You will see that every part of your hand describes the same sized circle, fingertips and palm alike.) Figure 13.2(c) shows the circles described by points A and B on the earth (Fig. 13.2(b)). For all points on the earth to move in a circle of this size it is necessary for there to be a centripetal acceleration (or force/unit mass) of the same magnitude everywhere and directed parallel to the earth/moon axis toward the moon. This centripetal acceleration is, on average, provided by the gravitational attraction of the moon, but as this attraction is not quite uniform over the surface of the earth, there is a slight excess of gravitational acceleration on the near side to the moon and a slight defect on the opposite side. (The gravitational acceleration is not uniform because the distance from the moon to points on the earth's surface varies from place to place.)

Figure 13.3(a) shows the earth/moon pair, with the moon close to the earth to emphasize some of the angles for convenience in showing components of forces (per unit mass, i.e. accelerations) later. (In the following, the term "force" will be taken to be per unit mass without stating so explicitly every time.) We will consider the forces at four points on the earth's surface, at A and C which are the nearest and farthest points from the moon, and at B and D which are equidistant from the moon. At points B and D, the gravitational force of the moon on unit mass on earth is shown by arrows F_b and F_d, respectively. These two forces will be equal in magnitude because of their equal distance from the moon. Because they occur at very nearly the same distance from the moon as is the centre of the earth (remembering that the angle BME (about $1°$ in fact) is grossly exaggerated in Fig. 13.3(a)), the value of this gravitational force will be almost exactly equal to the average value for the force of the moon on unit mass on earth (F/M_e). From Newton's Law of Gravitation, $F/M_e = G M_m/r^2$ ($\equiv CF = 3.32 \times 10^{-5}$ m s^{-2}), where G is the Gravitational Constant ($= 6.67 \times 10^{-11}$ N kg^{-2} m^2), M_e and M_m are the masses of the earth and moon respectively and r is the distance between their centres. The moon's gravitational force F_a at point A will be larger than the average value F/M_e because it is nearer to the moon, while the value at C (F_c) will be less than F/M_e. The differences are about $\pm 3\%$. The values of the four forces F_a, F_b, F_c and F_d are shown semi-quantitatively by the full-line arrows in Fig. 13.3(a). The total centripetal force needed to keep the earth and moon at their nearly constant distance apart while rotating is provided by their gravitational attraction and so must be equal to the value F, or F/M_e per unit mass ($= CF$), as is shown in Fig. 13.3(a) by the dashed arrows.

At D, by Newton's Law of Gravitation, the force/unit mass $F_d = GM_m/(r^2 + R^2)$ where R is the radius of the earth. Rewriting the denominator as $r^2(1 + R^2/r^2)$ and expanding in series gives $F_d = (GM_m/r^2)(1 - R^2/r^2 + \ldots)$. As $R/r \simeq 1/60$, F_d is nearly equal to CF in size, the difference being 1 part in 3600. Resolving F_d into components parallel and perpendicular to the line EM, i.e. F_{dx} ($= F_d r/(r^2 + R^2)^{1/2}$) and F_{dy} ($= F_d R/(r^2 + R^2)^{1/2}$) respectively, we get, again using series expansion

$$F_{dx} = (GM_m/r^2)[1 - (3/2)(R^2/r^2) + \ldots]$$
$$F_{dy} = (GM_m R/r^3)[1 - (3/2)(R^2/r^2) + \ldots].$$

To an accuracy of 1/2400, we may take $F_{dx} = CF$ and the residual to be $F_{dy} \simeq CF/60$ directed *inward*. (Note, however, that the size of $(F_{dx} - CF)$ is about 1/40 of F_{dy}.) Similarly at B there will be the residual force F_{by} of exactly the same

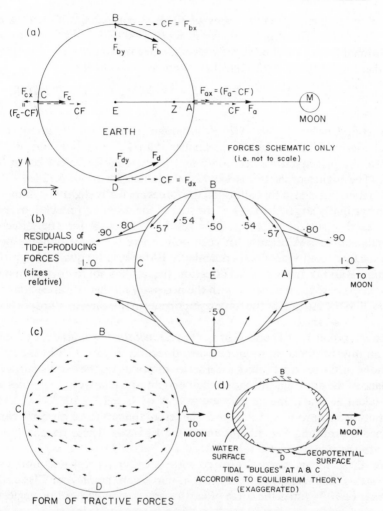

FIG. 13.3   (a) Directions of centripetal force per unit mass (CF) and moon's gravitational force per unit mass (F_a, F_b, F_c, F_d) at points on earth (not to scale), (b) directions and relative magnitudes of residuals of CF and F at various points on the earth's surface (correct to first order—see text), (c) form of the horizontal tractive forces over the earth's surface, (d) tidal "bulges" (much exaggerated) at A and C according to the equilibrium theory for an ocean covering the entire earth.

size as F_{dy} also directed *inward*. At A, $F_a = GM_m/(r-R)^2$ is directed *outward* (toward the moon). Expanding we have

$$F_a = (G M_m/r^2) (1 + 2R/r + 3R^2/r^2 + \ldots), \text{ and the residual}$$
$$F_{ax} = F_a - CF = (2G M_m R/r^3) [1 + (3/2) (R/r) + \ldots] \text{ directed } outward.$$

To first order (i.e. neglecting terms of $O(R/r)$ compared with unity) F_{ax} = $CF/30 = 2F_{by}$. The neglected term is about $1/40$ of F_{ax}. At C the moon's gravitational force F_c will be slightly less than needed to provide CF so that a mass there will behave as though there were an *outward* force

$$F_{cx} = F_c - CF = GM_m/(r+R)^2 - CF = -(2GM_m R/r^3)[1 - (3/2)(R/r) + \ldots]$$

Again to first order, $F_{cx} = -CF/30$ although its size is really about 2.5% smaller than $CF/30$ or about 5% smaller than F_{ax}. For our purposes, to illustrate the principles, we need only use the residual forces correct to first order. (The higher-order terms would be retained for more detailed calculations.) Remember also that although the forces are really slightly asymmetric, F_{ax} being slightly larger than $-F_{cx}$, the sum of the residuals (integral) over the whole earth's surface is zero. This result is obvious for the direction perpendicular to EM because for each point there is a corresponding point where the force component perpendicular to EM is exactly equal and opposite, e.g. points B and D. In the parallel direction, the result is not so obvious and it is left as an exercise for the reader with the necessary mathematical background to show that the integral of the force components in this direction proportional to R^2/r^4 does vanish.

Between points B or D and A or C, the residual force will gradually change from an inward direction to an outward direction. Figure 13.3(b) shows the directions of the residual forces for a series of points on the earth's surface in the plane of the drawing, and shows their magnitudes relative to the values at A or C taken as unity; the actual values at A or C $\simeq 1.1 \times 10^{-6}$ m s^{-2}. The distribution of the residual forces over the rest of the surface of the earth can be obtained by rotating Fig. 13.3(b) about the EM axis. These residual forces, then, are the *tide-producing forces*.

Now these forces can be resolved into (locally) vertical and horizontal components. The vertical component at its maximum (points A or C) is only of the order of 10^{-7} times the force of earth's gravitation and can be neglected. The horizontal forces are also small in absolute magnitude but are of the same magnitude as other horizontal forces in the ocean and are therefore significant. The distribution of these horizontal components is of the form shown in Fig. 13.3(c). They are zero along a meridian through the poles (B and D) and at the far and near points (A and C) from the moon; they are maximal along circles at $45°$ from A and C and are toward the moon on the near side of the earth and away from it on the far side. These horizontal components are the effective components of the tide-producing forces and are called the "tractive forces".

This pattern of tractive forces is tied to the earth–moon axis, but the earth has so far been assumed not to be rotating. If we now restore the rotation of the earth about its polar axis, each point on the earth's surface will, in one day, pass through a complete pattern of tractive forces, as in Fig. 13.3(c), with *two* passages of forces toward the moon and *two* away from it. That is, the tractive forces will have a period of one-half day even though the earth has a rotation period of one day. This is the basic reason for the existence of semi-diurnal tides for a diurnal rotation of the earth. (The day here is the lunar day of about 24.8 h because the moon is also moving with a 27.3-day period around the earth compared with the earth's rotation in 24.0 h relative to the sun.)

13.22   *Components of the tide-producing forces*

The earth/sun pair sets up a similar pattern of forces to that for the earth/moon pair but it differs from the latter in two respects: the maximum effect due to the sun is only about one-half of that due to the moon (because its greater distance outweighs its greater mass), and as the sun and moon do not rotate in synchronism the force patterns rotate independently and hence give rise to a rather complicated resultant.

The facts that the paths of rotation of the sun and moon about the earth are not circles but ellipses, and that the planes of rotation are not always in the equatorial plane but move north and south with an annual cycle for the sun and a monthly cycle for the moon add further complications to the resultant tide-producing forces. In addition, longer period variations of up to 19 years period occur due to other motions (such as the 18.6-year period due to the regression of the lunar nodes, the points where the moon crosses the plane of the ecliptic). However, the motions of the sun and moon are known very exactly and it is possible to express the resultant tide-producing forces as the sum of a number of simple harmonic (sine or cosine wave) *constituents*, each of which has its own characteristic (constant) period, phase and amplitude. These constituents are usually divided into three "species", with periods respectively of about one-half day (semi-diurnal), about one day (diurnal) and long period (longer than one day, generally a fortnight or longer). Some of the more important constituents with their size relative to the largest, the M_2 or principal lunar constituent, taken as 100 are given in Table 13.1. There are up to sixty-five constituents which are recognized as significant in some circumstances, e.g. in describing tides in river estuaries. The approximately 19-year period constituents are allowed for by multiplying the other constituents by a "node factor" which is changed annually.

13.3   Ocean responses to the tide-producing forces—tidal theories

There are two tidal theories, the equilibrium theory and the dynamical theory, both of which use the (horizontal) tractive forces to drive the oceans.

TABLE 13.1. *Characteristics of some of the principal tide-producing force constituents*

Species and name	Symbol	Period (solar hours)	Relative size
Semi-diurnal:			
Principal lunar	M_2	12.42	100
Principal solar	S_2	12.00	47
Larger lunar elliptic	N_2	12.66	19
Luni-solar semi-diurnal	K_2	11.97	13
Diurnal:			
Luni-solar diurnal	K_1	23.93	58
Principal lunar diurnal	O_1	25.82	42
Principal solar diurnal	P_1	24.07	19
Larger lunar elliptic	Q_1	26.87	8
Long period:			
Lunar fortnightly	M_f	327.9	17
Lunar monthly	M_m	661.3	9
Solar semi-annual	S_{sa}	4383	8

13.31 *The equilibrium theory of the tides*

The equilibrium theory was put forward by Newton as a first attempt to investigate the response of the oceans to the tide-producing forces and has been investigated by many workers since, regarding it as a standard of reference for more realistic theories. In the equilibrium theory, the entire earth is assumed to be covered with water of uniform depth and density. The tractive forces (Fig. 13.3(c)) then tend to cause the water to converge toward A and C and diverge from the meridional belt through B and D. If an infinite time is allowed, an equilibrium will be established when the hydrostatic pressure forces resulting from the slope of the water surface relative to geopotential surfaces (Fig. 13.3(d)) balance the tractive forces. Then if the solid earth rotates, a sequence of two high and two low waters each lunar day will be observed (except along the two zonal rings about B and D where there is no resultant rise or fall). This theory explains the semi-diurnal nature of the tides and the inequality of height of successive high and of low waters (when the moon is not in the equatorial plane). However, the predicted ranges of tidal rise and fall, about 0.55 m for the lunar and 0.24 m for the solar tides, with a combined maximum of 0.79 m, are far less than is observed for the real oceans. Other failings of the equilibrium tide are that the responses of the real ocean to the various force constituents (Table 13.1) are not in proportion to the sizes of these constituents but vary from place to place, and that high water does not always occur at the time of the moon's transit (zenith passage) but may be hours before or after this.

This theory is clearly very artificial with its assumptions of an earth completely covered with water and of infinite time allowed for equilibrium to become established between the hydrostatic and the tractive forces. (An alternative way to state the latter assumption is to say that the mass of the water is subject to gravitational forces but has no inertia, so that the equilibrium shape can be established instantly, friction being ignored.)

13.32   *The dynamical theory of the tides*

About a hundred years after Newton, Laplace put forward the dynamical theory. In this, a homogeneous ocean is still assumed to cover the whole earth at constant depth but the periodic tractive forces are considered to generate waves with periods corresponding to the constituents, i.e. forced waves. Coriolis acceleration and vertical particle acceleration are neglected. In subsequent developments, considerable effort was expended in seeking analytic solutions to the Laplace equations for water bodies of regular shape, such as narrow canals (to represent channels or long, narrow seas) or sectors of the surface of a sphere (to represent larger ocean areas). Most of these studies were for water of constant depth and they generally neglected any free oscillations which might be generated in a real ocean by the forced motions interacting with ocean bottom topography, although this aspect and some effects of density stratification have been studied lately. Other influences on the tides which have been shown to be significant are the effects of self-attraction in the water mass and of deformation of the solid earth by the ocean tides, while the Coriolis terms and vertical accelerations must be taken into account, particularly for internal tidal waves.

It has not been possible to obtain analytic solutions to the equations of motion for basins which realistically represent the actual oceans. However, it is possible to solve the equations numerically and promising results have been obtained for somewhat simplified models of the ocean basins (see Section 13.64). The limitations to obtaining detailed solutions lie in the large amount of computer time required if the ocean shapes and depths are to be represented without too much simplification.

13.4   The practical approach to tidal analysis and prediction

13.41   *Harmonic analysis—the classical method*

If present theory is inadequate to predict tides from dynamic principles, how is it that one can buy, for a modest sum, books of *tide tables* (predictions of times and heights of the tides and in some places for tidal currents) for a large proportion of the ports of the world? In effect, one uses the ocean itself empirically as a computer to solve the equations of motion. One records the rise and fall of water as a function of time at a particular location for a period of time, analyzes the resulting tide-height curve for its simple harmonic constituents, and uses them to perform the prediction into the future. This procedure was used empirically even before the development of the present tidal theory.

The recorded curve is a complex harmonic, i.e. the sum of many simple harmonics of different periods, phases and amplitudes. It is resolved into its simple harmonic constituents by mathematical procedures which are straight-forward (but tedious if done by hand). Each constituent can be represented by a sine curve with its own period and phase, whose amplitude represents its contribution (relative to mean sea level) to the total tide. These constituents can be calculated as far into the future as we wish, and then for any future time we can determine the expected tidal height simply by adding together all the constituents for that time, with appropriate corrections for the long-period variations associated with nodal regression. In practice, our observations of the tide curve have a limited accuracy so there is a limit to the accuracy with which we can determine the constituents and our predictions into the future become less accurate the further ahead in time that we go. Naturally, the longer the series of field observations which we have, the more accurate are likely to be the constituents and therefore our predictions.

Although tidal theory itself cannot yet predict the tides to a satisfactory accuracy, it does tell the practical man what constituents to look for in his analysis of the tide records, i.e. those of which the constituents in Table 13.1 are a selection (and in addition, perhaps, constituents at sum and difference frequencies which arise when non-linear effects become important, usually in coastal areas and estuaries). It was this contribution from tidal theory which changed tidal prediction from a purely empirical procedure to one based on sound physical principles.

Until the recent past, the *harmonic analysis* of the recorded tide curve into its constituents was carried out by paper and pencil methods on tabulations of hourly water heights, but it is now done by digital computer. The prediction into the future was carried out by an analogue device, invented by Lord Kelvin, in which the constituents were represented by rotating eccentrics (cams) whose throw was proportional to the amplitude and whose rate of rotation was inversely proportional to the period. A steel tape passing over all the cams totalled their displacements and this total was recorded on a paper strip, the result being a curve of tidal height versus time for the future. This step is now carried out by digital computer, which is programmed to compute the times and heights of high and low waters and to assemble them into tabular form with appropriate headings, etc., ready for printing.

An important point to note is that the tide *height* constituents obtained by the analysis for a particular locality do not necessarily have the same relative proportions as the tide-producing *force* constituents. The particular shape of the ocean basin in the vicinity causes the water to respond more readily to some constituents than to others, which is the main reason for the differences between tides in different parts of the oceans. The purpose of recording the actual tide for a period is effectively an analogue procedure for determining the local response, when our theoretical techniques are not adequate to do so mathematically.

The number of analysed constituents used for prediction depends on the accuracy required. Often, the use of the first seven in Table 13.1 (i.e. M_2, S_2, N_2, K_2, K_1, O_1 and P_1) will be sufficient to predict the tide to within about 10%, but generally twenty to thirty constituents are used for predicting two or three years ahead for ports close to the ocean, and sixty or more for those in river estuaries where the tide is rendered more complicated by bottom topography and non-linear effects.

It is usual to use hourly values for a continuous period of 369 days for analysis into twenty to thirty constituents, and to repeat the analysis for several such "years" to improve accuracy. Analysis for fewer constituents may be done from continuous records for as few as 29 or even 15 days. Records of 369, 29 or 15 days have approximately an integral number of cycles of the main constituents which are to be calculated and such lengths of records are required for the old "paper and pencil" methods. With digital computer analysis, any record length can be used—the longer the better. Also, with digital computer methods, observations at equally spaced times (e.g. hourly) are not required, although because they are generally available equally spaced observations are usually used. The number of constituents which can be calculated with sufficient accuracy (or "resolved" as the process is often termed) depends on the record length. The reader needing more detailed information on the practice and limits of tidal analysis is referred to *The Analysis of Tides* by Godin (1972).

It should be remembered that sea level is affected by other factors as well as the tide-producing forces, e.g. atmospheric pressure, wind set-up, solar heating, etc. These so-called meteorological tides are left as a residual by the harmonic analysis procedure and cannot be predicted (unless they have a tidal period, e.g. land–sea breeze effects or radiational tides (Section 13.42) which will be included as part of the S_2 tide). In the absence of such disturbances, the tide predictions are generally accurate to about ± 3 cm and ± 5 min. Meteorological effects, leaving out extremes such as hurricanes, may cause differences of tens of centimetres and tens of minutes.

13.42   *Fourier analysis—the response method*

The harmonic analysis procedure only extracts from the tide record the response to the specified gravitational force constituents and, as mentioned above, there is always a residual left, sometimes substantial. An alternative analysis procedure, termed the *response method*, regards the tide record as just another time-series and carries out a Fourier spectral analysis, i.e. determines the amplitudes and phases at equal frequency intervals ($\Delta f = 1/T$) from f_1

$= 1/T$ to $f_n = 1/(2\Delta t)$ where T is the record length and Δt is the sampling interval. Munk and Cartwright (1966) pioneered this approach to tidal analysis and revealed, in addition to the astronomical constituents, a spectrum of non-gravitational ones as well as a noise continuum (which arises from non-periodic effects such as meteorological fluctuations). The non-gravitational constituents of solar semi-diurnal, diurnal and annual periods are referred to as *radiational tides* and are attributed to the effects of solar radiation, both direct (heating) and indirect (wind effects). In addition, interaction with the atmospheric tides is indicated. (In places, these radiational tides constitute up to 30% of the gravitational S_2 constituent.) The inclusion of the non-gravitational constituents can give improved prediction but the improvement is usually small.

The harmonic analysis does include a good part of the radiational tides but the advantage of the response method is that it can separate gravitational from non-gravitational effects. For prediction of the normal tidal effects, the harmonic method seems to be sufficient. The response method is potentially useful for storm-surge studies since it allows for non-periodic effects. However, for prediction it requires long, reasonably continuous, records of at least a year, preferably of several years' duration. Analysis of shorter records, which is practical (for the major constituents) by the harmonic method, is not possible with the response method.

13.5   The measurement of tides

The simplest procedure for recording tide heights is to mount a vertical scale on a pier or wharf and to note visually the height of the water at, say, hourly intervals for long enough to obtain a record suitable for analysis. This is a tedious procedure.

The great proportion of tide records have been obtained with float-type recorders. A "stilling well" is mounted in the water with a recorder on top of it. The stilling well is a vertical pipe with a small hole at the bottom, below the lowest water level, so that the effects of waves of periods much shorter than tidal are damped out and the rise and fall of water in the well follows chiefly the tidal rise and fall. (Longer period waves, such as tsunamis, will be recorded to some degree.) A float on the water in the well is connected by a wire to a pulley which drives, through a reducing mechanism, a pencil which then moves back and forth parallel to the axis of a drum, carrying paper, which is rotated uniformly by a clockwork drive. The pencil then records on the paper a graph of tide height versus time.

In other instruments, a pressure sensor is mounted in the water below lowtide level and connected to a shore-mounted instrument which records the variations of hydrostatic pressure with time. The pressure can then be converted to water depth. Suitable damping is used to make the sensor or the recorder insensitive to waves. In the bubbler-type gauge, a tank of air under pressure is connected through a pressure-reducing mechanism to a pipe whose open end is fixed in the water below low tide. The shore instrument then measures the air pressure needed to just cause air to bubble out of the open end of the pipe, thus measuring the water pressure there and hence the water depth. Again, a record of pressure versus time yields the desired tide-height curve. The advantage of this type of tide gauge over the remote pressure sensor type is that all the instrumentation is on shore and accessible for servicing. All that is in the water is a length of hose.

For use in cold regions, where sea ice might damage any structure mounted through the water surface, the pressure sensor type may be used with shore recording, the connecting cable being buried in a trench. For severe ice conditions, a self-contained pressure/time recorder can be mounted on the sea bottom, being placed in position during one ice-free season and recovered during the next. This procedure is also useful in relatively shallow water, e.g. coral reefs, where there is no emergent land or structure on which to mount a recorder, or where there may be ship traffic. These self-recording instruments are now also being used to record tides over the continental shelf and slope, on sea mounts and even in the deep sea, but the amount of information which has been obtained away from the shore is still very small.

The tidal *ranges* (vertical differences between successive high and low waters) to be measured vary from almost zero, e.g. in places in the Faeroe Islands, to about 15 m in the Bay of Fundy in Canada.

The only promising technique for obtaining tidal height information for large areas of the open oceans is satellite altimetry. The microwave altimeter in *Seasat* in 1978 had a precision of ± 7 cm while the next generation of altimeters for TOPEX (ocean dynamics Topography Experiment of NASA) is being designed for a precision of ± 2 cm, although extraction of tide-height data from the records is somewhat more complicated than from shore-mounted tide gauges because allowance has to be made for the shape of the satellite orbit, for the geoid (the shape of the earth) and for the intervals between successive passages over the selected ocean area.

The instruments and methods used for measuring tidal currents (to be discussed in Section 13.7) are the same as those for ocean currents and are described in *Descriptive Physical Oceanography* (Pickard and Emery, 1982).

13.6   Tides in typical ocean regions

13.61   *Tides at the coast*

In the introduction to this chapter it was mentioned that there is considerable variety among observed tides (Fig. 13.1); this arises from the difference in response locally to the semi-diurnal and diurnal force constituents and to their relative phases. The simplest classification of tides uses, as the distinguishing feature, the response to the diurnal or to the semi-diurnal constituents of the tide-producing forces. In Fig. 13.4 are shown the two basic types with varieties of the second. For *diurnal* tides (Fig. 13.4(a)) there is one high water and one low water in each lunar day (about 24.8 h), while for the *semi-diurnal* tides (Fig. 13.4(b, c)) there are two high and two low waters in the same time interval. For semi-diurnal tides, in some regions two successive high waters will have nearly the same height and two successive low waters will have nearly the same (lower) height (Fig. 13.4(b), called *semi-diurnal equal tides*); in other regions, successive high waters and successive low waters will each have different heights (Fig. 13.4(c), *semi-diurnal unequal tides*). In some locations, a predominantly semi-diurnal tide becomes diurnal for a short time each month during neap tides.

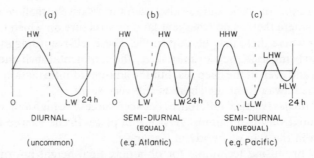

FIG. 13.4   *Simple classification of tides as: (a) diurnal, (b) semi-diurnal (equal), (c) semi-diurnal (unequal). HW = high water, LW = low water, HHW = higher high water, LLW = lower low water, LHW = lower high water, HLW = higher low water.*

As the forces due to the sun and moon come into phase, the range of the tide increases to a maximum (*spring tides*). This maximum occurs when the sun and moon are both on the same side of the the earth or both on opposite sides (syzygy). When the sun and moon are nearest to 90° to each other (quadrature) the resultant force has its minimum value and the tides have their minimum range (*neap tides*). Successive spring or neap tides occur at intervals of about 15 days. A feature of semi-diurnal tides is the so-called *age* of the tide whereby spring tides lag (occasionally lead) the time of syzygy by a day or two while neap tides lag (or lead) quadrature by about the same interval. It is related to the difference in phase of the M_2 and S_2 constituents and is attributed to tidal friction.

A more systematic classification of tidal types uses the "form ratio" $F = (K_1 + O_1)/(M_2 + S_2)$ of the sum of the amplitudes of the two main diurnal constituents of the actual tide (obtained from the analysis of the tide record) to that of the two main semi-diurnal amplitudes as follows (examples being shown in Fig. 13.1):

$F = 0$ to 0.25 : Semi-diurnal tides; high waters and low waters of about the same height each day, mean spring tide range $= 2(M_2 + S_2)$.

$F = 0.25$ to 1.5 : Mixed, mainly semi-diurnal tides; large inequalities in range and time between the highs and lows each day, mean spring tide range $= 2(M_2 + S_2)$.

$F = 1.5$ to 3.0 : Mixed, mainly diurnal tides; frequently only one high water per day, mean spring tide range $= 2(K_1 + O_1)$.

$F > 3.0$: Diurnal tides; generally only one high water per day, mean spring tide range $= 2(K_1 + O_1)$.

(The dividing values for F are somewhat arbitrary.)

Often the same type of tide is found for long distances along a coast so that a tide record at one port in the region will be sufficient to determine the type of tide for the whole region. The differences to be expected are in the relative phase and amplitude of the tide at other points in the region. It is therefore sufficient to collect long-term records at a few (principal) points, usually ports, to determine the important constituents and then to make shorter-term observations at subsidiary points to determine the relative phases (times of high or low water relative to those at the principal ports) and relative tidal ranges in order to prepare tide tables for the region. This procedure works for an open coast with simple bottom topography. Along a complicated coast, such as that of British Columbia or southern Chile, or in an island archipelago, it may be necessary to have the principal ports for long-term observations closer together. Only measurements in the field can determine just how close together or far apart need be the recording stations.

For practical reasons, almost all of our information about tidal rise and fall is for the coast, because only here are there fixed structures to which one can mount tide gauges of a simple mechanical type which have good reliability and will record unattended for long periods of time. In the last 10 years or so, some effort has been made to obtain tide records in deeper water but most of these have been on the continental shelf to depths of about 200 m and for a month or so, with only a few measurements at much greater depths. This is no more than a start in the direction of obtaining an adequate description of deep water tides.

13.62 Tides in estuaries

The ocean tides at the mouths of river estuaries cause tide-height variations to progress up the estuary and sometimes up the rivers themselves—as far as 800 km up the Amazon River. The tide wave penetrating up the estuary is modified as a result of change of width and depth, of increased friction and of river flow to seaward.

We look at the effects of changes of width and depth first. If the estuary becomes narrower and shallower away from the sea, the tidal range will increase up the estuary ("funnelling effect"). We assume that the changes in dimensions are very gradual so that no reflection of energy takes place and the wave can be treated as a plane wave. Then if we neglect frictional losses, the wave energy flux will remain constant as the wave progresses up the estuary. As the wave energy density for a plane wave is $E = (\rho g A^2)/2$ where A = amplitude of the tidal wave, the energy per unit length of the estuary = (Eb) where b = width of estuary. Then the energy flux up the estuary = (EbC) will be constant (C being the phase and group or energy propagation speed in shallow water), i.e.

$$\frac{(\rho g A^2) b (gh)^{1/2}}{2} = \text{constant or } A \propto b^{-1/2} h^{-1/4}.$$

so that narrowing of the estuary is more important than shoaling in determining the tidal amplitude increase. This funnelling effect is probably the reason for the increase in tidal amplitude up the St. Lawrence River shown in Figs. 13.5 and 13.7.

In addition, as the wave frequency f remains constant, the wavelength Λ will decrease with decrease of depth because $C = f\Lambda$,

$$\therefore \Lambda = \frac{C}{f} = \frac{(gh)^{1/2}}{f}, \qquad \text{i.e. } \Lambda \propto h^{1/2}.$$

So, if the depth decreases up the estuary, the amplitude increases and the wavelength decreases, i.e. the wave becomes steeper as $A/\Lambda \propto h^{-3/4}$.

The above is a very simple treatment neglecting friction and non-linear effects which can modify the wave so that the shape of the tide height/time curve becomes unsymmetrical, e.g. in Fig. 13.5, for the St. Lawrence estuary, compare the curve for Father Point toward the mouth with those for Quebec City which is 350 km up-estuary. In this estuary, the average times of rise and fall at Father Point are equal at 6 h 12 min but at Quebec the rise lasts for only 5 h 2 min compared with the fall for 7 h 23 min. In other locations, the difference is much greater, e.g. 2 h 10 min rise against 10 h 8 min fall at the limit of tidal influence in the Gironde River. The difference is largely due to the greater depth of water and hence greater speed of travel toward high water than near low water in these river estuaries. The river outflow also contributes to the asymmetry.

In some estuaries, the incoming tide reaches a critical speed and the front rises steeply, sometimes almost to the vertical (called a *bore*). In the Petitcodiac River in eastern Canada, the water may rise by 1 m in 10 seconds and continue to rise a further 2 m in 20 minutes. Much higher bores have been reported in

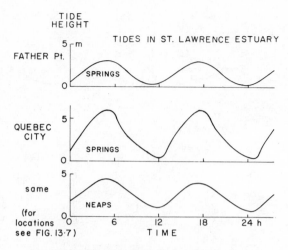

FIG. 13.5 *Showing development of asymmetrical form of tide curve as tide progresses up an estuary (from Father Point to Quebec City in the St. Lawrence Estuary, Canada). Note the greater asymmetry at spring that at neap tides.*

the Amazon River and in China near Shanghai. If the water depths on the deep and shallow sides of the bore are h_1 and h_2 respectively, then the speed of propagation (C_b) is given (Doodson and Warburg, 1941, section 27.9) by

$$C_b = [1 + (h_1 - h_2)/(h_1 + h_2)] [g(h_1 + h_2)/2]^{1/2}. \qquad (13.1)$$

For $h_1 = h_2 = h$, i.e. $(h_1 - h_2) = 0$ or no bore, $C_b = (gh)^{1/2} = C_l$, the long-wave speed, while if the bore height $(h_1 - h_2)$ is finite, $C_b > C_l$, tending to the limit $C_b = (2gh_1)^{1/2}$ as $h_2 \to 0$. The behaviour of a bore in an estuary is essentially the same as that of waves forming surf as they approach the shore.

13.63   *Tides in bays—resonance*

In some bays, the tidal range is very large compared with the range in the ocean near the mouth of the bay. This phenomenon is often attributed to *resonance*—the water in the bay having a natural period of oscillation close to that of the astronomical tide and therefore accumulating energy from it. The Bay of Fundy in eastern Canada is a frequently quoted example.

Let us examine the conditions necessary for resonance to occur. First, consider a long, narrow body of water (Fig. 13.6(a)) of length L, of depth h when the water is still, and of constant width. For simplicity we assume that the bottom is flat and the ends vertical. Such a body of water can be caused to oscillate, and the simplest mode is one in which the water at the ends (A, E) goes

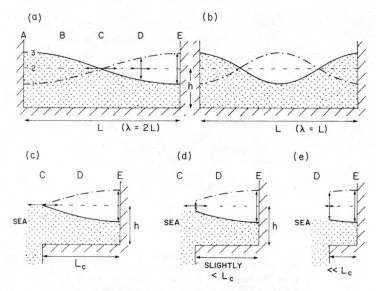

FIG. 13.6   *Tidal resonance in bays of different lengths.*

up and down parallel to the end walls (anti-nodes), that at the middle (C) goes back and forth with no vertical motion (node), while that in between, as at B and D, moves both up and down and horizontally. The lines 1, 2 and 3 show three successive positions of the water surface. It is quite easy to demonstrate this phenomenon in an ordinary household bath, partly filled with water, by moving one's hand back and forth through a few centimetres near the middle (C) or up and down at one end. It can also be done in a swimming pool, though here it needs a concerted effort by several people moving their bodies to get the water to oscillate satisfactorily. In either case it will be found that it is necessary to apply the stimulus at a specific frequency or period to generate and maintain the oscillations. For the household bath the period will be of the order of 2 to 3 seconds, while for the pool it would be of the order of 10 seconds. In the bath you could also determine that the period of oscillation depends on the depth of water, becoming less as the depth increases. Application of a more frequent periodic stimulus could cause the next mode of oscillation as in Fig. 13.6(b). The condition of maximum response to an applied periodic force is known as "resonance" and the waves are referred to as "standing waves" in contrast to the progressive waves discussed in Chapter 12.

The reason for these standing waves or *seiches* is that progressive waves travelling along the body of water are reflected at the far end and the two sets of waves travelling in opposite directions can interfere constructively with each other, i.e. their amplitudes will add together, if the wave speed and the length of the water body are such that the time for a wave to travel from one end to the other and back is a whole number of wave periods. This time $= 2L/C = nT$ where C = wave speed, n = a positive integer and T = period of oscillation, i.e.

$$T = \frac{2L}{nC} = \frac{2L}{n(gh)^{1/2}} \text{ which is known as Merian's formula.} \quad (13.2)$$

For the simplest oscillation (the fundamental), as in Fig. 13.6(a), the fundamental period, i.e. for $n = 1$, is

$$T_f = \frac{2L}{(gh)^{1/2}}. \quad (13.3)$$

Because the length L in this case is one-half of the length (Λ) of the travelling waves which interfere $(L = \Lambda/2 = C/2f)$, this arrangement is called a "half-wave oscillator". A few values for the period T for various values of L and h are shown in Table 13.2. This formula (13.3) may be used as a first approximation to calculate the fundamental period of oscillation of a lake; higher-order oscillations will have periods of $T_f/n = T_f/2, T_f/3$, etc. (To calculate the periods of oscillation of a real lake it would be necessary to take into account variations of depth and width along the length of the lake.)

TABLE 13.2.   *Values of $T_f = 2L/(g\,h)^{1/2}$ (hours)*
for combinations of L (km) and h (m)

h (m)	$L = 10$	100	500	1000 km
			Period	
50	0.25	2.5	12.6	25.1 hours
100	0.18	1.8	8.9	17.7
200	0.13	1.3	6.3	12.6
500	0.08	0.8	4.0	7.9
1000	0.06	0.6	2.8	5.6

Now suppose that instead of having a body of water closed at both ends (a "closed basin") we have one which is open to a tidal sea at one end (an "open basin") as in Fig. 13.6 (c) so that water can flow in during the flood tide and out during the ebb. It would be possible for the shorter length bay, CE, to behave like the right-hand half of that in Fig. 13.6 (a) and oscillate with a specific natural period. This body of water would be called a "quarter-wave oscillator" as its length L_c is one-quarter of the length of the travelling wave. Then $L_c = 0.5L$ so that the natural period of oscillation

$$T_c = \frac{4L_c}{(g\,h)^{1/2}}. \quad (13.4)$$

The situation in Fig. 13.6(c) is rather artificial because we have imagined water flowing in and out at the node C without any vertical motion. This arrangement might be set up in the laboratory but is not likely in nature. A more probable situation is that in Fig. 13.6 (d) where the end of the bay is inside the node C so that the sea goes up and down as well as flowing in and out at the same time. The important feature of this arrangement is that the vertical amplitude of surface motion (tidal range) is greater at the closed (head) end E than at the open (mouth) end of the bay near C; in other words, amplification of the tidal range occurs. For such amplification, the resonant length of the bay (L_c) is related to the depth by equation (13.4), and for a semi-diurnal tidal period of 12.4 h some related values of L_c and h are given in Table 13.3.

TABLE 13.3.   *Related values of length L_c and depth h for the fundamental period of oscillation of an open bay to be 12.4 h*

h =	50	100	200	500	1000 m
L_c =	247	350	495	782	1110 km

Real bays, etc., of course do not have flat bottoms of uniform depth nor flat, vertical ends, but it is possible to calculate the resonant length for an irregular bay to reasonable accuracy by a step-by-step method to allow for the varying depth. However, we can compare the ideal calculations of Table 13.3 with the dimensions of fjords (as in the coasts of British Columbia/Alaska, Norway or Chile) using their mean depth and actual length. We find that for these "bays", a typical mean depth is about 500 m with a length of 100 km. They are therefore much shorter than the critical length for large amplification (Table 13.3). Their dimensions are more like those of Fig. 13.6 (e) and only a small amplification occurs (about 5%).

Furthermore, friction has been neglected in the above discussion but in real bodies of water, particularly if shallow, its effect cannot be ignored. It can be shown that the effect of introducing friction is to make the response less sharp, e.g. the point nodes at C (Fig. 13.6 (a, c)) and sharp change of phase (180°) between the two sides of C in Fig. 13.6(a) or at the two nodes in (b), would extend over a finite length of the basin, the extension increasing as friction increases.

The Bay of Fundy in eastern Canada is much shallower ($\sim$ 100 m) than the fjords and its length ($\sim$ 300 km for the Bay itself) is nearer to the critical length for resonance to the semi-diurnal tide components. It has often been quoted as an example of resonant amplification because tidal ranges of some 15 m occur near the head compared with 5 m near the mouth. Some investigators have disputed this hypothesis, calculating a resonant period as low as 9.0 h which would be too far from the semi-diurnal constituents for significant amplification to occur. It was suggested instead that the large ranges at the head were simply due to the funnelling effect of the bay toward the head (Section 13.62). We should point out that one difficulty in making a step-by-step calculation of the resonant period is to decide where to take the mouth of the bay. For a long narrow fjord which opens suddenly into the ocean, there is not much uncertainty but for a wide-mouthed bay system, such as the Bay of Fundy together with the Gulf of Maine (for which system the axis is not perpendicular to the shoreline), there may be a temptation to select a length for the bay system which suits one's preconceived ideas! However, C. Garrett (1972) compared the ratios of the major semi-diurnal tide characteristics inside and outside the bay system, i.e. compared the response characteristics of the system with the forcing characteristics of the ocean tide. From this calculation he concluded that the resonant period for the system is about 13.3 h which is close enough to the probably forcing period (12.0, 12.4 and 12.7 h) for the resonant response explanation to be acceptable. In this example, the mouth of the bay system is determined by where the depth increases rapidly on the continental shelf, causing a sharp change in response, rather than by the position of the horizontal opening to the ocean.

13.64   *Tides of the open ocean*

The discussions in Sections 13.61 to 13.63 above have been about tides in relatively shallow waters where field observations are available. Until recently, the only tidal records available were from such shore observation points along the continents and at islands, and even now, as mentioned earlier, there is very little direct information for the open (deep) ocean.

Although it has little direct practical value, there has long been speculation about the character of tides in the open sea. In principle, Laplace's dynamical theory should provide a means for calculating the tides anywhere in the ocean; in practice, the actual ocean boundaries and bottom topography are so complex and the character of the equations is such that analytical solutions for the real oceans are not possible. Therefore, the theoretical studies of tides, based on Laplace's dynamical theory, have been of idealized oceans of relatively simple shape (e.g. channels, sectors of a spherical surface) and generally of constant depth. However, the solutions obtained did yield a description of the tides over the whole area studied, not just at the edge (i.e. the coast). It was found that while in long, narrow channels the tide wave progresses as a plane wave along the channel, in wider bodies of water the form is often as a wave rotating about a centre or nodal point like a spoke of a wheel, i.e. as a Kelvin wave (Section 12.10.3). The node or *amphidrome* is a location of zero rise and fall with cotidal lines rotating usually anti-clockwise in the northern hemisphere and clockwise in the southern. (*Cotidal* or *cophase lines* join points of equal phase at any instant. For a plane wave progressing along a channel they will be straight lines perpendicular to the direction of propagation; for an amphidromic system of waves rotating about a point they will have the appearance of spokes of a wheel, though generally not straight spokes. *Corange lines* join points of equal tidal range.) Such then were the results of purely theoretical studies.

The second approach has been an empirical one. The observed tides at the coast generally show a progression of constant phase along a coast and often a progressive change of range. Since the early 1800s investigators have empirically prepared ocean tide charts by constructing cotidal and corange line patterns for ocean areas. They essentially extrapolated from the known coastal tide characteristics into the open ocean, using ideas and patterns from the theoretical studies referred to above as guides. Such charts are generally prepared for the main constituents of the tide, e.g. M_2, S_2, K_1, O_1, etc. Figure 13.7 shows a recent example (Godin, 1980) of such a chart for the M_2 constituent in the St. Lawrence Estuary, Canada. The solid lines are cophase lines connecting points of equal phase lag expressed in degrees ($360° \equiv 24$ h). (Physically the phase lag is the delay between the time that a tide-producing force constituent reaches its maximum and the time that it causes a maximum contribution to the local vertical tide.) The dashed lines are for equal tidal amplitudes (cm) for the area. A wave enters from the Atlantic through Cabot Strait and there is an amphidromic system rotating anticlockwise centred at about 47°N, 62.5°W. The tide wave then progresses west and southwest along the St. Lawrence River estuary with its amplitude increasing as it does so.

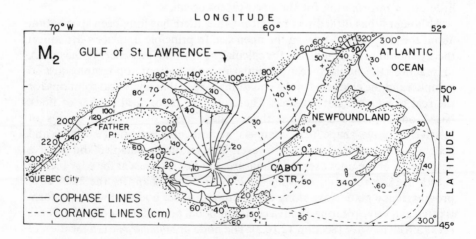

FIG. 13.7   *Cotidal charts for M_2 constituent in St. Lawrence Estuary, Canada. Full lines = cophase lines in degrees ($360° \equiv 24$ h, see text); dashed lines = corange lines in cm. (From Godin, 1980.)*

The third approach to the study of the open ocean tides has been to use numerical methods to solve the Laplace equations. The method is simple in principle but requires much calculation if realistic ocean shapes and bottom topography and the relevant interactions (e.g. with the solid earth tides) are to be taken into account. It was the development of automatic computers in the 1950s which offered the possibility of doing the calculations for large areas. A number of numerical computations of world ocean tides have been made, initially using very simplified coastal shapes and bottom topography but more recently with fairly realistic shapes and topography. A recent example by Accad and Pekeris (1978) is shown in Fig. 13.8 for the M_2 tide. This solution was derived solely from the known tide potential and ocean topography and includes effects of self-attraction (between tidal "bulges") and tidal loading (of the solid earth against which the tidal rise and fall is measured). It shows quite good correspondence with observed (coastal) tides in most of the oceans. Other investigators have constrained their solutions, during calculation, to agree with known tidal characteristics along the coast and therefore these solutions cannot be tested against present tide records.

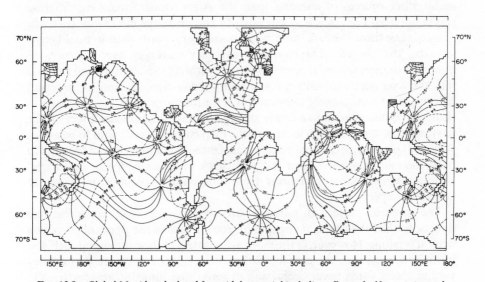

FIG. 13.8   *Global M_2 tide calculated from tidal potential including effects of self-attraction and of tidal loading, 2° grid. Full lines = cophase lines (Greenwich hours), dashed lines = corange lines (cm). (From Accad and Pekeris, 1978.)*

These numerical solutions also reveal "anti-amphidromes" or regions of local maximal tidal amplitude with very little phase change, as seen in Fig. 13.8 in the centre of the Indian Ocean and in the western equatorial Pacific. The various solutions show many features which are similar but others which differ between solutions or do not agree with known tide data, so that much work remains to be done. Serious tests of the open ocean aspects of the solutions await suitable measurements from satellites.

Finally, although the character of the open ocean tides may have limited *direct* practical application, it is of significance for other studies. For earth tide studies and in satellite geodesy it is necessary to know the distribution of the ocean water mass over the globe. Also, as will be discussed in Section 13.9, the results are needed for studies of global tidal energy.

13.7 Tidal currents

Although tidal currents are the basic phenomenon, they have been less studied than the tidal rise and fall. There are several reasons. The instrumentation required for current measurement is more complicated and less likely to operate satisfactorily for long periods unattended, and the logistic problems of laying and recovering current meters and the costs involved are much greater than those associated with measuring the rise and fall. In addition, there are many other sources of currents than the astronomical forces, e.g. internal waves, internal tides, seiches, wind-driven and estuarine circulations, etc. In consequence there is much more noise in current records than in tidal height records. Also current characteristics may vary markedly over distances of hundreds or even tens of metres horizontally (rarely the case for tidal rise and fall) and over depths of only a few metres. In consequence, and for practical reasons, most tidal current information is for locations in narrow passages in shipping routes where currents are strong and have a significant effect on the navigation of ships. In such locations, the tidal current speeds tend to be about 90° out of phase with the tidal rise and fall, i.e. the maximum current speeds will be at the middle of the rise or fall, with slack water near high or low water, although considerable variations from this pattern occur near coasts of complex shape.

The common pattern in narrow waterways is of a *flood* current in one direction while the tide is rising and an *ebb* current in the opposite direction while it is falling. However, in the more open waters of the continental shelf and in shallow open seas, the characteristics of the tidal currents are that as they vary in speed, often never decreasing to zero, their direction rotates, usually with a semi-diurnal period dominating. Such currents may be represented by a current vector hodograph, i.e. the figure traced out by the tip of a vector representing the current at, say, hourly invervals over the tidal cycle. The figure will be an ellipse where tides are purely semi-diurnal or diurnal but more complicated for more general tides. Typical examples would be those in Fig. 13.9.

Tidal currents, as for tidal heights, may be separated by harmonic analysis into constituents for the purpose of prediction. For tides in channels, where the

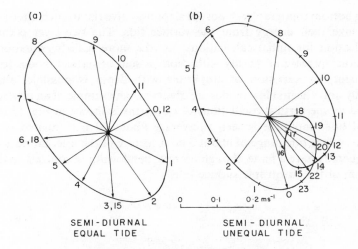

FIG. 13.9   *Form of tidal current hodograph for (a) semi-diurnal equal tide, (b) semi-diurnal unequal tide. Arrows represent current velocities for lunar hours from high tide (t = 0 h). (To avoid crowding, only even hour vectors are shown in (b)).*

current is essentially parallel to the centre-line of the channel, the total current would be analysed; for rotary currents it would be usual to resolve the current into rectangular components before analysis.

Away from the coast, tidal current speeds are generally less than about 0.1 m s^{-1} but much higher values are common in straits and passages—a maximum of 8 m s^{-1} for Seymour Narrows in western Canada is frequently quoted as one of the highest speeds measured, certainly in regularly used shipping channels.

13.8   Internal tides

In all of the previous discussion, it was tacitly assumed that the ocean was of uniform density, i.e. that the tides were barotropic with currents uniform from top to bottom. However, in most ocean regions the water is actually stratified. The first approximation to reality is to consider a two-layer ocean with an upper layer (of tens to hundreds of metres thickness) of lower density water separated by a sharp pycnocline from the deeper water of greater density. Then interaction of the barotropic motions with bottom topography may give rise to vertical oscillations of the pycnocline of tidal period, i.e. internal tides similar to the internal waves of Chapter 12. For instance, onshore tidal motions impinging on a sloping shore or continental shelf will acquire a vertical motion which may temporarily displace the pycnocline above its equilibrium position and then relaxation oscillations will take place. It has been shown that either sloping bottom topography or abrupt steps may give rise to such internal tides which take their energy from the surface tide. The two-layer ocean is a simplification for initial calculations; for the more realistic situation of a continuous increase of density with depth, a sum of vertical modes for each constituent, i.e. variation of amplitude with depth is possible, although generally only the first few modes are observed. An example of an internal tide revealed by the vertical oscillations of isotherms is shown in Fig. 13.10. The internal tide amplitude is seen to average about 20 m compared with the average surface tide range of about 3 m. A feature of the internal tides is that they generally have a finite, though narrow, band width in contrast to the line spectrum of the barotropic surface tide.

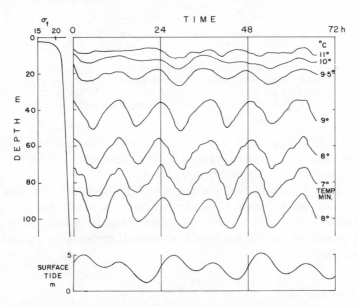

FIG. 13.10   *Internal tide, Bute Inlet, British Columbia, Canada, July 1953, showing vertical motions of isotherms and of surface tide. (Note the difference between the depth scale and the surface tide scale; the surface tide is magnified by a factor 4 compared with the internal tide.) On left, smoothed profile of σ_t.*

Although the internal tides may have large amplitudes compared with the surface tides, their energy content is much less because of the small density differences within the water column compared with the air/water density difference at the sea surface. In addition, the phase speed (speed of progress) of internal tidal waves is smaller by a factor of about 100 than the phase speed of the surface tides (which is of the order of 200 m s^{-1} = 700 km h^{-1}). However, the water particle speeds associated with internal tides (baroclinic and therefore varying with depth) are of the same order as those for the barotropic tides (independent of depth) and it may be difficult to separate the two when analysing current records.

Internal tide records are generally fairly symmetrical in form, like semi-diurnal equal tides (Fig. 13.4(b)) but numerous cases of asymmetrical ones have been observed, e.g. Fig. 13.11, and even cases where the internal tidal wave has the appearance of surf breaking near the shore. Some cases were in shallow water where the cause was probably the influence of the shallow bottom. In other cases, in deep water, Defant (1961) has argued that the breaking is a consequence of dynamical instability induced by velocity shear, e.g. in straits or by interference between internal tides and inertial motions.

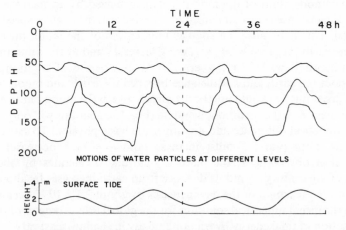

FIG. 13.11 *Asymmetric shape of internal tide, Myrmidon Reef near the continental shelf-break off the Great Barrier Reef, 18°19'S, 147°21'E, November 1981, showing motion of water particles at different levels and the surface tide. (Note the difference between the depth scale and the surface tide height scale; the surface tide is magnified by a factor 12.5 compared with the internal tide.) (Data from E. Wolanski, Australian Institute of Marine Science.)*

Since internal tides are baroclinic they exhibit velocity shear and are considered to have a significant effect on mixing in the ocean and so affecting the thermohaline structure to an extent not expected from the barotropic tide. In regions where the internal tides break, mixing will be intense. A further effect, which has been discussed by Defant (1958) is that the influence of the internal motions "can be very disconcerting when attempting to obtain the mean distribution of the oceanographic elements" (i.e. water properties, using bottle cast or CTD techniques).

13.9 Tidal friction and the moon's motion

It is interesting, although to a physicist remembering Newton's Third Law of Motion it should not be unexpected, to note that while the moon generates tides on earth, these tides also affect the motion of both moon and earth. This provides a second reason for interest in the global tides.

The reason for the interaction is as follows. The motion of the ocean waters relative to the bottom gives rise to friction which, on average, causes the axis of the tidal "bulges" to lag behind the direction from earth centre to moon centre. In consequence, there is a gravitational torque on the earth tending to retard its rotation so that the period of the earth's rotation, the day, gets progressively longer. Because the angular momentum of the earth/moon system must remain constant, loss of angular momentum by the earth must be compensated by gain of angular momentum of the moon. This is achieved by an increase in the earth/moon distance (r) accompanied by a decrease in the angular speed (ω) of the moon. This is because the angular momentum of the moon (mass M_m) about the earth, $M_m r^2 \omega = M_m r^2 (2\pi/T)$, increases and at the same time, by Kepler's Third Law, $T^2 \propto r^3$, where $T = 2\pi/\omega$ = period of rotation of the moon about the earth and r = distance between the earth and moon centres. (Strictly speaking we should use the distance between the moon centre and the centre of mass Z of the earth/moon system (Fig. 13.2) but this will only make a small arithmetical difference and not affect the basic physics.) To satisfy both conditions, both r and T must increase (and $\omega = 2\pi/T$ decrease). From astronomical observations, the moon's longitude decelerates by about 27 seconds of arc century^{-2} and its distance from earth increases by about 6 cm year^{-1} while the length of the day increases by about 2.4×10^{-3} s century^{-1}.

From the recent numerical solutions for the global tide, estimates of the rate of dissipation of tidal energy by friction (mostly in shallow seas) vary from 2.2 to 2.8×10^{12} watts, while the energy input from the astronomical forces is estimated at up to about 5×10^{12} watts, so that there is a discrepancy, at present not accounted for, between input and loss. Part of this is attributable to energy transfer from the surface tide to internal tides but this is probably only about 10% of the whole. The mechanical energy of the ocean tides has been estimated at 5.1×10^{17} joules for the M_2 constituent, equally divided between potential and kinetic energies.

13.10 Storm surges

These are mentioned here to emphasize that unusual rises of sea level may occur due to other causes than exceptionally high tides (and tsunamis as discussed in Chapter 12). Storm surges are the result of the frictional stress of strong winds blowing toward the land and causing the water level to be "set up" by as much as several metres. This effect has caused severe flooding of low-lying areas at the south end of the North Sea during strong northerly winds and in areas at the north end of the Bay of Bengal during cyclones. Particularly in the latter case, the low atmospheric pressure at the centre of the cyclone can additionally raise the sea level (the "inverted barometer effect"). A drop of pressure of 3 kPa ($\equiv$ 30 mbar) can raise sea level by 0.3 m. This value is the response to an atmospheric pressure decrease for a stationary system after a long enough time for equilibrium to be reached. The actual response may be greater or lesser than this amount depending on the bottom topography of the area and the speed at which the storm centre moves.

In addition to the rise in water level itself, another factor must be considered. On a low coast, a rise of mean sea level of 2–3 metres may result in disastrous erosive effects of the large surface gravity waves associated with the storm.

In bounded bodies of water, such as small seas or lakes, after the wind stress decreases relaxation oscillations may take place. These are standing waves or *seiches* and they die out as the mechanical energy of the water is dissipated by friction.

13.11 Epilogue

This chapter has been mainly a descriptive review of some aspects of the history of the study of tides and of aspects of practical interest. More details of the mathematical aspects, from the oceanographic point of view, may be found in Dietrich *et al.* (1980) and in Defant (1961, Vol. 2); Cartwright (1977) presented an excellent review of tidal studies to that date and Hendershott's Chapter 10 in *Evolution of Physical Oceanography* (Warren and Wunsch, Eds., 1981) brings the presentation up to date with discussions of many aspects of long waves and ocean tides.

14

Some Presently
Active and Future Work

We have described many models, both analytic and numerical, which attempt to demonstrate the importance of particular dynamic effects. A model is always a simplification of the real system and it attempts to include only the effects which are important in producing a particular phenomenon. Such models are useful in improving our understanding even when they do not represent the details of the real ocean. For example, Stommel's rectangular ocean model, discussed in Chapter 9, is clearly not intended to represent a real ocean; however, it does demonstrate the importance of the variation of the Coriolis parameter with latitude. The more complicated and detailed numerical simulation models described in Chapter 11 attempt to represent the ocean more realistically.

Because of the complicated geometry of the oceans and the importance of non-linear effects, analytical modelling has not and probably cannot produce a complete ocean model. While the numerical models can be more representative and show promise, none of them, as yet, has produced a quantitatively correct model of the general ocean circulation. Both types of modelling have been used to demonstrate possible mechanisms to explain certain features of the ocean circulation. It is clear that much work remains to be done before we can be certain that we have an adequate quantitative understanding of ocean dynamics. It is also quite clear that our observational data base is inadequate. This is unfortunate as it is difficult to understand a system theoretically when one cannot describe it sufficiently well.

Exploration of possible parameter ranges has been limited by the speed of available computers, particularly in models attempting to simulate the actual ocean. But the new, faster machines which have become available should help to reduce this problem. By running models of sufficient resolution that all dynamic effects are permitted, and comparing them critically with observations, it may be possible to determine which effects are important. The necessary verification observations will probably have to be collected specially for the comparisons, and obtaining them will be a non-trivial problem, to say the least. (Successful models of smaller regions such as rivers, estuaries and coastal seas provide examples of how such models may be adjusted to simulate a geophysical flow adequately, that is, to predict the information which we want, such as tidal currents, height of storm surges or the effects of engineering structures on the circulation. We are still a long way from having the necessary data base to follow this sort of procedure in the open sea.) Once an adequate model of a reasonably large ocean region (for example the North Atlantic) is achieved, parameterization schemes that allow simulation of the desired larger-scale features with lower resolution can be explored and larger regions on longer time scales can be studied. Until the fine resolution to allow all important dynamic effects is achieved, and adequate detailed verification data are obtained allowing modellers to work back up to limited resolution, it will be difficult to devise parameterization schemes that will allow adequate simulations of the real ocean for large regions and long time-scales. Work toward these goals is continuing.

There is also the question of whether or not a complex non-linear system, such as the ocean or the atmosphere or the combined ocean–atmosphere system, has a unique statistically steady solution. Examples of much simpler non-linear systems which are always oscillating can be constructed. The decade-to-decade variations in climate that occur with apparently the same external solar and tidal forcing suggest that the combined system may be able to exist in a number of quasi-steady states with jumps from one to another when a large perturbation occurs. There are variations in the forcing, however (for example, variations in the solar wind with the sun-spot cycle and the earth's orbital parameters), so the system may be predictable given the right data. There is some evidence that very long-term climatic fluctuations are associated with the periodicities of the earth's orbital variations. Much more work will have to be done before this question of uniqueness can be answered with any assurance.

Much work has been done on the time-dependent motions, which we have mentioned only briefly as our concern has been with the large-scale average circulation. This work starts with considerations of planetary or Rossby waves mentioned in Chapter 12, and includes such things as effects of topography on these waves and their interaction with one another as non-linear effects become more important. It leads ultimately, using numerical simulations, to an investigation of the behaviour of the strongly non-linear case of quasi-two-dimensional or geostrophic turbulence. The mesoscale eddies, as they are termed by oceanographers, are phenomena of this type. Because these eddies may interact with the mean flow, this work on time-dependent motions may be very helpful in our search for a fuller understanding of how the general circulation works.

Much more information is needed about this kind of turbulence and also the smaller-scale three-dimensional turbulence which is responsible for frictional and mixing effects. At present, parameterization of these effects is crude, largely due to the lack of knowledge of these phenomena in the ocean and how they interact with the large-scale flow. Further work examining various turbulence models and how these different turbulence models affect the large-scale circulation is needed. Then observational programmes need to be conducted to gather data to determine which type of turbulence model is most representative.

Because of the small spatial scale and the long time scale, the logistics of getting good observations of such phenomena as the ocean eddies are extremely difficult. Large efforts have been made by groups from the U.S.S.R (the POLYGON experiment) and the U.S.A.–U.K. (MODE, the Mid-Ocean Dynamics Experiment). Each has observed one eddy fairly well and fringes of others. A further experiment, POLYMODE, a combination of U.S.S.R.– U.S.A.–U.K. efforts was completed in 1978 but extensions of it are continuing. As part of these projects a considerable amount of theoretical work, both analytical and numerical, has been done to help to design the sensor arrays to optimize the data that have been and are being collected. In addition to these detailed array studies using current meters, XBT's and CTD's, historical BT (= bathythermograph, a temperature-depth recorder) data and new, more detailed, XBT sections are being examined to get a better idea of the statistics of the existence and intensity of such eddies in various ocean regions: satellite observations are also being used.

The relatively large projects like POLYMODE, involving many scientists and strong interactions between theoretical and observational oceanographers, are likely to be a common feature for future oceanographic investigations. In the past, most oceanographic studies have involved rather smaller groups of people working independently on particular problems. While this traditional method will continue to be used, many oceanographic problems will require the much larger project approach to solve them. These large programmes of direct observations of deep ocean circulation are likely to continue for many years. The results of theory, both analytical and numerical, must be used to design the observational programmes so that data can critically test the theory. The observational results will no doubt lead to more modifications of our theoretical ideas until we achieve a better understanding of the ocean than we have at present.

Studies of the upper mixed layer of the ocean, mentioned in Chapter 10, will continue. As noted there, understanding of the development and evolution of this layer, e.g. the formation and breakdown of the seasonal thermocline at mid-latitudes, is important for oceanographic purposes, both physical and biological, and for meteorological purposes.

On the longer time-scale, the ocean must play an important role in determining the earth's climate and its variations. Thus, a better understanding of the ocean's general circulation is required if we are to obtain a sufficient understanding of the physical basis of climate to be able to predict climatic variations. Because of the effects of climate variations on human activities, particularly agriculture, the study of climate and the role which the ocean plays in it has become an active area of research which is likely to continue for many years to come.

The problem of the generation of surface gravity waves by the wind continues to be an area of active research. We still do not have an accurate quantitative understanding of this process. However, the new work described briefly in Chapter 12 has suggested some interesting new possibilities and further investigation of these may in future lead to a better understanding of this process. Also, because of their effects on the distributions of water properties as well as their time-dependent effects on observations of them and of currents, internal waves will continue to be studied.

While the oceanography of coastal and estuarine regions is beyond the scope of this book it should be noted that the study of dynamical oceanography in these regions is also an active area of research. Studies of coastal upwelling regions and estuaries have important applications because they are such biologically important regions. The need for better navigation and for better knowledge of the transport of material both natural (e.g. silt) and man-made (e.g. sewage, oil spills) are other applications for the study of the physical oceanography of coastal regions.

Calculations of the tidal response of the oceans from the basic equations of motion are continuing. Such studies for the deep ocean are being aided by the data obtained with internally recording deep-sea tide gauges, although the amount of data from this source is still far too small to test even the present open ocean tide calculations. Calculations of tides and, more particularly, tidal currents are also being made in coastal areas. While we can always make tidal height calculations empirically as described in Chapter 13, modelling which allows prediction of tidal currents is very important because measurement of these currents in detail is so difficult, time-consuming and expensive.

Development of new instruments to assist the dynamical oceanographer in his investigations is also continuing. The continued development of better moored current meters is an example. They will provide longer-term measurements more reliably. The data provided by satellites is another example. Sea-surface temperatures may be obtained over far larger areas than would ever be possible from surface platforms and the temperatures, together with infrared photographs, help to delineate the paths of ocean currents. Radar altimeter techniques for obtaining sea-surface elevations with a relative accuracy of about 10 cm are anticipated in the near future. Such observations would give information on the tides of the open ocean to permit the calculations mentioned above to be tested adequately. With suitable averaging, the sea surface elevation data would yield information about the stronger currents of the general circulation. If an accuracy of 1 cm could be achieved then such observations along with geostrophic calculations would provide a powerful tool for determining the general ocean circulation for comparison with theory. Microwave observations from satellites can give information about wave heights and surface winds. The obvious advantage of these observations from satellites is the total ocean area coverage which is potentially possible together with frequent repetition to enable time-dependent characteristics to be studied.

The study of finestructure (vertical variations with scales of one to a few metres) and of microstructure (scales of millimetres to decimetres, i.e. the small-scale turbulence) is another area of active research. Such studies should eventually lead to a better description and understanding of the details of the mixing and friction processes and better parameterizations for vertical eddy viscosity and diffusivity. Better parameterization for the horizontal eddy processes for small scales should result too. For large scales, such parameterization may well come from the study of mesoscale eddies.

The study of the Antarctic region and the Antarctic Circumpolar Current which we mentioned briefly in Chapter 11 received renewed attention in the International Southern Ocean Study (ISOS) which was part of the International Decade of Ocean Exploration (IDOE) during the 1970s sponsored by the National Science Foundation of the United States of America. (MODE and POLYMODE were partly supported by IDOE as was the Coastal Upwelling Ecosystems Analysis Programme (CUEA), another large-scale programme with physical oceanographic aspects and the North Pacific Experiment (NORPAX) which was essentially physical.) The Antarctic Circumpolar Current is the main feature of the only ocean circulation system which has no complete meridional (north–south) boundary, and is therefore particularly interesting. While it is agreed that wind-driving is very important, there was a very wide range of transport estimates for the current in the past and controversy over the importance of other possible driving mechanisms. With the extensive moored current meter data collected during ISOS, as well as other observations, a better description and understanding of this circulation system has been achieved. Further work will no doubt be indicated as these and continuing observations are analysed.

An interesting review of results obtained from the many large programmes included in the IDOE is available in *Oceanus*, Volume 23, Number 1, 1980.

Clearly our knowledge of the sea is very incomplete at present and much remains to be done. Thus for the student interested in observing and interpreting the oceans there are still many opportunities.

Appendix 1

Mathematical Review with Some Elementary Fluid Mechanics

A.1.1 Introduction

The purpose of this Appendix is to provide a brief review of some symbolism, mathematical procedures and some aspects of fluid mechanics relevant to the text, primarily for students whose field is not in the physical sciences.

The main symbols used to represent physical quantities in equations in this text have been given at the beginning of the book (List of Symbols).

For coordinate axes we use a right-handed system (Fig. A.1(a)) with the positive x-axis directed horizontally to the east, the positive y-axis horizontally to the north, and the positive z-axis positive vertically upward. The corresponding velocity components are then u positive to the east, v positive to the north and w positive upward. (It should be noted that *The Oceans* (Sverdrup *et al.*, 1946) uses a left-handed system with the z-axis positive downward.)

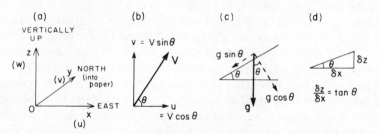

FIG. A.1 (a) *Orientation of axes used in this text with corresponding velocity components* (u, v, w), (b) *rectangular components of a vector* **V**, (c) *components of* **g** *on a slope*, (d) *gradient* $\delta z / \delta x$ *of a slope*.

A.1.2 Scalars and vectors

Scalar quantities are those which are expressed by a number and a unit only (e.g. temperature, mass) while *vector* quantities possess direction in addition (e.g. velocity, acceleration, force). Vector quantities are indicated in this text by the use of bold type, e.g. **V**. If we use the same symbol in ordinary type it means

287

the magnitude or size of the vector quantity regardless of direction. (For velocity **V**, the magnitude V is called the "speed". In some texts "velocity" is used where speed would be appropriate; we have tried not to do this.) It is often convenient to split a vector quantity into components, e.g. using axes as in Fig. A.1(a); in Fig. A.1(b) the vector **V**, which is directed at an angle θ to the x-axis can be resolved into components: $V \cos \theta = u$ to the east and $V \sin \theta = v$ to the north. When we wish to be specific about direction we write $V \cos \theta$ as iu where **i** is a "unit vector" directed to the east (positive x-direction). Then $-$iu would represent a component u directed to the west (negative x-direction). Similarly **j** and **k** are unit vectors in the north $(+y)$ and upward $(+z)$ directions respectively. Another example (Fig. A.1(c)) is where the acceleration due to gravity **g**, which acts vertically, is resolved into a component $g \sin \theta$ down the sloping surface which is at angle θ to the horizontal and a component $g \cos \theta$ perpendicular to the surface. In oceanographic applications of this resolution into components, the angle θ is generally small so that $g \sin \theta$ is a small quantity while $g \cos \theta$ is nearly equal to g.

A.1.3   Derivatives

The abbreviation δz means "a small distance in the z-direction" and δx means "a small distance in the x-direction". Then the quotient $\delta z / \delta x$ is physically a measure of the slope of the surface (e.g. Fig. A.1(d)) relative to the horizontal, i.e. $\delta z / \delta x = \tan \theta$. In the limit, when δx is assumed to approach zero $(\delta x \rightarrow 0)$ so that we are looking at a point in space, $\delta z / \delta x$ is written as dz/dx which is called the *derivative, gradient* or *rate of change* of z with respect to x. This is the case if the variable (z in this case) is a function of x only. If the variable is a function of other parameters as well, e.g. $S = S(x, y, z, t)$, then the derivative of S with respect to x *only* is written as $\partial S / \partial x$, called the *partial derivative* of S (with respect to x), assuming that y, z and t remain constant.

One special case is when the derivative $\partial / \partial t$ of a quantity is zero, e.g. $\partial S / \partial t = 0$, $\partial u / \partial t = 0$. There is no change of the quantity with time, but it does not imply that the quantity itself is zero, e.g. $\partial u / \partial t = 0$ means that there is no change of u with respect to time at any point in the region under study but u need not be zero, i.e. the water may be moving but at a constant (steady) speed at each point. This situation is referred to as the *steady state*. Note also that the combined statements $S = S(x, y, z)$, $\partial S / \partial t = 0$ imply that S does not change with time anywhere but the value of S may differ from one point to another.

One rule which is often needed is that for derivatives of products, i.e. $\partial(\rho u)/\partial x = \rho(\partial u / \partial x) + u(\partial \rho / \partial x)$.

A.1.3.1   *The total (or individual) derivative*

In fluid dynamics a special case arises when a quantity q varies with (x, y, z) and with time (t), i.e. $q = q(x, y, z, t)$. Then we will show that the time derivative dq/dt, called the *total* (or *individual*) *derivative* because it is the time change of q with respect to both time and space following a particular piece of the fluid, is given by

$$\frac{dq}{dt} \quad = \quad \frac{\partial q}{\partial t} \quad + \quad \left(u\frac{\partial q}{\partial x} + v\frac{\partial q}{\partial y} + w\frac{\partial q}{\partial z} \right). \tag{A.1}$$

$$\begin{array}{ccc} \text{Total} & \text{Local} & \text{Advective} \\ \text{derivative} & = \quad \text{term} \quad + & \text{terms} \end{array}$$

Physically this equation states that q may vary with time ($\partial q/\partial t$) at a position (x, y, z) and also vary as the fluid moves from this point to another point $(x + \delta x, y + \delta y, z + \delta z)$. The first term on the right of equation (A.1) is then called the *local* term and the other three are the *advective* terms because they are related to the flow (advection) components u, v and w.

To derive this expression, consider first a case in which the value of q at all points does not change with time—the steady state. Mathematically we write $\partial q/\partial t = 0$. However, q may still change with position. In this case, a small "parcel" of fluid or "fluid element" moving through the field must undergo changes, i.e. dq/d$t \neq 0$ unless q is the same everywhere. Initially we will suppose that there is motion only in the x-direction and variations only in the x-direction so that at time t a parcel is at point x with property $q(x)$ while at a slightly later time $(t + \delta t)$ it is at $(x + \delta x)$ with property $q(x + \delta x)$. Now using Taylor's series expansion we can write

$$q(x + \delta x) = q(x) + (\partial q/\partial x)\delta x + \text{terms of order } (\delta x)^2 \text{ or smaller.}$$

These latter terms can be neglected (since in the limit as $\delta x \to 0$ they will be negligible). The property change from x to $(x + \delta x)$ is therefore $(\partial q/\partial x)\delta x$ and the rate of change following the motion is

$$\frac{\text{property change}}{\text{time change}} = \frac{\left(\dfrac{\partial q}{\partial x}\right)\delta x}{\delta t} = \frac{\partial q}{\partial x}\frac{\delta x}{\delta t}.$$

In the limit, as $\delta t \to 0$, $\delta x/\delta t \to u$, and the rate of change $= u(\partial q/\partial x)$ in the x-direction. In the more general case when there are also v and w components of velocity and variations in all three component directions, the rate of change associated with the motion of the fluid is

$$u\frac{\partial q}{\partial x} + v\frac{\partial q}{\partial y} + w\frac{\partial q}{\partial z}.$$

This is the advective component of the property change.

If we now include changes with time ($\partial q/\partial t$) at the point itself in the fluid, we will have the total derivative of equation (A.1) above. The term "total" derivative then means the sum of the local and the advective or "field" terms and is the same as the "individual" derivative which means "following the motion of an individual parcel of fluid".

If the quantity q is a scalar property of the fluid, e.g. salinity (S) or temperature (T), then equation (A.1) is the complete expression of the total derivative. If the quantity q is a vector quantity, e.g. velocity (**V**), there will in general be three components to the total derivative as

$$x\text{-direction:} \quad \frac{du}{dt} = \frac{\partial u}{\partial t} + u\frac{\partial u}{\partial x} + v\frac{\partial u}{\partial y} + w\frac{\partial u}{\partial z},$$

$$y\text{-direction:} \quad \frac{dv}{dt} = \frac{\partial v}{\partial t} + u\frac{\partial v}{\partial x} + v\frac{\partial v}{\partial y} + w\frac{\partial v}{\partial z},$$

$$z\text{-direction:} \quad \frac{dw}{dt} = \frac{\partial w}{\partial t} + u\frac{\partial w}{\partial x} + v\frac{\partial w}{\partial y} + w\frac{\partial w}{\partial z}. \qquad \text{(A.2)}$$

A.1.3.2 The total derivative—a physical picture

To get a physical picture of the total derivative, consider the following situation. Fluid is flowing along a pipe ABCD whose dimension in the z-direction is constant (Fig. A.2(a)) but which narrows smoothly between B and C. Then there will be no z-component of velocity, and close to the centre-line v and $\partial u/\partial y$ will be negligible. Then the x-component from equations (A.2) becomes

$$\frac{du}{dt} = \frac{\partial u}{\partial t} + u\frac{\partial u}{\partial x},$$

$$\text{total} = \text{local} + \text{field acceleration.}$$

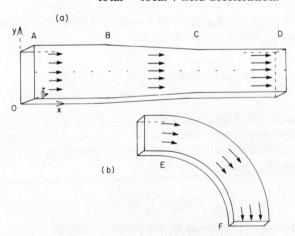

FIG. A.2 *Flow patterns for discussion of total or individual derivative, (a) linear flow, (b) curved flow.*

(a) If the volume flow is constant:
    (i) at every point, u = constant, $\therefore \partial u/\partial t = 0$,
                i.e. local acceleration = 0;
    (ii) at every instant between AB and CD, u = constant, so $u(\partial u/\partial x)$ = 0,
           i.e. field acceleration = 0, $\therefore$ total acceleration = 0;
    (iii) at every instant between BC, u increases with x, so $u(\partial u/\partial x) = +$,

$$\text{i.e. } \frac{du}{dt} = u\frac{\partial u}{\partial x} = + \text{ (but a steady state),}$$

        or total = field acceleration only.

(b) If the volume flow is increasing with time:
    (i) at every point, u is increasing with time, $\therefore \partial u/\partial t = +$;
    (ii) at every instant between AB and CD, u is increasing but $\partial u/\partial x = 0$, $\therefore u(\partial u/\partial x) = 0$,

$$\text{i.e. } \frac{du}{dt} = \frac{\partial u}{\partial t} \text{ or total = local acceleration only;}$$

    (iii) at every instant between BC, u increases with x and $\partial u/\partial x = +$, $\therefore u(\partial u/\partial x) = +$,

$$\text{i.e. } \frac{du}{dt} = \frac{\partial u}{\partial t} + u\frac{\partial u}{\partial x} = +,$$

        $\therefore$ total = local + field acceleration.

(c) If the volume flow is decreasing with time:
    (i) at every point, u is decreasing with time, $\therefore \partial u/\partial t = -$;
    (ii) at every instant between AB and CD, u is decreasing

$$\text{but } (\partial u/\partial x) = 0, \therefore \frac{du}{dt} = - = \text{total = local acceleration;}$$

    (iii) at every instant between BC, u increases with x and $\partial u/\partial x = +$, $\therefore u(\partial u/\partial x) = +$,

$$\text{i.e. } \frac{du}{dt} = \frac{\partial u}{\partial t} + u\frac{\partial u}{\partial x} = (-\text{ term}) + (+\text{ term}),$$

        $\therefore$ total = local + field accelerations
        and it would be possible for du/dt to be $+$, zero or $-$.

For an alternative flow pattern (Fig. A.2(b)) along a pipe of constant z-dimension but curved in the xy-plane, where at E, $v = 0$, $u \neq 0$ and at F, $u = 0$, $v \neq 0$ and for constant volume flow, $(\partial u/\partial t) = (\partial v/\partial t) = 0$,

and
$$\frac{du}{dt} = 0 + u\frac{\partial u}{\partial x} + v\frac{\partial u}{\partial y}; \frac{dv}{dt} = 0 + u\frac{\partial v}{\partial x} + v\frac{\partial v}{\partial y},$$

i.e. total = field accelerations only, i.e. a steady state.

A.1.4   Integrals

Integration essentially means summation. For instance, in Fig. A.3(a) we may want to determine the value of the single-shaded area between the curve, the x-axis and the vertical lines at $x = x_1$ and $x = x_2$. We can do so by dividing the area into narrow strips, such as the double-shaded one whose height is y_1 and width δx, and whose area is therefore $\delta A = y_1 \delta x$ and then summing the areas of all the strips so that the total area

$$A = \sum \delta A = \sum_{x_1}^{x_2} y\delta x.$$

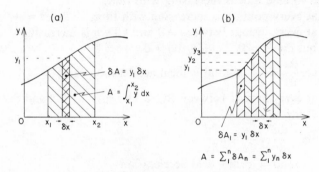

FIG. A.3   (a) *Significance of an integral*, (b) *area under a curve obtained by summation of areas of strips.*

Alternatively this can be written as

$$A = \int_{x_1}^{x_2} y\,dx,$$ i.e. the *integral* of y with respect to x.

If y is a simple function of x there are standard rules for integrating y with respect to x, i.e. determining the value of the integral, but if y is a complicated or irregular function of x, as in Fig. A.3(b), there may be no known rule. It will then be necessary to go back to the basic procedure

$$A = \sum_{x_1}^{x_2} y\,\delta x$$

by dividing the total area into a number of narrow strips and adding the individual areas so that $A = \delta A_1 + \delta A_2 + \delta A_3 +$, etc. $= y_1\delta x + y_2\delta x + y_3\delta x +$, etc. This procedure is used in Chapter 8 in geostrophic current calculations.

A.1.5   Fields

Physicists use the word *field* for the distribution of a quantity in space, e.g. geographical distributions of temperature or salinity are scalar fields while current distribution patterns represent vector fields, e.g. Fig. A.4(b). It must be noted that in such field patterns, isopleths (lines of equal value) of the property generally cannot cross or touch because this would mean that the quantity had two different values at the same point.

A.1.6 Descriptions of fluid flow

There are two ways of describing the pattern of flow in a region: *Lagrangian* in which we describe or plot the path (trajectory) followed by each fluid particle, specifying when each particle reaches each point in its path, and *Eulerian* in which we describe the velocity (speed and direction) of the fluid at every point in the fluid at every instant of time. (In practice, of course, we can only plot a selection of paths in a Lagrangian description or velocities at a selection of points for an Eulerian one.)

The two types of plots are illustrated in Fig. A.4 using the North Pacific Ocean as an example. If the circulation does not change with time, then these two figures will be related very simply—each arrow of the Eulerian pattern will be a tangent to a continuous line of the Lagrangian pattern. Therefore, if current measurements are made at a series of points to obtain initially an Eulerian picture, the Lagrangian one may be constructed by drawing continuous lines through the arrows so that the latter are tangents to the lines. In this case the Lagrangian lines will also be true *streamlines* (lines which at any given time are everywhere tangent to the Eulerian velocity). However, if the currents are changing with time, then the streamline pattern, which is an instantaneous feature, will be continually changing and the Lagrangian pattern, which shows the history of the flow, will be a step-by-step summation of the Eulerian pattern and will not be the same as the streamline pattern for any one time.

It may be noted that in the North Pacific example which we have chosen, the main (*long-term average*) character of the clockwise circulation probably does not change much, and the difference between the Lagrangian trajectories and the individual streamline patterns may not be large. However, in other places, such as the Western Coral Sea or the Equatorial Indian Ocean, the direction of the currents reverses during its annual cycle and there is a considerable difference between streamline and trajectory patterns.

The Lagrangian method is commonly used for descriptions of ocean circulation, using separate patterns for the different seasons if there is a marked seasonal variation. It is the form in which information comes from measurements made by following drifting floats or buoys. However, it is difficult to handle mathematically. The Eulerian description is obtained by putting current meters at a number of fixed points and recording speed and direction at every point simultaneously. It is more tractable mathematically and we use it when investigating the equations of motion in the sea. In principle one can

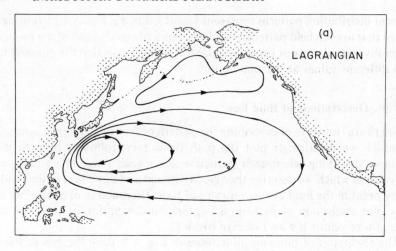

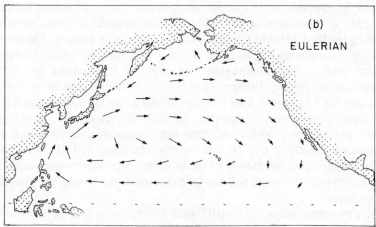

FIG. A.4   *Flow pattern described in (a) Lagrangian manner (flow lines),*
(b) Eulerian manner (velocities at points).

convert from one system to the other; in practice one rarely has enough observations to do so confidently.

A.1.7 Convergences and divergences

A common feature of flow patterns at the sea surface is one where the flow converges toward a point or line (Fig. A.5(a, c)) or diverges from a point or line (Fig. A.5(b, d)). In the convergent patterns, because the water cannot just disappear it must flow downward at the point or line of *convergence*. For the

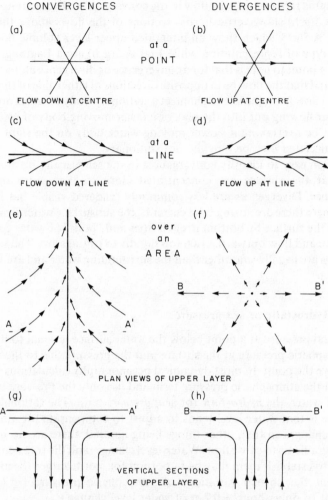

FIG. A.5 *Flow patterns at convergence or divergence at (a, b) a point, (c, d) a line, (e, f) an area—plan view, (g, h) an area—side view.*

divergences, the water must come up from below the surface and then flow outward. The patterns in the figure are mathematical idealizations because the down or up flow has infinitesimal cross-section, which is not possible in the real world where water is, to a first approximation, incompressible and where the flow is not just in a mathematically thin surface layer but actually has a significant depth. In reality, convergence or divergence generally occurs over an area as shown in Figs. A.5(e, f). These are plan views of the surface with the arrow lengths representing the flow in the convergent or divergent upper layer. Figures A.5(g, h) show vertical cross-sections of the flows above them. Note that Figs. A.5(e f), which show an integrated upper layer volume flow, are an Eulerian type of representation while Figs. A.5(g, h) are a Lagrangian form.

Another point to note is that for a convergence or divergence along a line, it is not essential that the flow be in opposite directions at either side of the line. For instance, a flow from the left might meet a stationary body of water on the right (e.g. a river flowing out into the sea), or a faster-moving body of water on the left might be overtaking a slower-moving water body on the right. Also the convergence area may be moving, not stationary.

It is often easy to identify convergences in the sea visually because objects floating at the surface are concentrated close to the centre or line of convergence. Divergences are less commonly rendered visible but in coastal waters where there are strong tidal currents, the subsurface water is sometimes forced to the surface by bottom irregularities and "boils" of water come up to the surface and flow outward, often too rapidly to row against. The surfaces of these boils are usually smoother than the surrounding water and are visible for this reason.

A.1.8 Hydrostatic or sea pressure

The total pressure at a point below the water surface is equal to the sum of the atmospheric pressure at the surface and the pressure due to the weight of fluid above the point. In most dynamical oceanography calculations it is usual to neglect the atmospheric pressure term and use only the pressure due to the weight of water, the *hydrostatic* (or *sea*) *pressure* term. The rationale for this procedure is that water level tends to adjust to compensate for atmospheric pressure changes so that, other things being ignored, there are no horizontal pressure gradients to cause flow. Water-level changes and the transient currents required to establish them may be important for storm surges (Section 13.10) (Note that the normal variations of atmospheric pressure (± 2 kPa) are equivalent to only about ± 0.2 m of water level change.)

Referring to Fig. A.6(a), if the water is uniform in density ρ (ignoring the compressibility effect) the hydrostatic pressure at level z (depth $h = -z$) is $p_z = -\rho g z$ where g is the acceleration due to gravity which is here assumed to be independent of depth. This simple situation is rarely found in the sea where the density generally increases with depth. If the water column were made up, for example, of three layers each of uniform density as in Fig. A.6(b), then the pressure at $z = z_1 + z_2 + z_3$ would be $p_z = -(\rho_1 z_1 + \rho_2 z_2 + \rho_3 z_3)g$. More generally in the ocean, the density increases with depth in a manner similar to that shown at the right in Fig. A.6(c) with an upper mixed layer of nearly uniform density, then a zone of increasing density (the *pycnocline zone*) grading into a deeper zone of more slowly increasing density. In this case, if we write

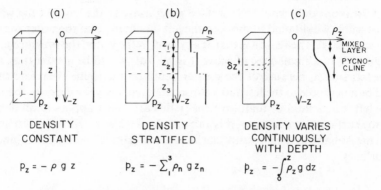

(a)   (b)   (c)

DENSITY   DENSITY   DENSITY VARIES
CONSTANT   STRATIFIED   CONTINUOUSLY
                                               WITH DEPTH

$p_z = -\rho g z$ $p_z = -\sum_1^3 \rho_n g z_n$ $p_z = -\int_0^z \rho_z g \, dz$

FIG. A.6 *Pressure in a fluid: (a) density (ρ) constant, (b) density (ρ_n) varying in steps, (c) density (ρ_z) varying smoothly with depth.*

$dp = -\rho_z g dz$ for the pressure due to a small layer of water of thickness dz where the density is ρ_z, then the total hydrostatic (sea) pressure will be

$$p_z = \int_0^z -\rho_z g \, dz.$$

The minus signs occur because z increases upward; with $z = 0$ at the surface, z is negative below the surface while p_z must be positive. If the density happens to vary with depth according to some mathematical function which can be integrated, then this integral may be evaluated directly. However, it is usually necessary to use the procedure of Fig. A.6(b), dividing the water column into a sufficient number of thin layers, each of essentially uniform density, to represent closely the real distribution with depth. For analytical studies, the shape of density variation of Fig. A.6(c) has sometimes been represented by an upper mixed layer of depth h of uniform density ρ_1 and a lower layer of density $\rho_2 = \rho_d - (\rho_d - \rho_1) \exp(1 + z/h)$ for $z < -h$, so that in the lower layer the density approaches asymptotically the deep water value ρ_d.

A.1.9   Slope effects

If a block of wood of mass m were placed on a sloping surface (Fig. A.7(a)) it is quite likely that it would remain stationary because the component $mg \sin\theta$ of its weight (mg) down the slope would be less than the possible friction force up the slope. However, if *water* were poured on the slope (Fig. A.7(c)), the frictional force between it and the slope would be much smaller than the weight component down the slope initially since there is no fluid friction until motion begins. The water would flow down the slope with increasing speed until the friction associated with the flow balanced the weight component down the slope. If we had a container of water and established a slope on the surface

(Fig. A.7(b)) not only would the surface water move to the left, but the whole body of water would tend to move to the left (assuming that the water were of uniform density). The reason is that at any level in the water, the pressure p_R on the right side of a small cube of water (Fig. A.7(d)) would be greater than that on the left side p_L because of the greater depth on the right. Therefore there would be a net force to the left and a consequent tendency for the cube to move to the left. For a fluid of uniform density this argument applies at all depths from the surface to the bottom. It is an example of a *barotropic* force system, in which the isobaric (equal pressure) surfaces are all parallel to each other (and to the surface).

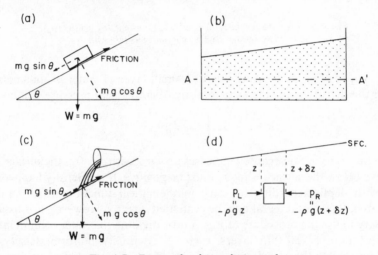

FIG. A.7 *Forces related to a sloping surface.*

The above statements apply to the situation when the water is initially stationary. It is shown in the text that if the water is moving appropriately (into the paper in the northern hemisphere in this case) the Coriolis force (see Chapter 8) to the right may be sufficient to balance the hydrostatic pressure force to the left and the slope may be maintained.

A situation in which the net horizontal hydrostatic force does not penetrate to the bottom could occur if the density of the water in the upper layer above level AA′ in Fig. A.7(b) decreased progressively to the right so that the hydrostatic pressure

$$p_A = \int_0^{z_A} \rho g \, dz$$

remained constant along AA′. Then there would still be a tendency for the water above AA′ to move to the left but if the density were uniform below AA′

there would be no resultant hydrostatic pressure force and no tendency for the deep water to move. In the layer above AA', where the density varies with horizontal position, the isobaric surfaces will be inclined to each other, rather than parallel, and the force system is then described as *baroclinic*.

A.1.10 Compressibility

The compressibility β of a fluid is defined as $\beta = -(1/V)(dV/dp)$ where V is volume and p is pressure, and for a fluid to be incompressible $\beta = 0$. Now we can write

$$\beta = -\frac{1}{V}\frac{dV}{dp} = -\left(\frac{1}{V}\frac{dV}{dt}\right)\left(\frac{dt}{dp}\right) = -\left(\frac{1}{V}\frac{dV}{dt}\right) \div \left(\frac{dp}{dt}\right)$$

where dV/dt and dp/dt are changes with time of volume and pressure respectively, following a small parcel (in the limit infinitesimal) of fluid. If the pressure is changing but $\beta = 0$, we conclude that $(1/V)(dV/dt) = 0$. Now for a constant mass of fluid of volume V and hence density $\rho = m/V$

$$\frac{1}{\rho}\frac{d\rho}{dt} = \frac{V}{m}\frac{d}{dt}\left(\frac{m}{V}\right) = -\frac{1}{V}\frac{dV}{dt} = 0,$$

so that we can regard incompressibility as meaning either

$$\frac{1}{V}\frac{dV}{dt} = 0 \quad \text{or} \quad \frac{1}{\rho}\frac{d\rho}{dt} = 0.$$

A.1.11 Equation of state

Real fluids are compressible, i.e. $\beta \neq 0$, and the relation between the change of volume and the applied pressure is called the *compressibility equation*. It is the pressure component of the full equation of state. If a fixed mass of liquid has volume V_0 at $p = 0$ and V at $p = p$, then the compression curve is of the form shown in Fig. A.8(a). For low pressures (tens of bars) the relation is linear, for hundreds of bars pressure it is quadratic and for thousands of bars it is cubic. (The bar has been the usual unit for pressure in liquid compressibility studies and, even though it is not an SI unit, it has been adopted for use with the 1980 international equation of state for sea water, as mentioned in Chapter 2.) The relations between p and V may be expressed through the *tangent bulk modulus* (K_T) (the reciprocal of compressibility, i.e. $K_T = 1/\beta$) which is defined as

$$K_T = -V\frac{dp}{dV}$$

so that K_T is the slope of the tangent to the curve at (p, V). For very low pressures K_T may, for practical purposes, be regarded as constant but for

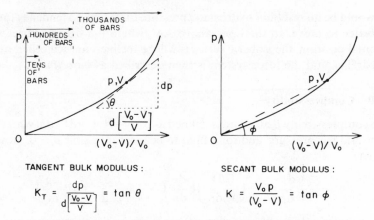

TANGENT BULK MODULUS :

$$K_T = \frac{dp}{d\left[\dfrac{V_0-V}{V}\right]} = \tan\theta$$

SECANT BULK MODULUS :

$$K = \frac{V_0\,p}{(V_0-V)} = \tan\phi$$

FIG. A.8   *Schematic compression curve for a liquid to show definitions of (a) tangent bulk modulus (K_T), (b) secant bulk modulus (K).*

higher pressures it will be proportional to some power of p (assuming isothermal conditions for the moment).

Hayward (1967) reviewed the great variety of compressibility equations which had been used and showed that a modified compressibility equation using the *secant bulk modulus*

$$K = \frac{V_0 p}{V_0 - V} = K_0 + mp \qquad (A.3)$$

is easier to derive from experimental data and just as accurate in practice as the tangent bulk modulus. The quantity K is the average value of the bulk modulus over the range 0 to p, i.e. it corresponds to the slope of the chord or *secant* cutting the compression curve at the origin and at (p, V) (Fig. A.8(b)) and K_0 is the value of the bulk modulus at $p = 0$. Note that V_0 and V are proportional to the specific volume of the liquid at $p = 0$ and p.

In a recent study of the full equation of state for sea water, Millero, Chen, Bradshaw and Schleicher (1980) showed that the experimental data could be best described using a secant bulk modulus equation of the form

$$K(s,t,p) = \frac{\alpha_0 p}{\alpha_0 - \alpha_p} = K_0 + Ap + Bp^2, \qquad (A.4)$$

where α_0, α_p = specific volume of sea water at $p = 0$ and p respectively, K_0, A and B include temperature dependent parameters for pure water and also temperature and salinity-dependent parameters for sea water.
Then the equation of state is expressed in the forms

$$\alpha(s,t,p) = \alpha(s,t,0)[1 - p/K(s,t,p)]$$

or
$$\rho(s,t,p) = \rho(s,t,0)/[1 - p/K(s,t,p)]. \qquad (A.5)$$

Details of the constants in this equation of state are given in Appendix 3.

It is interesting to note that Tait (1888), when studying the compressibility of sea water samples from the *Challenger* Expedition (from 1872 to 1876), proposed a compressibility equation for what he called "average compressibility" which, when rearranged, is of identical form to equation (A.4) above. However, as Hayward (1967) pointed out, the "Tait's equation" which has been used since by "two generations" of workers in the field of liquid compressibility is a misquotation of Tait's original equation and is, in fact, more complicated, less easy to use and no more accurate than the genuine one. The origin of the misquotation is not known.

A.1.12   Centripetal and centrifugal forces

Acceleration is defined as the rate of change of velocity of a body and may consist of a rate of change of speed and/or of direction. If the body has mass, a resultant force is required to cause either change. If the direction of motion is changing, the body must be travelling on a curved path and there must be a resultant force directed inward toward the centre of curvature of the path. This force, which acts on the body itself, is called the *centripetal* force. Now Newton's Third Law of Motion states that to every acting force there must be a reacting force, equal in magnitude, opposite in direction and *acting on another body*. This reaction is called the *centrifugal* force and it is important to note that it does not act on the body itself but on something else.

A simple example of this force pair is provided when a small mass attached to a string is whirled around in a circle, the inner end of the string being held in the hand. Then the inward force of the string provides the *centripetal force on the mass*, while the outward pull of the string *on the hand* is the centrifugal force.

Some authors study the dynamics of such a system by using a fictional outward force on the mass (equal to its mass × its acceleration) and call this a "centrifugal force". While the procedure of using an outward "mass acceleration" is a legitimate (and often convenient) device for solving the dynamics, it is incorrect to call this the "centrifugal force". We prefer to treat problems in rotation in terms of the *inward* directed centripetal force on the mass itself.

In the example above, the tension in the string acting on the orbiting mass provides the centripetal force while in the astronomical case, the gravitational attraction of the earth on the moon, for example, provides the centripetal force needed to maintain the moon in orbit. Likewise, the reciprocal gravitational attraction of the moon on the earth provides the centripetal force needed to keep the earth revolving about the centre of mass of the earth–moon system.

Appendix 2

Units Used in Physical Oceanography

A.2.1 Introduction

Hitherto, physical oceanographers have used a mixed system of units which has its advantages for those thoroughly familiar with it but which offers some difficulties for the beginner in the field. Because the International System of Units (abbreviated SI) is now coming into general use we have used it in this text. In SI there are base units and derived units (from the base units) and a number of temporary units which are accepted but will be phased out of use eventually. In the following we will first list the base, derived and temporary units used in dynamical oceanography and then will relate them to the older mixed units. Our basic reference for SI practice is the Metric Practice Guide published by the Standards Council of Canada (reference CAN-3-001-02-73, CSA Z234.1-1973).

A.2.2 Base units

The base units used in this text, their abbreviations and physical dimensions are:

Quantity	Base unit	Abbreviation	Dimension
Length	metre	m	[L]
Mass	kilogram	kg	[M]
Time	second	s	[T]
Thermodynamic temperature	kelvin	K	[K]

A.2.3 Derived and temporary units

The units are given with their abbreviations and under the name of the unit is given the symbol used in equations in this text together with (in square brackets) the physical dimensions of the unit in terms of length [L], mass [M], time [T] and thermodynamic temperature (above absolute zero) [K].

Quantity (Symbol) [Dimension]	Unit and abbreviation
Length (L) $[L]$	1 centimetre (cm) $= 10^{-2}$ m, 1 decimetre (dm) $= 10^{-1}$ m, 1 kilometre (km) $= 10^{3}$ m, 1 international nautical mile (n ml) $= 1852.0$ m (temporary unit).
Mass (m, M) $[M]$	1 gram (g) $= 10^{-3}$ kg, 1 tonne (t) $= 10^{3}$ kg.
Time (t, T) $[T]$	1 minute (min) $= 60$ s, 1 hour (h) $= 60$ min, 1 day (d) $= 24$ h $= 86\,400$ s (mean solar day), $\qquad\qquad\qquad$ $(86\,164$ s $= 1$ sidereal day), 1 year (a) $=$ not defined in SI, $\qquad\qquad\qquad$ here taken as 365 d.
Area (A), $[L^{2}]$	1 m^2.
Volume (V), $[L^{3}]$	1 m^3, 1 litre $\simeq$ 1 dm^3 $= 10^{-3}$ m^3, (not SI, not used for high precision measurements).
Speed; components $(V; u, v, w; C)$ $[LT^{-1}]$	1 m s^{-1} $= 10^{2}$ cm s^{-1}, 1 knot (kn) $=$ 1 n ml h^{-1} $= 0.514$ m s^{-1} (temporary unit).
Acceleration (a), $[LT^{-2}]$	1 m s^{-2}.
Density (ρ), $[ML^{-3}]$	1 kg m^{-3} $= 10^{-3}$ g cm^{-3}.
Relative density (d), $[none]$	Value given relative to pure water for liquids and to air for gases—formerly "specific gravity".
Specific volume (α), $[M^{-1}L^{3}]$	1 m^3 kg^{-1} $= 10^{3}$ cm^3 g^{-1}.
Force (F), $[MLT^{-2}]$	1 newton (N) $=$ 1 kg m s^{-2} $=$ force required to give 1 kg mass an acceleration of 1 m s^{-2} $(= 10^{5}$ dynes—not SI).

Pressure (p), $[ML^{-1}T^{-2}]$	1 pascal (Pa) = 1 N m^{-2} (= 10 dyne cm^{-2}—not SI), (= 10^{-5} bar—not SI)
Energy (E, W), $[ML^2T^{-2}]$	1 joule (J) = 1 N m (= 10^7 ergs—not SI).
Dynamic viscosity (μ), $[ML^{-1}T^{-1}]$	1 pascal second (Pa s) (= 10 poise—not SI).
Kinematic viscosity $(v = \mu\rho^{-1})$, $[L^2T^{-1}]$	1 m^2 s^{-1} = 10^4 cm^2 s^{-1} (= 10^4 stokes—not SI).
Kinematic diffusivity (κ), $[L^2T^{-1}]$	1 m^2 s^{-1}.
Temperature (T), $[K]$	The Celsius temperature (°C) is the difference between the thermodynamic temperature T and the temperature $T_0 = 273.15$ K, i.e. 0°C = 273.15 K, unit = 1 K).

A.2.4 Units used in dynamical oceanography and some numerical values

Some of these units are peculiar to physical oceanography; numerical values for some of the quantities are given for illustration.

Quantity (Symbol) [Dimensions]	Description, typical values
Sverdrup: (Sv) $[L^3T^{-1}]$	a unit of volume flow = 10^6 m^3 s^{-1}; (not SI). Typical values for major currents: Pacific equatorial currents, 10 to 70 Sv; Gulf Stream, 50 to 150 Sv; Antarctic Circumpolar Current to 290 Sv.
Salinity: (S_A, S) [none]	*Absolute salinity* (S_A) is defined as the ratio of the mass of dissolved material in sea water to the mass of sea water. *Practical salinity* (S) is defined in terms of the ratio K_{15} of the electrical conductivity of the sea water sample at the temperature of 15°C and the pressure of one standard atmosphere to that of a potassium chloride (KCl) solution in which the mass fraction of KCl is 32.4356×10^{-3} at the same temperature and pressure. The K_{15} value exactly equal to 1 corresponds, by definition, to a practical salinity exactly equal to 35. The practical salinity is defined in terms of the ratio K_{15} by the following equation:

$$S = a_0 + a_1 K_{15}^{1/2} + a_2 K_{15} + a_3 K_{15}^{3/2} + a_4 K_{15}^2 + a_5 K_{15}^{5/2}$$

where $a_0 = 0.0080$, $a_1 = -0.169\,2$, $a_2 = 25.385\,1$, $a_3 = 14.094\,1$, $a_4 = -7.026\,1$, $a_5 = 2.708\,1$, $\Sigma a_i = 35.000\,0$, $2 < S < 42$.

The mean value for the world ocean is about $S = 34.7$, with values in the Red Sea over 42 and values close to 0 in coastal regions near rivers.

As it may not be possible to make measurements at precisely 15°C, in practice one uses Standard Sea Water (of known S, *circa* 35, which is labelled with its measured K_{15} value) to calibrate the conductivity instrument and measures the ratio R_t between the unknown sea water sample conductivity and the standard sea water conductivity at some constant temperature t°C. Then one uses R_t in place of K_{15} in the equation for S and, if $t \neq 15$°C, one adds a correction of

$$\Delta S = \frac{(t-15)}{1+k(t-15)}(b_0 + b_1 R_t^{1/2} + b_2 R_t + b_3 R_t^{3/2} + b_4 R_t^2 + b_5 R_t^{5/2})$$

where $b_0 = 0.0005$, $b_1 = -0.0056$, $b_2 = -0.0066$, $b_3 = -0.0375$, $b_4 = 0.0636$, $b_5 = -0.0144$, $k = 0.0162$.

In situ conductivity instruments measure R, the ratio of the *in situ* conductivity to the standard conductivity ($\equiv 1$ at $S = 35$, $t = 15$°C, $p = 0$). The ratio R is factored into three parts as $R = R_p r_t R_t$ where R_p is the ratio of the *in situ* conductivity to the conductivity of the same sample at the same temperature but at $p = 0$; r_t is the ratio of the conductivity of standard sea water of $S = 35$ at temperature t to its conductivity at $t = 15$°C. Knowledge of R_p and r_t allows calculation of $R_t = R/(R_p r_t)$. With t in °C and p in bars (10^5 Pa).

$$r_t = c_0 + c_1 t + c_2 t^2 + c_3 t^3 + c_4 t^4$$

where $c_0 = 0.676\,609\,7$, $c_1 = 2.005\,64 \times 10^{-2}$, $c_2 = 1.104\,259 \times 10^{-4}$, $c_3 = -6.969\,8 \times 10^{-7}$, $c_4 = 1.003\,1 \times 10^{-9}$,

$$R_p = 1 + \frac{p(e_1 + e_2 p + e_3 p^2)}{1 + d_1 t + d_2 t^2 + (d_3 + d_4 t)R}$$

Salinity (*Contd.*)

where $e_1 = 2.070 \times 10^{-4}$, $e_2 = -6.370 \times 10^{-8}$, $e_3 = 3.989 \times 10^{-12}$, $d_1 = 3.426 \times 10^{-2}$, $d_2 = 4.464 \times 10^{-4}$, $d_3 = 4.215 \times 10^{-1}$, $d_4 = -3.107 \times 10^{-3}$.

(A formal statement about salinity and the equation of state is given in Unesco *Technical Papers in Marine Science*, No. 36, 1981, together with further references.)

Density:
(ρ)
$[ML^{-3}]$

Is a function of S, T and p, i.e. $\rho = \rho(s,t,p)$.
For sea water of salinity $S = 35.00$, temperature $T = 10.00°C$ at standard atmospheric pressure (= zero hydrostatic pressure) $\rho(35,10,0) = 1026.954$ kg m^{-3}. Strictly speaking, sea water density is operationally measured relative to pure water as the standard and is therefore "relative density". However, in physical equations it must be treated dimensionally as density. See Appendix 3 for further information.

Sigma-*t*:
(σ_t)
$[ML^{-3}]$

This quantity is introduced for convenience and is defined as: $\sigma_t = (\rho(s,t,o) - 1000.00)$. For the sample of sea water above, the value of $\sigma_t = 26.95$. (It is usual to omit the units (kg m^{-3}) when stating values for σ_t, which is usually used for descriptive purposes rather than as a component in equations.) In the mixed units system, with ρ in g cm^{-3}, sigma-*t* is defined as $\sigma_t = (\rho - 1) \times 10^3$, which gives the same numerical value as the SI unit definition given above although the units are different (i.e. mg cm^{-3}).

Note that although sea water density may be stated to six significant figures, and σ_t to four figures, the absolute values are not known to this accuracy. Differences between densities for typical oceanic waters may be accurate to the second decimal place (0.01 kg m^{-3}) but absolute values only to the first decimal place (0.1 kg m^{-3}), i.e. five significant figures in ρ or three in σ_t, using the older tables. The IES 80 is claimed to have a standard error of 3.6×10^{-3} kg m^{-3}, i.e. of 0.0036 in σ_t.

Specific volume:
($\alpha = 1/\rho$)
$[M^{-1} L^3]$

Is a function of S, T and p, i.e. $\alpha(s,t,p)$. For sea water of salinity $S = 35.00$, temperature $T = 10.00°C$ at standard atmospheric pressure, $\alpha(35,10,0) = 0.973\,753 \times 10^{-3}$ m^3 kg^{-1}.

Specific volume anomaly: (δ), $[M^{-1} L^3]$

Defined as $\delta = \alpha(s,t,p) - \alpha(35,0,p)$ (units $m^3 kg^{-1}$).

Thermosteric anomaly: $(\Delta_{s,t})$ $[M^{-1} L^3]$

A term in the expression of δ (see text, Section 2.23). SI units are $m^3 kg^{-1}$; for $\sigma_t = 26.95$, the value of $\Delta_{s,t}$ $= 109.5 \times 10^{-8} m^3 kg^{-1}$. In the old mixed units system the corresponding value is $109.5 \times 10^{-5} cm^3 g^{-1}$, or $109.5 cl t^{-1}$ ($=$ centilitres tonne^{-1}) to avoid having to write the power of 10. ($1 cl t^{-1} = 10^{-8} m^3 kg^{-1}$.)

Pressure: (p) $[ML^{-1} T^{-2}]$

$1 bar = 10^3 mbar$ ($= 10^6 dyn cm^{-2}$) $= 10^5 Pa = 100 kPa$,
$1 decibar = 1 dbar = 10^4 Pa = 10 kPa$.
Standard atmospheric pressure $= 1013.25 mbar$ $= 101.325 kPa$. In the open ocean, the pressure at a geometrical depth of 1000 m is about 1010 dbar $= 10100 kPa$

Geopotential: (Φ) $[L^2 T^{-2}]$

The work done per unit mass to raise a body vertically through a small distance z in the vicinity of the earth $= g z$ joules kg^{-1}. In SI, the unit of Φ is $1 J kg^{-1}$ $= 1 m^2 s^{-2}$, and using $g = 9.80 m s^{-2}$ for the acceleration due to gravity, then for a vertical lift of 1 m, $\delta\Phi$ $= 9.80 J kg^{-1}$.

In the past, in dynamical oceanography, a quantity called "dynamic height" (D) has been used. This is geopotential expressed in units such that 1 dynamic metre $= 10.0 J kg^{-1} = 10.0 m^2 s^{-2}$. Then the dynamic height difference between the sea surface and 1000 m geometrical depth $= -980 dyn m$ for $g = 9.80 m s^{-2}$.

Viscosity: (μ) $[ML^{-1} T^{-1}]$ $(v, A_x,$ etc.$)$, $[L^2 T^{-1}]$

For sea water of $S = 35$ at $T = 10°C$
Molecular viscosity –
    Dynamic: $\mu \simeq 1.4 \times 10^{-3} kg m^{-1} s^{-1}$,
    Kinematic: $v = \mu\rho^{-1} \simeq 1.4 \times 10^{-6} m^2 s^{-1}$.
Eddy viscosity –
    Kinematic: A_x, A_y up to $10^5 m^2 s^{-1}$
                     A_z up to $10^{-1} m^2 s^{-1}$.

Diffusivity: $(\kappa_T, \kappa_S)[L^2 T^{-1}]$

For sea water, kinematic:
Molecular diffusivity:
    for heat, $\kappa_T \sim 1 \times 10^{-7} m^2 s^{-1}$.
    for salt, $\kappa_S \sim 1 \times 10^{-9} m^2 s^{-1}$.

$(K_S, K_T)[L^2 T^{-1}]$

Eddy diffusivity: same ranges as for eddy viscosity above.

Appendix 3

Sources of Information for the Estimation of Specific Volume and Density from Values of Salinity, Temperature and Pressure

A.3.1 The table method

Sources of tables of specific volume, etc., data using the old mixed units system, salinity (S') prior to PSS 78 and based on the Knudsen–Ekman equation of state (i.e. pre-IES 80) are as follows, using primed symbols to identify mixed units values (references are in the Bibliography, Section B.2):

$\alpha'(35,0,p)$ – Sverdrup *et al.*, Table 1   For $p = 0$ to 9900 dbar
(99 000 kPa).

           Neumann and Pierson,
           Table IV   ditto,
           N.O.O. 614, Table IV   ditto.

$\Delta'_{(s,t)}$ – Sverdrup *et al.*, Table III   for $\sigma_t = 23$ to 28,
           Neumann and Pierson, Table I   ditto,
           N.O.O. 614, Table V   For $T = -1.9$ to 29.9°C,
                               $S' = 21.0$ to 37.9.

$\delta'_{t,p}, \delta'_{s,p}$ – Sverdrup *et al.*, Tables IV and V $\left. \begin{array}{l} p = 0 \text{ to } 10\,000 \text{ dbar} \\ S' = 30 \text{ to } 40 \\ T = -2 \text{ to } 30°C. \end{array} \right.$
           Neumann and Pierson, Tables
           II and III, N.O.O. 614, Tables   for
           VI and VII

			T	S'
σ'_t	– Knudsen's Tables	for	-2 to 33°C	0 to 40,
	N.O.O. 614		-2 to 30°	30 to 38,
	N.O.O. 615		-2 to 30°	0 to 40,
	Fleming (1939) (S' and T values			
	for unit values of σ'_t)		-2 to 30	22 to 41.

$\rho'(s,t,p)$ – reciprocal of $\alpha'(s,t,p)$.

The relations between SI numerical values (unprimed symbols) and the mixed units values (primed) are:

$$\alpha = \alpha' \times 10^{-3}, \quad \rho = \rho' \times 10^3,$$
$$\Delta_{s,t} = \Delta'_{s,t} \times 10^{-3}, \quad \sigma_t = \sigma'_t,$$
$$\delta_{t,p} = \delta'_{t,p} \times 10^{-3}, \quad p \text{ (kpa)} = p' \text{ (dbar)} \times 10,$$
$$\delta_{s,p} = \delta'_{s,p} \times 10^{-3}, \quad T \text{ and } S' \text{ are the same, using the older}$$
$$\text{definition of salinity, i.e. not PSS 78.}$$

To indicate their order of magnitude, Table A.3.1. gives a selection of values for the specific volume anomaly terms. Blanks in the table indicate that the combinations of parameters do not occur in the sea.

TABLE A.3.1 *A selection of values for specific volume anomaly terms in units of $10^{-8}\, m^3\, kg^{-1}$*

	Temperature	Salinity:	30	32	34	35	36
$\Delta_{s,t}$:	$-2°C$	$\Delta_{s,t} = 378$	224	70	-7	-84
	0	383	230	76	0	-76
	10	481	332	183	110	36
	20	682	536	390	318	245

	Temperature	Pressure:	0	1	2	5	10×10^4 kPa
		(Depth $\simeq$	0	1000	2000	5000	10 000 m)
$\delta_{t,p}$:	$-2°C$	$\delta_{d,p} =$	0	-6	-11		
	0		0	0	0	0	0
	2		0	6	10	22	38
	10		0	21	41		
$\delta_{s,p}$:	Salinity						
	30	$\delta_{s,p} =$	0	-8			
	34		0	-2	-3	-7	
	34.8		0	0	-1	-1	-2
	35		0	0	0	0	0
	36		0	2	3		

N. B. The above values are for S according to PSS 78 and using IES 80.

Table A.3.2 shows in the first column the current accuracy of routine measurement of the three parameters, salinity (from conductivity), temperature and pressure, and in the second and third columns respectively the accuracy of density and of specific volume corresponding to the variation of each parameter individually. For comparison the values for density, etc., for water of salinity 35.00, temperature 10°C and zero hydrostatic pressure ($p = 0$) are given at the bottom of the table.

TABLE A.3.2   *Accuracy of measurement of salinity, temperature and pressure and related accuracies of density and specific volume*

Accuracy of measurement	Related accuracies of	
	Density ($kg\,m^{-3}$)	Specific volume ($m^3\,kg^{-1}$)
$\Delta S = \pm 0.003$	$\Delta\rho = \Delta\sigma_t = \pm 0.002$	$\Delta\alpha = \Delta(\Delta_{s,t}) = \mp 0.2 \times 10^{-8}$
$\Delta T = \pm 0.02°C$	$\Delta\rho = \Delta\sigma_t \mp 0.002$ ($T = 2°C$, $S = 35$) to ∓ 0.006 ($T = 25°C$, $S = 35$)	$\Delta\alpha = \Delta(\Delta_{s,t}) = \pm 0.3 \times 10^{-8}$
$\Delta p = \pm 50$ kPa ($\equiv \Delta z = \pm 5$ m for upper 1000 m)	$\Delta\rho = \pm 0.24$*	$\Delta\alpha = \mp 2.2 \times 10^{-8}$

For sea water of:
$$S = 35.00 \quad \text{then} \quad \rho = 1026.95\ kg\,m^{-3}$$
$$T = 10.00°C \qquad \sigma_t = 26.95\ kg\,m^{-3}$$
$$p = 0 \qquad \alpha = 0.973\,76 \times 10^{-3}\ m^3\,kg^{-1}$$
$$\Delta_{s,t} = 109.7 \times 10^{-8}\ m^3\,kg^{-1}$$

* The uncertainty due to pressure differences may be greater in deep water because of greater uncertainty in pressure measurement.

A.3.2   International Equation of State of Sea Water, 1980

The new equation of state is presented by Millero and Poisson (1981) and also given in Unesco Technical Paper in Marine Science Number 36 (Unesco, 1981). The density of sea water as a function of practical salinity (S), temperature ($T°C$) and sea pressure (p bars) is given by

$$\rho(s,t,p) = \rho(s,t,o)/[1 - p/K(s,t,p)]$$

where $K(s,t,p)$ is the secant bulk modulus (see Section A.1.11). The specific volume is given by

$$\alpha(s,t,p) = \alpha(s,t,o)\,[1 - p/K(s,t,p)].$$

The polynomial expressions for $\rho(s,t,o)$ and $K(s,t,p)$ are given below.

For the IES 80, the density of sea water at one standard atmosphere pressure ($p = 0$) is given by:

$$\rho(s,t,o) =$$

$$
\begin{aligned}
&+999.842\,594 && +6.793\,952 \times 10^{-2} \times T \\
&- \ \ 9.095\,290 \times 10^{-3} \times T^2 && +1.001\,685 \times 10^{-4} \times T^3 \\
&- \ \ 1.120\,083 \times 10^{-6} \times T^4 && +6.536\,332 \times 10^{-9} \times T^5 \\
&+ \ \ 8.244\,93 \ \times 10^{-1} \ \ \times S && -4.089\,9 \ \ \times 10^{-3} \times T \ \times S \\
&+ \ \ 7.643\,8 \ \ \times 10^{-5} \times T^2 \times S && -8.246\,7 \ \ \times 10^{-7} \times T^3 \times S \\
&+ \ \ 5.387\,5 \ \ \times 10^{-9} \times T^4 \times S && -5.724\,66 \ \times 10^{-3} \qquad \times S^{3/2} \\
&+ \ \ 1.022\,7 \ \ \times 10^{-4} \times T \ \ \times S^{3/2} && -1.654.6 \ \ \times 10^{-6} \times T^2 \times S^{3/2} \\
&+ \ \ 4.831\,4 \ \ \times 10^{-4} \qquad \times S^2.
\end{aligned}
$$

For the IES 80, the secant bulk modulus is given by:

$$K(s,t,p) =$$

$$+\,19\,652.21$$

+	$148.420\,6$	$\times T$	$-2.327\,105$	$\times T^2$	
+	$1.360\,477 \times 10^{-2} \times T^3$		$-5.155\,288 \times 10^{-5} \times T^4$		
+	$3.239\,908$	$\times p$	$+1.437\,13 \times 10^{-3} \times T \times p$		
+	$1.160\,92 \times 10^{-4} \times T^2 \times p$		$-5.779\,05 \times 10^{-7} \times T^3 \times p$		
+	$8.509\,35 \times 10^{-5} \times p^2$		$-6.122\,93 \times 10^{-6} \times T \times p^2$		
+	$5.278\,7 \times 10^{-8} \times T^2 \times p^2$				
+	$54.674\,6$	$\times S$	$-0.603\,459$	$\times T$	$\times S$
+	$1.099\,87 \times 10^{-2} \times T^2$	$\times S$	$-6.167\,0 \times 10^{-5} \times T^3$		$\times S$
+	7.944×10^{-2}	$\times S^{3/2}$	$+1.648\,3 \times 10^{-2} \times T$		$S^{3/2}$
−	$5.300\,9 \times 10^{-4} \times T^2$	$\times S^{3/2}$	$+2.283\,8 \times 10^{-3}$	$\times p$	$\times S$
−	$1.098\,1 \times 10^{-5} \times T \times p \times S$		$-1.607\,8 \times 10^{-6} \times T^2 \times p \times S$		
+	$1.910\,75 \times 10^{-4} \times p \times S^{3/2}$		$-9.934\,8 \times 10^{-7}$	$\times p^2 \times S$	
+	$2.081\,6 \times 10^{-8} \times T \times p^2 \times S$		$+9.169\,7 \times 10^{-10} \times T^2 \times p^2 \times S$		

The above polynomials are taken from Unesco *Technical Papers in Marine Science* No. 36, 1981.

The following values may be used for checking the correct use of the IES 80, ρ being in $\mathrm{kg\,m^{-3}}$ and K in bars:

S	$T\,°C$	p bars	$\rho(s,t,p)$	$K(s,t,p)$
0	5	0	999.966 75	20 337.803 75
		1000	1044.128 02	23 643.525 99
	25	0	997.047 96	22 100.721 06
		1000	1037.902 04	25 405.097 17
35	5	0	1027.675 47	22 185.933 58
		1000	1069.489 14	25 577.498 19
	25	0	1023.343 06	23 726.349 49
		1000	1062.538 17	27 108.945 04

In the above polynomials, the pure-water terms are those not containing salinity (S). Procedures for converting salinities estimated by earlier methods to the Practical Salinity Scale, 1978, for use with the above polynomials are given by Lewis and Perkin (1981).

Tables of sea-water properties derived from the International Equation of State 1980 are promised as future volumes of the International Oceanographic Tables (Unesco, 1981).

Bibliography

B.1 Introduction

This bibliography is divided into three sections: (a) suggestions for further reading in the form of texts which enlarge on or supplement the material of the present text and a short list of journals which are largely devoted to marine science, (b) a list of data, etc., sources, and (c) the specific journal references mentioned in the text for the convenience of the reader wishing to go into a topic in more detail. (Note that a few of the text references will be found in Section B.2.1)

More comprehensive lists of journal references may be found in *The Oceans* (1946), by Sverdrup, Johnson and Fleming, for material published before about 1942, and *Evolution of Physical Oceanography* (1981), edited by Warren and Wunsch, for later material.

B.2 Suggestions for further reading

B.2.1 Texts

BARBER, N. F. (1969) *Water Waves,* Wykeham Publishing Company, p. 142. Descriptions of wave characteristics and behaviour with only a little mathematics.

BASCOM, W. (1964) *Waves and Beaches,* Anchor, Doubleday, p. 267. A lively account of experimental studies of ocean waves, particularly near the shore, and of their relations to beach form.

BATCHELOR, G. K. (1967) *Introduction to Fluid Mechanics,* Cambridge University Press, p. 615. A review of the field; graduate level.

BJERKNES, V. *et al.* (1933) *Physikalische Hydrodynamik,* Springer, Berlin, p. 797. For the transformations from fixed to rotating axes for the equations of motion, and an immense quantity of other material in physical hydrodynamics.

CSANADY, G. T. (1982) *Circulation in the Coastal Ocean,* Riedel, p. 274. The dynamics of the coastal ocean and shallow seas; graduate level.

DARWIN, G. H. (1911) *The Tides and Kindred Phenomena in the Solar System,* Houghton Mifflin; reprint Freeman, San Francisco, 1962, p. 378. The text of a series of public lectures with excellent non-mathematical discussions/expositions of topics in the field.

DEFANT, A. (1958) *Ebb and Flow, the Tides of Earth, Atmosphere and Water,* University of Michigan Press, p. 121. A very interesting descriptive account with illustrations.

DEFANT, A. (1961) *Physical Oceanography,* Pergamon Press, p. 1319. An advanced-level text. Volume 1, Pt. 1, is descriptive while Pt. 2 and all of Vol. 2 are dynamical.

DIETRICH, G., K. KALLE, W. KRAUSS and G. SIEDLER (1980) *General Oceanography—an Introduction,* Second Edition, p. 626 (translated by S. and H.U.Roll). A fairly comprehensive text on descriptive and dynamical oceanography with summaries of other disciplines; relatively little mathematics, good illustrations.

DOBSON, F., L. HASSE and R. DAVIS (Eds.) (1980) *Air—Sea Interaction; Instruments and Methods,* Plenum p. 801. A series of papers reviewing methods for measuring velocity, temperature, etc., in air and water, air—sea interaction techniques and platforms from which measurements are made. Reviews the theoretical bases as well as the mechanics of the instruments. The best reference available on instrumentation.

DOODSON, A. T. and H. D. WARBURG (1941) *Admiralty Manual of Tides*, His Majesty's Stationary Office, p. 270. A review of tidal descriptions, theory and harmonic analysis.

DYER, K. R. (1973) *Estuaries—a Physical Introduction*, Wiley, p. 140. A summary of the descriptive and dynamical oceanography of estuaries.

FOMIN, L. M. (1964) *The Dynamic Method in Oceanography*, Elsevier, p. 212. Devoted to the use of the geostrophic method and dynamic topographies for current determination.

FREELAND, H. J., D. M. FARMER and C. D. LEVINGS (Eds.) (1980) *Fjord Oceanography*, Plenum, 1980, p. 715. Proceedings of a symposium on the physical and biological oceanography of fjord estuaries.

GILL, A. E. (1982) *Atmosphere–Ocean Dynamics*, Academic Press, p. 662. A unified, comprehensive approach to the study of oceanic and atmospheric circulations, basically mathematical but with excellent verbal descriptions.

GODIN, G. (1972) *The Analysis of Tides*, University of Toronto Press, p. 264. A detailed, advanced description of the various techniques.

GOLDBERG, E. D. *et al.* (Eds.) (1977) *The Sea: Ideas and Observations*, Wiley-Interscience, Vol. 6, *Marine Modelling*, p. 1048. A collection of papers on modelling of physical, geological, chemical and biological systems in the sea.

GOWER, J. F. R. (Ed.) (1981) *Oceanography from Space. Marine Science*, Vol. 13, Plenum Press, p. 978. A collection of papers on techniques of and some results from satellite observations.

GROSS, M. GRANT (1982) *Oceanography: a View of the Earth*, Third Edition, Prentice-Hall, p. 498. A good descriptive account, well illustrated.

HILL, M. N. (Ed.), *The Sea: Ideas and Observations*, Wiley-Interscience: Vol. 1, 1962, *Physical Oceanography*, p. 864. A collection of advanced papers on dynamical oceanography. Vol. 2, 1963, *The Composition of Sea Water; Comparative and Descriptive Oceanography*, p. 554. A series of mainly descriptive papers on the chemistry, biology and physics of the ocean.

HINZE, J. O. (1975) *Turbulence*, McGraw-Hill, 2nd Edition, p. 790. A comprehensive, advanced text.

IPPEN, A. T. (Ed.) (1966) *Estuary and Coastline Hydrodynamics*, McGraw-Hill, p. 744. A collection of papers summarizing wave and tide theory, harbour resonance and many aspects of estuarine dynamics.

KINSMAN, B. (1965) *Wind Waves, their Generation and Propagation on the Ocean Surface*, Prentice-Hall, p. 676. A detailed, mathematical review of wind-wave generation and characteristics.

KRAUSS, E. B. (Ed.) (1977) *Modelling and Prediction of the Upper Layers of the Ocean*, Pergamon Press, p. 325. Papers by a number of authors on these subjects, advanced.

LACOMBE, H. (1965) *Course d'Océanographie Physique*, Gauthier-Villars, p. 392. Covers many of the subjects in the present text (not tides or modelling) but in more mathematical detail.

LAMB, H. (1932) *Hydrodynamics*, Dover (reprint), 6th Edition, p. 738. The basic reference on classical hydrodynamics.

LEBLOND, P. H. and L. A. MYSAK (1978) *Waves in the Ocean*, Elsevier, p. 602. A comprehensive, up-to-date text, advanced.

MCLELLAN, H. J. (1965) *Elements of Physical Oceanography*, Pergamon Press, p. 150. An introduction to descriptive and dynamical physical oceanography.

MELCHIOR, P. (1982) *Tides of the Planet Earth*, Pergamon Press, 2nd Edition, p. 630. An advanced text on tidal theory.

MURTY, T. S. (1977) *Seismic Sea Waves–Tsunamis*, Department of Fisheries and the Environment, Fisheries and Marine Services, Ottawa, p. 337. A review of the character and theory of tsunamis.

NEUMANN, G. and W. J. PIERSON (1966) *Principles of Physical Oceanography*, Prentice-Hall, p.545. Moderately advanced text on dynamical oceanography with one chapter on descriptive oceanography.

OFFICER, C. B. (1958) *Introduction to the Theory of Sound Transmission, with Applications to the Ocean*, McGraw-Hill, p. 284. A mathematical treatment.

OFFICER, C. B. (1976) *Physical Oceanography of Estuaries and Associated Coastal Waters*, Wiley, p. 456. A moderately advanced account of the physical theory with applications to typical areas around the world.

PEDLOSKY, J. (1979) *Geophysical Fluid Dynamics*, Springer Verlag, p. 624. An advanced text of interest to graduate students in physical oceanography.

PHILLIPS, O. M. (1966) *The Dynamics of the Upper Ocean*, Cambridge University Press, p. 261. A graduate level text on surface and internal waves and on oceanic turbulence.

PICKARD, G. L. and W. J. EMERY (1982) *Descriptive Physical Oceanography*, Pergamon Press, 4th Edition, p. 249. An introduction to descriptive (synoptic) physical oceanography for science undergraduates and graduates.

PIERSON, W. J., G. NEUMANN and R. W. JAMES (1955) *Practical Methods for Observing and Forecasting Ocean Waves*, U.S. Naval Oceanographic Office, Publ. 603, p. 284. A technical treatise with applications and a chapter on wave refraction plotting.

PROUDMAN, J. (1953) *Dynamical Oceanography*, Methuen, p. 409. A basic mathematical treatise with much detail on waves and tides; everything done from first principles.

REID, R. O. (Ed.) (1975) *Numerical Models of Ocean Circulation*, National Academy of Sciences, Washington, D.C., p. 364. Proceedings of a symposium on numerical modelling. Advanced level but does include several review papers on the character of ocean circulation and some discussion on directions for future study.

RICHARDS, F. A. (Ed.) (1981) *Coastal Upwelling*, Coastal and Estuarine Sciences, Vol. 1, American Geophysical Union, p. 529. A series of papers on recent studies in coastal upwelling particularly from the Coastal Upwelling Ecosystems Analysis programme.

ROBINSON, A. R. (Ed.) (1963) *Wind-driven Ocean Circulation*, Blaisdell, p. 161. A collection of reprints of papers on the theory of this subject.

ROLL, H. U. (1965) *Physics of the Marine Atmosphere*, Academic Press, p. 426. An advanced description of the influence of the sea on the atmosphere above it, characteristics and turbulent flow of the atmosphere, thermodynamics.

RUSSELL, R. C. M. and D. M. MACMILLAN (1952) *Waves and Tides*, Hutchinson, p. 348. Mainly descriptive, with illustrations.

STERN, M. E. (1975) *Ocean Circulation Physics*, Academic Press, p. 246. Geophysical fluid dynamics for graduate physical oceanographers—mathematical.

STOMMEL, H. (1965) *The Gulf Stream*, University of California Press, 2nd Edition, p. 248. Both a description of this ocean feature and a review of theoretical studies of it. An excellent introduction to physical oceanography for upper year undergraduates and graduate students in physics.

SVERDRUP, H. U., M. W. JOHNSON and R. H. FLEMING (1946) *The Oceans, their Physics, Chemistry and General Biology*, Prentice-Hall, p. 1087. A comprehensive reference text on all aspects of oceanography to that date.

TENNEKES, H. and J. L. LUMLEY (1972) *A First Course in Turbulence*, MIT Press, p. 300. A first course for graduate students.

THOMSON, R. E. (1981) *Oceanography of the British Columbia Coast*, Department of Fisheries and Oceans, Ottawa, p. 291. An excellent account of most aspects of the physical oceanography, descriptive and dynamical, of a coastal region. Well illustrated, not mathematical.

TOLSTOY, I. and C. S. CLAY (1966) *Ocean Acoustics*, McGraw-Hill, p. 293. Another mathematical treatment of the subject.

TRICKER, R. A. R. (1965) *Bores, Breakers, Waves and Wakes*, Elsevier, p. 250. An interesting account of waves near the shore and of bores in rivers.

TURNER, J. S. (1973) *Buoyancy Effects in Fluids*, Cambridge University Press, p. 367. Discusses various consequences of gravity acting on small density differences in fluids, e.g. internal waves, instability of shear flow, buoyant convection, double diffusion and mixing. Graduate level.

URICK, R. J. (1975) *Principles of Underwater Sound*, McGraw-Hill, Second Edition, p. 384. A summary of principles, effects and phenomena of underwater sound with quantitative data and some theory; illustrated and with practical applications.

VARIOUS (1971) *Oceanography—Readings from Scientific American*, Freeman, p. 417. A collection of stimulating articles on many aspects of oceanography.

VARIOUS (1977) *Ocean Science—Readings from Scientific American*, Freeman, p. 307. Contains articles additional to those in the previous collection.

VERONIS, G. (1973) "Large-scale Ocean Circulations" in *Advances in Applied Mechanics*, **13**,

1–92. A review of analytic theory with careful attention to approximations used; some discussion of attempts at laboratory simulation of oceanic flows.

VON ARX, W. S. (1962) *An Introduction of Physical Oceanography*, Addison-Wesley, p. 422. A stimulating introduction to physical oceanography, with special emphasis on current measurements and the use of physical scale models.

WARREN, B. A. and C. WUNSCH (Eds.) (1981) *Evolution of Physical Oceanography*, MIT Press, p. 623. An excellent series of survey articles on the general ocean circulation, physical processes, techniques and ocean/atmosphere interaction, tracing the development of these fields and laying out the state of our present knowledge. Prepared as a tribute to Henry Stommel. Advanced but essential to the library of any serious physical oceanographer.

WIEGEL, R. L. (1964) *Oceanographical Engineering*, Prentice-Hall, p. 532. Some of the applications of physical oceanographic knowledge, particularly in relation to coastal structures and with an emphasis on waves.

B.2.2   Journals

Recent papers in dynamical oceanography may be found in these journals (and in many others as the reader may find on examining the journal references).

Continental Shelf Research.   Pergamon Press (from 1982).
Deep-Sea Research.   Pergamon Press (from 1953).
Estuarine and Coastal Marine Science.   Academic Press (from 1973).
Journal of Geophysical Research—Oceans and Atmosphere.   American Geophysical Union, Washington, D.C. (from 1959).
Journal of Marine Research.   Sears Foundation for Marine Research (from 1939).
Journal of Physical Oceanography.   American Meteorological Society (from 1971).
Oceanus.   Woods Hole Oceanographic Institution (from 1952). (Has short, up-to-date, non-mathematical accounts of recent developments in most aspects of oceanography.)
Tellus.   Stockholm (from 1949).
Transactions (later *EOS*), American Geophysical Union (from 1920).

Two annual reviews of various aspects of oceanography are:
Oceanography and Marine Biology—an Annual Review.   M. BARNES (Ed.), Aberdeen University Press (from 1963).
Progress in Oceanography.   M. V. ANGEL and J. O'BRIEN (Eds.), Pergamon Press (from 1964).

B.3   Sources of Data

Sets of numerical data, etc., useful in the numerical practice of dynamical oceanography may be found in the following references.

B.3.1   Salinity defined in terms of chlorinity; Knudsen-Ekman equation of state

FLEMING, R. H. (1939) Tables for Sigma-t, *Journal of Marine Research*, **2**, 9–11. (Tables of values of temperature and salinity for whole number values of sigma-*t*. Useful for plotting T–S diagrams.)

Handbook of Oceanographic Tables (1966) U.S. Naval Oceanographic Office, Washington, D. C., Special Publication 68, p. 712. (A collection of tables of use to oceanographers.)

Instruction Manual for Obtaining Oceanographic Data, 3rd Edition, (1968.) U.S. Naval Oceanographic Office, Washington, D.C., Publication 607, p. 210. (A description of routine oceanographic procedures and of standard instruments (not CTD's nor STD's).)

International Oceanographic Tables, Vol. 1 (1966), Unesco, Paris and National Institute of Oceanography, England, p. 128. (For the conversion of conductivity ratio to salinity.)

KNUDSEN, M. (1901) *Hydrographische Tabellen*, G.E.C. Gad, Copenhagen, p. 63. (Tables for the calculation of sigma-*t* from values of salinity and temperature.)

LAFOND, E. C. (1951) *Processing Oceanographic Data*, U.S. Naval Oceanographic Office, Washington, D. C., Publication 614, p. 114. (A compilation of tables needed for correcting reversing thermometers, calculating density, specific volume, etc.)

NEUMANN, G. and W. J. PIERSON, (1966) (see Section B.2.1). (Tables for α $(35,0,p)$, Δ (s,t), $\delta(t,p)$, (t,p), $\delta(s,p)$, (s,p) and stability (E).)

SVERDRUP, H. U., M. W. JOHNSON and R. H. FLEMING, (1946) (see Section B.2.1). (Tables for $\alpha(35,0,p)\Delta(s,t)$, $\delta(t,p)$, $\delta(s,p)$.)

Tables for Sea Water Density, (1952) U.S. Naval Oceanographic Office, Washington, D.C., Publication 615, p. 265. (Tables for calculating sigma-t from values of salinity and temperature.)

B.3.2   Practical Salinity Scale 1978 and International Equation of State 1980

MILLERO, F. J. and A. POISSON (1981) International one-atmosphere equation of state of seawater. *Deep-Sea Research*, **28A**, 625–629. (IES 80 for one atmosphere pressure but also gives the full formula for the pressure effect, i.e. $\alpha(s,t,p)$.)

Unesco, Ninth Report of the Joint Panel on Oceanographic Tables and Standards (1979). Unesco *Technical Papers in Marine Science*, No. 30, p. 31 (PSS 78).

Unesco, Tenth Report of the Joint Panel on Oceanographic Tables and Standards (1981). Unesco *Technical Papers in Marine Science*, No. 36, p. 24. (Contains the definitions of the PSS 78 and IES 80.)

A group of papers reviewing the background to and the development of the Practical Salinity Scale 1978 (PSS 78) were published in the IEEE *Journal of Oceanic Engineering*, Vol. 5, No. 1, 1980.

B.4   Journal references

ACCAD, Y. and C. L. PEKERIS (1978) Solution of tidal equations for M_2 and S_2 tides in the world ocean from a knowledge of the tidal potential alone. *Philosophical Transactions, Royal Society of London A*, **290**, 235–266.

BAKER, D. J. (1981) Ocean instruments and experiment design. Chap. 14, pp. 396–433, in *Evolution of Physical Oceanography*, B. A. WARREN and C. WUNSCH (Eds.), MIT Press.

BEHRINGER, D. W. (1979) On computing the absolute geostrophic velocity spiral. *Journal of Marine Research*, **37**, 459–470.

BEHRINGER, D. W. and H. STOMMEL (1980) The beta spiral in the North Atlantic subtropical gyre. *Deep-Sea Research*, **27A**, 225–238.

BREKHOVSKIKH, L. M., K. N. FEDEROV, L. M. FOMIN, M. N. KOSHYAKOV and A. D. YAMPOLSKY (1971) Large-scale multi-buoy experiment in the tropical Altantic. *Deep-Sea Research*, **18**, 1189–1206.

BRYAN, K. and M. D. COX (1972) The circulation of the world ocean: a numerical study. 1. A homogeneous model. *Journal of Physical Oceanography*, **2**, 319–335.

BRYAN, K. and L. J. LEWIS (1979) A water mass model of the world ocean. *Journal of Geophysical Research*, **84**, 2503–2517.

CARTWRIGHT, D. E. (1977) Oceanic tides. *Reports on Progress in Physics*, **40**, 665–708.

CHENEY, R. E. and J. G. MARSH (1981a) *Seasat* altimeter observations of the dynamic topography in the western North Atlantic. *Journal of Geophysical Research*, **86**, 473–483.

CHENEY, R. E. and J. G. MARSH (1981b) Oceanic eddy variability measured by *GEOS 3* altimeter crossover differences, *EOS. Transactions of the American Geophysical Union*, **62**, 743–752.

CORIOLIS, G. (1835) Mémoire sur les équations du mouvement relatif des systèmes de corps. *Journal de l'Ecole Royale Polytechnique*, **15**, p. 142.

COX, M. D. (1970) A mathematical model of the Indian Ocean. *Deep-Sea Research*, **17**, 47–75.

COX, M. D. (1975) A baroclinic numerical model of the world ocean. In *Numerical Models of Ocean Circulation*, pp. 107–120, National Academy of Sciences, Washington, D.C.

DAVIES, R. E., R. DE SZOEKE and P. NIILER (1981) Variability in the upper ocean during MILE.

Part 2: Modeling the mixed layer response. *Deep-Sea Research*, **28**, 1453–1476.

DEFANT, A. (1950) Reality and illusion in oceanographic surveys. *Journal of Marine Research*, **9**, 120–138.

DONELAN, M. (1982) The dependence of the aerodynamic drag coefficient on wave parameters. In *Proceedings, First International Conference on Meteorology and Air–Sea Interaction of the Coastal Zone* (in press).

EKMAN, V. W. (1905) On the influence of the earth's rotation on ocean currents. *Arkiv för Matematik, Astronomi och Fysik*, **2**, No. 11, p. 52.

EKMAN, V. W. (1908) Die Zusammendrückbarkeit des Meereswassers nebst einigen Werten für Wasser und Quecksilber. *Publications de Circonstance, Conseil Permanent International pour l'Exploration de la Mer*, No. 43, 1–47.

EMERY, W. J. (1983) On the geographical variability of the upper level mean and eddy fields in the North Atlantic and North Pacific. *Journal of Physical Oceanography*, **13**, 269–291.

FOFONOFF, N. (1962) Dynamics of ocean currents. Section III, pp. 323–395, in *The Sea-Ideas and Observations*, M. N. Hill (Ed.), Wiley Interscience, Vol. I.

FORRISTAL, G. Z. (1978). On the statistical distribution of wave heights in a storm. *Journal of Geophysical Research*, **83**, 2353–2358.

GARRATT, J. R. (1977) Review of drag coefficients over oceans and continents. *Monthly Weather Review*, **105**, 915–929.

GARRETT, C. (1972) Tidal resonance in the Bay of Fundy and Gulf of Maine. *Nature*, **238**, 441–443.

GARWOOD, R. W. (1977) An oceanic mixed-layer model capable of simulating cyclic states. *Journal of Physical Oceanography*, **7**, 455–468.

GODIN, G. (1980) Cotidal charts for Canada. *Marine Sciences and Information Directorate, Department of Fisheries and Oceans, Ottawa*, MSS Report Series No. 55, p. 93.

HANEY, R. L. (1979) Numerical models of ocean circulation and climate interaction. *Reviews of Geophysics and Space Physics*, **17**, 1494–1507.

HASSELMAN, K. *et al.* (1973) Measurements of wind-wave growth and swell decay during the North Sea wave project (JONSWAP). *Deutsche Hydrographische Zeitschrift*, Supplement A, **8**, (12), p. 95.

HAYWARD, A. T. J. (1967) Compressibility equations for liquids; a comparative study. *British Journal of Applied Physics*, **18**, 965–977.

HELLERMAN, S. (1967) An updated estimate of the wind stress on the world ocean. *Monthly Weather Review*, **95**, 606–626.

HELLERMAN, S. (1968) Correction. *Monthly Weather Review*, **96**, 63'–74.

HOLLAND, W. R. (1973) Baroclinic and topographic influences on the transport in western boundary currents. *Geophysical Fluid Dynamics*, **4**, 187–210.

HOLLAND, W. R. (1978) The role of mesoscale eddies in the general circulation of the ocean—numerical experiments using a wind-driven, quasi-geostrophic model. *Journal of Physical Oceanography*, **8**, 363–392.

HOLLAND, W. R. and A. D. HIRSCHMAN (1972) A numerical calculation of the circulation of the North Atlantic Ocean. *Journal of Physical Oceanography*, **2**, 336–352.

HOLLAND, W. R. and L. B. LIN (1975) On the origin of mesoscale eddies and their contribution to the general circulation of the ocean. I. A preliminary numerical experiment; II. A parameter study. *Journal of Physical Oceanography*, **5**, 642–669.

KNAUSS, J. A. (1960) Measurement of the Cromwell Current. *Deep-Sea Research*, **6**, 265–286.

KNUDSEN, M. (1901) *Hydrographische Tabellen*, G. E. C. Gad, Copenhagen, p. 63.

LARGE, W. G. and S. POND (1981) Open ocean momentum flux measurements in moderate to strong winds. *Journal of Physical Oceanography*, **11**, 324–336.

LEETMAA, A. and A. F. BUNKER (1978) Updated charts of the mean annual wind stress, convergences in the Ekman layers and Sverdrup transports in the North Atlantic. *Journal of Marine Research*, **36**, 311–322.

LEETMAA, A., J. P. MCCREARY and D. W. MOORE (1981) Equatorial currents; observation and theory. Chap. 6, pp. 184–196, in *Evolution of Physical Oceanography*, B. A. WARREN and C. WUNSCH (Eds.), MIT Press.

LEETMAA, A., P. NIILER and H. STOMMEL, (1977) Does the Sverdrup relation account for the Mid-Atlantic circulation? *Journal of Geophysical Research*, **72**, 4959–4975.

LEWIS, E. L. (1980) The Practical Salinity Scale 1978 and its antecedents. *Journal of Oceanic Engineering*, **5**, 3–8.

LEWIS, E. L. and R. G. PERKIN (1981) The Practical Salinity Scale 1978: conversion of existing data. *Deep-Sea Research*, **28A**, 307–328.

MCWILLIAMS, J. C. (1979) A review of research on mesoscale ocean currents. *Reviews of Geophysics and Space Physics*, **17**, 1548–1568.

MEEHL, G. A., W. M. WASHINGTON and A. J. SEMPTER, Jr. (1982) Experiments with a global ocean model driven by observed atmospheric forcing. *Journal of Physical Oceanography*, **12**, 301–312.

MILES, J. W. (1961) On the stability of heterogeneous shear flows. *Journal of Fluid Mechanics*, **10**, 496–508.

MILLERO, F. J., C-T. CHEN, A. BRADSHAW and K. SCHLEICHER (1980) A new high-pressure equation of state for sea-water. *Deep-Sea Research*, **27A**, 255–264.

MILLERO, F. J. and A. POISSON (1981) International one-atmosphere equation of state of sea-water. *Deep-Sea Research*, **28A**, 625–629.

MODE Group (1978) The Mid-Ocean Dynamics Experiment. *Deep-Sea Research*, **25**, 859–910.

MUNK, W. H. (1950) On the wind-driven ocean circulation. *Journal of Meteorology*, **7**, 79–93.

MUNK, W. H. (1981) Internal waves and small-scale mixing processes. Chap. 9, pp. 264–291, in *Evolution of Physical Oceanography*, B. A. WARREN and C. WUNSCH (Eds.), MIT Press.

MUNK, W. H. and G. F. CARRIER (1950) The wind-driven circulation in ocean basins of various shapes. *Tellus*, **2**, 158–167.

MUNK, W. H. and D. E. CARTWRIGHT (1966) Tidal spectroscopy and prediction. *Philosophical Transactions, Royal Society of London A*, **259**, 533–581.

NIILER, P. (1975) Deepening of the wind-mixed layer. *Journal of Marine Research*, **33**, 405–422.

NIILER, P. and E. B. KRAUSS (1977) One-dimensional models of the upper ocean. Pp. 143–172 in *Modelling and Prediction of the Upper Layers of the Ocean*, E. B. KRAUSS (Ed.), Pergamon.

O'BRIEN, J. J. (1971) A two-dimensional model of the wind-driven North Pacific *Invest. Pesqu.* **35**, 331–349.

PHILANDER, S. G. H. (1973) Equatorial undercurrent: measurements and theories. *Reviews of Geophysics and Space Physics*, **11**, 513–570.

REID, R. O. (1948) The equatorial currents of the eastern Pacific as maintained by the stress of the wind. *Journal of Marine Research*, **7**, 75–99.

ROSSBY, C-G. and R. B. MONTGOMERY (1935) The layer of frictional influence in wind and ocean currents. *Papers in Physical Oceanography and Meteorology*, **3**, No. 3, p. 101.

SARKISIAN, A. S. and V. F. IVANOV (1971) The combined effect of baroclinicity and bottom topography as an important factor in the dynamics of ocean currents. *Izvestia Acad. Sci., USSR, Atmospheric and Oceanic Physics*, **1**, 173–188.

SAUNDERS, P. M. (1976) On the uncertainty of wind-stress curl calculations. *Journal of Marine Research*, **34**, 155–160.

SCHMITZ, W. J., Jr. and W. R. HOLLAND (1982) A preliminary comparison of selected numerical eddy-resolving general circulation experiments with observations. *Journal of Marine Research*, **40**, 75–117.

SCHOTT, F. and H. STOMMEL (1978) Beta-spirals and absolute velocities in different oceans. *Deep-Sea Research*, **25**, 961–1010.

SMITH, S. D. (1980) Wind stress and heat flux over the ocean in gale-force winds. *Journal of Physical Oceanography*, **10**, 709–726.

SNODGRASS, F. E. *et al.* (1966) Propagation of ocean swell across the Pacific. *Philosophical Transactions, Royal Society of London A*, **259**, 431–497.

STEWART, R. H. (1980) Ocean wave measurement techniques. Chap. 24, pp. 447–470 in *Air–Sea Interaction—Instruments and Methods*, F. DOBSON, L. HASSE and R. DAVIS (Eds.), Plenum.

STOMMEL, H. (1948) The westward intensification of wind-driven currents. *Transactions, American Geophysical Union*, **29**, 202–206.

STOMMEL, H. (1958) The abyssal circulation. *Deep-Sea Research*, **5**, 80–82.

STOMMEL, H., P. NIILER and D. ANATI (1978) Dynamic topography and recirculation of the North Atlantic. *Journal of Marine Research*, **36**, 449–468.

SVERDRUP, H. U. and W. H. MUNK (1947) Wind, sea and swell: theory of relations for forecasting. *United States Hydrographic Office, Publication 601*, p. 44.

SWALLOW, J. C. (1955) A neutral-buoyancy float for measuring deep currents. *Deep-Sea Research*, **3**, 74–81.

TAIT, P. G. (1888) Report on some of the physical properties of fresh water and sea water. *The Voyage of H.M.S. Challenger*, Vol. 2, Part 4, pp. 1–76.

TURNER, J. S. (1981) Small scale mixing processes. Chap. 8, pp. 236–262, in *Evolution of Physical Oceanography*, B. A. WARREN and C. WUNSCH (Eds.), MIT Press.

VERONIS, G. (1981) Dynamics of large-scale circulation. Chap. 5, pp. 140–183, in *Evolution of Physical Oceanography*, B. A. WARREN and C. WUNSCH (Eds.), MIT Press.

WARREN, B. A. (1977) Deep western boundary current in the eastern Indian Ocean. *Science*, **196**, 53–54.

WELANDER, P. (1971) The thermocline problem. *Philosophical Transactions, Royal Society of London A*, **270**, 69–73.

WILLIAMS, A. J. (1975) Images of ocean microstructure. *Deep-Sea Research*, **22**, 811–829.

WORTHINGTON, L. V. (1976) On the North Atlantic circulation. *Johns Hopkins Oceanographic Studies*, No. 6, p. 110.

WUNSCH, C. (1978) The North Atlantic general circulation west of 50°W determined by inverse methods. *Reviews of Geophysics and Space Physics*, **16**, 583–620.

WYRTKI, K. (1974) The dynamic topography of the Pacific Ocean and its fluctuations. Hawaii Institute of Geophysics, Report HIG-75-5, p.19 + 37 figs.

Index